Collecting ARIZONA

Collecting Arizona

State of Mines, Legacy of Minerals

recounted by
Tony L. Potucek, Les Presmyk
Richard Graeme and others

edited by Terry C. Wallace

with Gloria A. Staebler, Ray Grant
Suzanne Liebetrau and Tom Wilson

Lithographie • Denver

COPYRIGHT

PO Box 11613, Denver, CO 80211 USA

Edited by Terry C. Wallace
with Gloria A. Staebler, Ray Grant, Suzanne Liebetrau and Tom Wilson

Conceived by Les Presmyk and Gloria A. Staebler

Layout and design by Gloria A. Staebler

Printed in the United States of America

ISBN: 978-0-9836323-1-3

www.lithographie.org

IMAGE CREDITS

Front Cover
Wulfenite, 5 cm wide; Jobes stope, Rowley Mine, Maricopa County, Arizona
Mark Hay collection, Jeff Scovil photo

Frontispiece
Azurite on malachite, 5 cm tall; Morenci Mine, Greenlee County, Arizona
Stan Esbenshade collection, Jesse La Plante photo

Left
Copper twin, 16.4 cm tall; Ray Mine, Pinal County, Arizona
Paul S. Harter collection, Jeff Scovil photo

Facing page
Silver on calcite, 9.7 cm tall; Mineral Park, Mohave County, Arizona
Evan Jones collection, Jeff Scovil photo

Page ix
Wulfenite, 7.5 cm wide; Glove Mine, Cochise County, Arizona
University of Arizona collection (ex Richard Bideaux), Jeff Scovil photo

Page x
Lewis Page and Jason A Pickens, the 1908 Globe, Arizona Drilling Team, pose for a George H. Tbwoites 1908 photo, Tom Hughes collection

Page xii
Aerial view of the Ray Mine, Pinal County, Arizona
Marco van Middelkoop photo, March 29, 2009

Page xiv-xv
Monument Valley as seen from Hunts Mesa, Arizona
Dean Hueber photo, January 13, 2006

Back Cover
Copper Queen carbide lamp
Becksted collection, Jeff Scovil photo

THIS BOOK WAS MADE POSSIBLE THROUGH
THE GENEROUS SUPPORT OF

Jack W. Delbrook
Scott Rudolph
David W. Bunk
Crystal Classics & Kristalle
Terry C. Wallace
Gloria A. Staebler
Evan A. Jones
Meghan K. Funk
Daniel Trinchillo
Les and Paula Presmyk

AND ADDITIONAL SUPPORT FROM

Bruce Carter, Paul S. Harter, Jeffrey A. Scovil
Stuart and Donna Wilensky, George and Cindy Witters
John Lucking, William Larson and Ron Gladnick
Steve and Carol Maslansky, Peter Megaw, Mark Hay
Scott Werschky and The Flagg Mineral Foundation

DEDICATION

This book is dedicated to "Mr. Rockhound" **ARTHUR L. FLAGG**. The first curator of the Arizona Mineral Museum, Flagg mentored untold numbers of Arizona collectors with patience and a philanthropic spirit.

It is further dedicated to the father and son team of **GEORGE** and **RICHARD BIDEAUX**. Their spirit, foresight, and their dedication to field collecting and to the mineral collecting community have inspired generations of Arizona collectors.

JAS. A. DICKENS
PAGE
GLOBE, ARIZONA TEAM
Copyright 1908
by Geo. H. Thwaites

TABLE OF CONTENTS

PREFACE

The seed for this book was planted in the early 1990s when Dr. Carl Francis and his wife were visiting the Phoenix area after the Tucson Show. Carl spent an evening going through our collection and asked me about my efforts to preserve the history of the specimens in our collection. I assured him that the catalog contained as much prior collection history as was known. He went on to explain that the histories should include the stories about how the specimens were collected and by whom. Carl's point was driven home for me later while I was reading "The History of Mineral Collecting" by Wendell Wilson (*The Mineralogical Record*, 25:6).

In the fall of 2007, I broached the subject of putting together a book on Arizona mineral specimens with Lithographie publisher Gloria Staebler. Wayne Thompson had published his "Ikons" supplement to *The Mineralogical Record*, and he felt strongly that Paula and I should consider a book or supplement on our collection as a means of showcasing our collection and Arizona mineral specimens. While the idea was intriguing, the cost was prohibitive. Besides, I wanted the book to encompass more than just our collection. Gloria and I began discussing a broader approach, something that would showcase Arizona's mineral wealth and its strong mining history. I suggested that the book come out in conjunction with the 2012 Tucson show and Arizona's Centennial.

Those early discussions took shape in November 2009, when Gloria and I began to draft the format and project schedule. The standard organization for this sort of book would have been geographic—built around localities or mining districts. But Gloria envisioned a book with an historical framework. The idea resonated with me as I recalled C.H. Dunning's 1966 book *Rock to Riches*, which told the story of Arizona's mining history chronologically. As I excitedly explained how that book was constructed, the die was cast. But a project of this nature was going to require serious research—we were going to need a strong coauthor. The moment Gloria suggested him, there was no question as to the choice: my long-time friend Tony Potucek. Tom Hughes came onboard to supplement his research, retired USGS cartographer John Callahan agreed to draft our maps, and Dr. Ray Grant stepped up as technical editor—our team was taking shape!

Tony immediately got things moving, setting the book up to not only highlight Arizona's mineral collecting heritage but to celebrate the miners and collectors who had dug the specimens and preserved that heritage. Tony assumed the roles of coauthor, managing editor, and researcher for the first year of the project. He interviewed more than 100 living collectors and compiled documentation from many more who are no longer with us. Richard Graeme wrote the sections on Bisbee that are woven into the story while both Tony and I contacted collectors we knew and others that we knew of. The accounts retold here attest to Tony's tenacity and ability to persuade Arizona collectors to record their stories. Gloria took the reigns as managing editor and photo researcher in 2011, bringing former University of Arizona professor and curator Dr. Terry Wallace on to help to pull the collecting stories into a narrative and provide the historical context.

We venture here to relate the story of mineral collecting in Arizona and to share the rich and beautiful legacy of the state's mining history. While the stories of many of Arizona's fine specimens are left to our imaginations: What must it have been like to have been underground in Bisbee when a 40-foot-long tube of azurite was encountered? Or to have been with the first prospectors who stumbled across the vein outcrop at the Red Cloud? While not preserved in the literature, those events were momentous enough for miners to have taken the unusual step for the era and preserved them in the form of mineral specimens—testaments to their importance and clues to the circumstances of their formation and recovery.

–Les Presmyk, October 2011

ACKNOWLEDGEMENTS

Collecting Arizona has benefited from the great spirit of cooperation and desire that is alive in the mineral collecting community. The list of collectors and people who contributed to the story is long, and their contributions are gratefully acknowledged.

Significant finds were described by **Robert Allgood**, **Malcolm Alter**, **Andy Anderson**, **Garry Alexander**, **Brad Archer**, **Bruce and Debbie Barlow**, **Bob Bartsch**, **Bill Besse**, **Harry Brenton**, **Jim Bleess**, **Garth Bricker**, **Jon Brugman**, **John Callahan**, **Stan Celestian**, **Jack Crowley**, **Rock Currier**, **Ryan Dewitt**, **Bob Eveleth**, **Ron Gibbs**, **Ray Grant**, **Bernie Haag**, **Bill Hefferon**, **Ed Huskinson**, **Brian Huntsman**, **Bob Jackson**, **Mike Jensen**, **Walter Johnson**, **Roy Jones**, **Tony Jones**, **Mark Kielbaso**, **Stan Keith**, **Robert Kerr**, **Pete Knudsen**, **Joe Knue**, **William Larson**, **Ray Lasmanis**, **Wayne and Dona Leicht**, **Scott Lewis**, **Tristan Massart**, **John Mediz**, **Peter Megaw**, **Glen Miller**, **Dean Misantoni**, **Dick Morris**, **Barbara Muntyan**, **David New**, **Mike New**, **Marcus Origlieri**, **Neal Pfaff**, **Paul Pohwat**, **Neil Prenn**, **Jim Ricker**, **Ken Roberts**, **the Rogers Family**, **Jay Rosenbauer**, **Mike Sanders**, **Jeff Scovil**, **Mike Shannon**, **Frank Sousa**, **Mike and Sandra Sprunger**, **George Stevens**, **John Taylor**, **Layton Talbott**, **Richard Thomssen**, **Wayne Thompson**, **Frank Valenzuela**, **Les Wagner**, **Jim Walker**, **Mary Fong-Walker**, **Bill Williams**, and **Wendell Wilson**.

Continuous support, collecting documentation, and other information was provided by the following individuals as the authors and editors had to more than once go back to the well: **Andy Clark**, **Stan Esbenshade**, **George Godas**, **Mark Hay**, **Paul Harter**, **Tom Hughes**, **Mike and Mary Jaworski**, **Bob Jones**, **Evan Jones**, **Bobby Lane**, **John Lucking**, **John Mediz**, **Nyal Niemuth**, **Gene and Jackie Schlepp**, **Graham Sutton**, **Jimmy and Joyce Vacek**, and **William and Pamela Wilkinson**.

Additional contributions by **Anna Dimetrovic**, **Phil Richardson**, and **Bob Downs** improved the story and the book. To the many deceased collectors and professionals who contributed by written stories and articles, or otherwise spread the word knowingly or unknowingly, the authors are grateful, for that imparted knowledge greatly benefitted this endeavor.

Collecting Arizona has been likened to a tapestry woven of the stories and specimens of a centuries-old tradition of mining in the state. The diverse and often discontinuous threads of the "weft"—the named and unnamed

collectors and their stories both told and untold—form the visible patterns and textures. However were it not for the editorial team—the weavers and hidden "warp"—the collectors' stories would remain loose threads. If they have done their jobs well, our editors and their efforts will be all but invisible to our readers; yet they are vital to the strength of this fabric.

Our gratitude is thus extended to: **Ann Frazier**, who compiled and maintained the reference list; **Dr. Ray Grant** and **Richard Graeme** for the considerable knowledge and experience they have brought to their reading of the texts; **John Callahan** for carefully drawing and many times revising our maps; **Michael Bainbridge** and **Craig Wagner**, who have worked magic in moving our images from digital files to accurate, high-quality, color reproductions; **Dave Bunk**, who knowledgably and objectivity assisted with image selection and research; **Jeffrey A. Scovil**, who patiently and professionally worked with the community of collectors to get its mineral treasures and heirlooms photographed; **Tony Potucek**, who collected the bulk of the stories and compiled a great deal of essential background information; and to **Dr. Suzanne Liebetrau** and **Tom Wilson** who have tirelessly read and reread the many iterations of these chapters, supplying thoughtful, objective commentary, edits, and rewrites, all of which have improved this book immeasurably.

Finally the perspective, rewrites, and daily consultations with **Dr. Terry C. Wallace** have been indispensible in bringing *Collecting Arizona* to its fruition. His unmatched knowledge and experience, his dedication, as well as his passion for the minerals, geology, and history of the State of Arizona have made him a truly invaluable asset.

On A Personal Note

From Tony Potucek: When I agreed to take on this project, I realized that I would be working nights and weekends for nearly two years, and I needed a nudge. It was my life partner **Marcie Greenberg** who pointed out that I had participated in the halcyon years of mineral collecting and could provide unique perspective. She convinced me to "Write the story!" For her guidance and support through our lives, I offer my highest gratitude.

From Les Presmyk: I am indebted to all of the people whose stories are contained within these covers and to the many unnamed miners and collectors who have contributed to the rich history of mineral collecting in Arizona. But the one person who most deserves my personal acknowledgement is my wife **Paula**, my best friend and partner since 1970.

INTRODUCTION

A Collector's Paradise

Above: The U.S. one-cent was, with few exceptions, made mostly from copper until 1982 when the cost of the copper needed to make the coin rose above one cent. The Victor David Brenner likeness of President Abraham Lincoln has been on the obverse of the U.S. penny since 1909. This coin pictured was minted in San Francisco in 1912, the year that Arizona, the nation's greatest copper producer, became the 48th U.S. state.

Left: This 15.5 cm wide azurite on malachite was recovered in the 1890s from Bisbee's historic Czar Mine in Cochise County, Arizona. The specimen surfaced in the 1970s in the collection in Vista, California, and in 1971 was acquired by Bill MaGee, who sold it the following year to dealer/collector Bill Larson, its current owner.

Robert Weldon photo

From its earliest days as a territory when prospectors searched for gold and silver to the giant copper camps of the early twentieth century to today when Tucson is a mecca for mineral collectors worldwide, the historical fabric of Arizona is woven with minerals. The state's very name comes from an early silver rush. In 1736 silver nuggets were discovered near present day Nogales. The site of the discovery was near a Tohono O'odham village called *Ali-Shonak* or "place of small springs." The Spanish who rushed to retrieve the silver corrupted this strange word to *Arizonac.* And when Anglo-Americans arrived in the nineteenth century, the "c" was dropped, and the area became known as Arizona.

The mineral wealth of the state is profound: it has produced more than 90 billion pounds of copper, 500 million ounces of silver, and 16 million ounces of gold. As of 2011, there were some 11,000 abandoned or active mines and prospects scattered across the state. Arizona has been the nation's leader in annual copper production since 1910, having produced more copper than the other 49 states combined! Today Morenci is the largest copper mine in the United States and one of the largest in the world. All of this mining activity has produced an abundance of mineral specimens. Indeed, the state is blessed with a spectacular mineralogy and a stunning array of colorful minerals, many of which are simply the best in the world. Azurite, malachite, wulfenite, and vanadinite are the best known, but Arizona is also home to almost 900 mineral species. This long list of minerals contains 75 that were first discovered and described from mines within the boundaries of the state.

Arizona is a unique land and one of extremes. The state encompasses some 114,000 square miles and consists of three distinct physiographic provinces: the Colorado Plateau in the northern third of the state, characterized by relatively flat lands with elevations between 5,000 and 7,000 feet; the Basin and Range with mountain ranges and desert valleys in the southern half of the state; and a transition zone between the Colorado Plateau and the Basin and Range. The highest point in the state is Humphreys Peak (elevation 12,611 feet), north of Flagstaff, but the desert valleys near Yuma have elevations of less than 100 feet. The Basin and Range and to a lesser extent the Transition Zone are the home of the most of the state's mineral wealth. Much of this region is desert: annual precipitation is less than 10 inches, and fewer than 20 days per year see rainfall. In Yuma summertime temperatures can exceed 120 degrees Fahrenheit.

Arizona's inhabitants were mining for turquoise and salt more than 600 years ago, and there is strong evidence of Spanish and Mexican mining in the Santa Cruz Valley south of Tucson. However, the story of mining and Arizona's dramatic development from frontier to forty-eighth state really begins with the Gadsden Purchase in 1853.

Development in the state slowed with the onset of the Civil War as U.S. troops on the frontier were reassigned, and the forts that guarded the prospectors, ranchers, and miners were abandoned. But when the war ended, the forts were re-established and prospectors began to explore Arizona in earnest. Gold was first found along the Colorado River and then near modern-day Prescott. Tombstone, the greatest Arizona silver camp, was discovered in the mid-1870s. And in the 1880s, Arizona's most celebrated copper camp—**Bisbee**—was discovered.

The opening of these mining camps corresponded with a growing interest in the science of minerals and a nationwide interest in mineral collecting. Miners and mine management alike learned that collectors were willing to pay for colorful specimens and that a miner could supplement his income by loading his lunch pail with specimens. In many cases, these aesthetic and interesting snapshots of the earth's history are the only documentation of the rich veins and ores that were long ago mined, crushed, and smelted. Interestingly, much of the rich social, cultural, and industrial development of the state of Arizona—much of what links our lives to mining and to the earth's resources—is today reflected and preserved in public and private mining artifact, memorabilia, and mineral collections. The preserved specimens reflect the values and the circumstances of people, places, and times. Specimens that travel through collections can acquire veneers of history, which can increase their monetary value. A Red Cloud wulfenite collected by Ed Over and owned by Arizona mineralphile Richard Bideaux has an emotional appeal to many collectors. Every mineral specimen has a story; some have broad

Right: Ed Over enjoyed success collecting butter-yellow wulfenite on calcite at the Hilltop Mine, Cochise County, in the 1930s, and some of that material was later distributed by Schortmann's Minerals. Because collector Andy Toussaint purchased this 6.8 cm wide specimen from Schortmann's, it may have been collected by the legendary Over. This pretty miniature spent more than a decade in each of the collections of Andy Toussaint, Phil Scalisi, and Ken Hollman before being acquired by its present owner Mark Hay.

Jeff Scovil photo

Above: Yale-educated William Phipps Blake (1826–1910) was a geologist, mineralogist, and a director of the School of Mines at the University of Arizona

Facing page, upper: Pendant made from shell, mother-of-pearl, stone, and turquoise, 5.8 cm long
Tonto Basin Salado culture
(ca. 1150–1450 CE)

Private collection, Jeff Scovil photo

Facing page, lower: Native copper pseudomorph after hammerhead, 20.6 cm wide recovered from the 800 level of the Shattuck Mine in 1925

Bisbee Mining and Historical Museum collection, Jeff Scovil photo

implications, others express local impressions.

Over the years, a number of books have been written on the minerals, mineralogy, and mining history of Arizona. William Phipps Blake published the first of these in 1909 in which he recognized approximately 100 species. In 1944 Arthur Flagg, who worked in the Arizona Department of Mineral Resources, published *Rockhounds and Arizona Minerals*, and in 1958 he published *Mineralogical Journeys in Arizona*. By then the list of minerals found in Arizona had grown to nearly 400. And in 1995, when long-time curator of the University of Arizona Mineral Museum John Anthony and his coauthors published the latest edition of the *Mineralogy of Arizona*, the list had grown to more than 800 species.

Collecting Arizona endeavors to provide a sense of the people, places, and times that spawned the state's mineral discoveries. We begin with a geologic overview, but the story is more a history than a scientific treatise and is divided into chapters, each covering a distinct era, beginning with the pre-Territorial period when missionaries, soldiers, and prospectors made the earliest inroads into the uncharted Pimería Alta, which would become southern Arizona.

This book closes with the present era, one in which the world's great mineral classics have become collectibles on par with cars, coins, or art. The influx into the mineral specimen market of aesthetic collectors has created a pronounced rift between the ethos that serves those collectors and that which continues to serve more traditional collectors, to whom aesthetics are no more important than history, science, or their personal connections to a specimen. What mineral collecting will look like in the future is anyone's guess, but it will most certainly reflect contemporaneous collectors, values, and times.

Collecting Arizona is a Centennial tribute to the state's legacy of mineral specimens; to those who have spent time digging, documenting, and preserving that legacy; and to the great state of Arizona, a state built on its minerals and its mines.

Above: Deciduous trees in Oak Creek Canyon reflect some of the varied flora and fauna found in the Transition Zone.

Dean Hueber photo, November 2006

Left: Erosion has cut through the sedimentary layers of the Colorado Plateau to form the glorious Grand Canyon. This photo of a side canyon was taken by adventurer and photographer Steve Ramras in February 2011.

CHAPTER 1

A Tour of Arizona's Geology

The geology of Arizona is spectacular and extraordinarily varied. In the northern part of the state the Colorado River slices through the thick sandstone and limestone of the Colorado Plateau, forming the Grand Canyon and exposing the nearly two-billion-year-old Vishnu Schist. Less than 60 miles away, the San Francisco volcanic field contains hundreds of cinder cones, some nearly 1,000 feet tall; the youngest is only about 1,000 years old.

In the northeast corner of the state, the colorful Permian sandstone buttes and bluffs frame the surreal landscape of Monument Valley and are testament to the large, shallow, ancient sea, which captured the sediments that eroded from an ancestral Rocky Mountains to the east. The boundary between continent and sea was rich with forest and grasslands. As the sea level ebbed and flowed, the forests were flooded, and many of the trees were preserved as colorful fossilized wood.

In the southeastern part of the state, towering mountains are separated by wide valleys, which formed as the crust of Arizona was stretched and extended. These mountains and desert valleys define a unique ecological environment: the peaks are "sky islands," hosting unique flora and fauna that commonly have ranges of only a few tens of square miles. The wonderful geology and topography of Arizona is a product of its long and complex geologic history. That history is also responsible for the variety of mineral deposits throughout the state.

Topography, the Face of Geology

Geologists divide Arizona into three physiographic provinces: the broad, high plains of the Colorado Plateau in the northern third of the state; the mountains and valleys of the Basin and Range in the southern half of the state; and the Transition Zone, or Central Mountain Province, situated between the two.

The Colorado Plateau, a large series of intermountain plateaus, covers nearly 130,000 square miles. It is centered on the Four Corners (the point at which the states of Arizona, Colorado, Utah, and New Mexico meet) and is sometimes referred to as Red Rock Country. The geology of the Colorado Plateau is relatively simple, consisting mainly of horizontal sedimentary Paleozoic, Mesozoic, and Cenozoic rocks. The Colorado River and its various tributaries drain about 90 percent of the Plateau, which is nearly

UTAH
NEVADA
CALIFORNIA
NEW MEXICO
MEXICO
N
S
E
W
114°
113°
112°
111°
110°
37°
36°
35°
34°
33°
32°
Four Corners
Vermillion Cliffs
National Monument
Monument
Valley
Colorado
River
Canyon de Chelly
National Monument
COLORADO PLATEAU
PROVINCE
Grand Canyon
Grandview
Mine
San Francisco
volcanic field
Sunset Crater Volcano
National Monument
Mineral Park
Oatman
Flagstaff
Winslow
Holbrook
Petrified
Forest
National Park
United Verde
Mine
Jerome
Old Dick
(Bruce) Mine
Bagdad
Rawhide
Mine
Antler Mine
Mystery Mine
group (Heson Mine)
TRANSITION ZONE PROVINCE
Diamond Point
Phoenix
Inspiration
Mine
Old Dominion Mine
Magma Mine
79
Mine
Ray
Iron Cap Mine
Morenci
North and South
Geronimo Mines
Red Cloud Mine
BASIN AND RANGE PROVINCE
Grand Reef
Mine
Tiger
Yuma
Vekol Mine
New Cornelia
Mine
Tucson
Tombstone
Red Mountain
(Flux) Mine
Bisbee
0
50 MILES
0
50 KILOMETERS

Sunset Crater is a cinder cone created around 1064 CE during a series of eruptions dominated by explosive expulsion of scoria. It is part of the San Francisco Volcanic Field, which covers nearly 1,700 square miles in northern Arizona on the Colorado Plateau. The volcanic field was active for about 6 million years and produced more than 600 volcanoes or craters. Arizona's highest mountain, Humphreys Peak (elevation 12,633 feet), which towers over Flagstaff, is part of the field and is its only strata volcano. Sunset Crater is the youngest crater is the field, although it is highly likely that there will be a future eruption somewhere in the area. Sunset Crater is a national monument; the crater, viewed here from the southwest, is thought to have been named in the nineteenth century by John Wesley Powell, who likened the colorful cone to a spectacular Arizona sunset.

Dwight Hanson photo, March 2009

uniform in elevation (between 5,500 and 7,000 feet). Beyond the uplift of the rocks to their present elevation, there has been little tectonic deformation of the thick sequence of sedimentary rocks over the past 600 million years of geologic history. Erosion has carved spectacular landforms across the Plateau with names belying a pastel pallet: Vermillion Cliffs, Pink Cliffs, and Gray Cliffs. In Arizona the Plateau is home to 17 national parks and monuments, the Grand Canyon, Canyon de Chelly, and Monument Valley among them.

The Basin and Range province covers the southern half of the state and is part of a much larger geologic terrane, which stretches into northern Mexico, New Mexico, California, and most of Nevada. The Basin and Range is characterized by northwest- to southeast-trending mountain ranges separated by wide valleys. The rocks exposed in the mountains range in age from Precambrian to the Holocene. The Basin and Range province is the response to widespread crustal extension over the last 17 million years. As the crust is stretched, normal faults develop that separate the high mountain peaks and the deep intervening valleys. The extension and faulting have produced a topography that is like an accordion. Igneous activity is widespread across the Basin and Range. The majority of Arizona's ore deposits and mineral occurrences are in this province.

The Transition Zone, or Central Mountain Province, is located between the Colorado Plateau and the Basin and Range and covers a smaller area. The geology of the Transition Zone is characterized by a combination of horizontal sediments and faulted ranges; the area has a variable and complex topography. Precambrian to Holocene rocks are exposed throughout, but Precambrian rocks predominate. Most of Arizona's Precambrian ore deposits are found in this region.

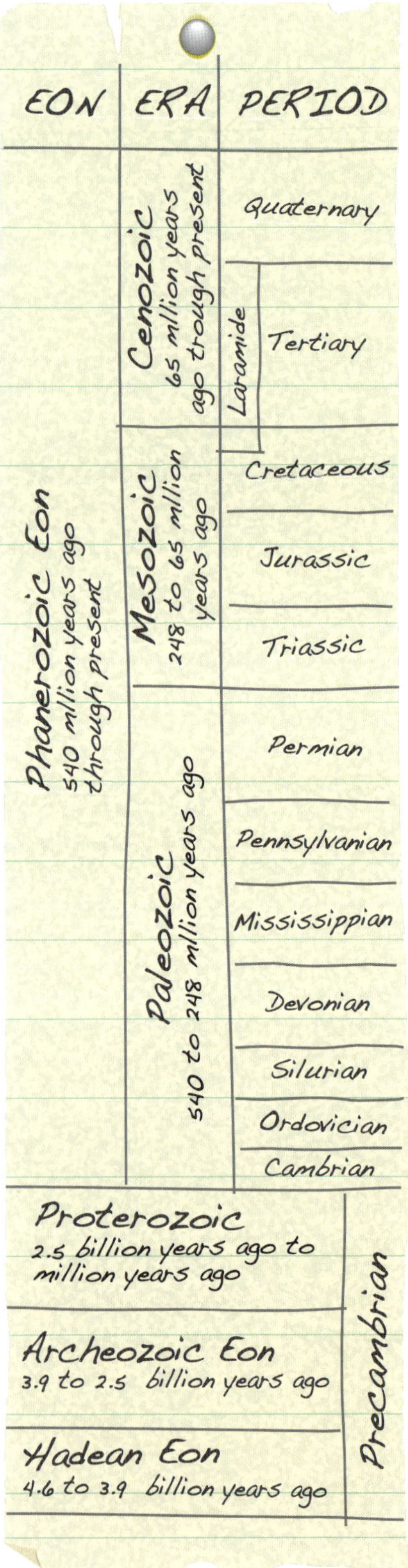

The Geologic History of Arizona

Two billion years ago a Precambrian ocean covered the area that is today Arizona. At the margin of the ocean, a series of volcanic arcs formed from the subduction of oceanic crust beneath other oceanic crust. Over time this subduction produced a large amount of igneous rock, which along with the associated sediments began to form continental crust. For more than 400 million years, these island arcs along with the associated deposition, igneous activity, and deformation moved southward, and by 1.6 billion years ago, the initial continental crust of Arizona was in place as part of North America.

During this period of continent formation (2 billion years to 1.6 billion years ago), a number of ore deposits formed. Magmatic fluids and hot seawater circulating through the volcanic rocks dissolved sulfur and metals such as iron, copper, zinc, lead, silver, and gold. These solutions vented through fractures onto the ocean floor, similar to the way that black smokers vent onto the bottom of the modern-day ocean. These black smokers are preserved in several mining localities, most prominently at the **United Verde Mine** in Jerome. This Precambrian activity deposited massive sulfides, in the form of pyrite, chalcopyrite, and sphalerite, onto the floor of the ancient ocean. The **Antler**, **Old Dick**, **Bruce**, and **United Verde** mines exploited just such ancient deposits. And indeed they are very rich: more than 3 billion pounds of copper were produced in the Verde mining district!

After the Precambrian crust formed, it must have then been thrust up into a series of mountains, after which there was a long period of erosion, which erased a huge section of time in terms of the rocks present. There is a gap of more than 700 million years between the oldest rocks exposed on the floor of the Grand Canyon—the Vishnu Complex—and the rocks deposited on top of them. The oldest rocks of the Vishnu Complex have been radiometrically dated at 1,950 million years before present (mybp). They are some of the earliest known rocks in the southwest and are representative of the processes that grew the bulk of the earth's continental crust. This gap is referred to as the Great Unconformity and was first observed by John Wesley Powell in 1869 as he explored the Colorado River.

By the beginning of the Paleozoic era, Arizona was firmly attached to the North American Craton. It was located at the edge of the continent and

Right: This polished slab of fossilized wood was cut from the trunk of a tree that was mineralized by silica containing enough chromium to cause the green color. It was found in the Chinle formation near Winslow in Navajo County.

Al and Sue Liebetrau collection
Jeff Scovil photo

Below: A silicified section of a conifer trunk preserved in the Petrified Forest National Park near Holbrook; these trees are related to tropical and subtropical trees such as those found today in Chile and Argentina and are a reflection of the climate of Arizona during the Triassic period when the trees were living.

was covered with a shallow, tropical sea. A thick sequence of Paleozoic sedimentary rocks was deposited above the Vishnu as the waters of the sea advanced and retreated numerous times. The shales, sandstones, and limestones that characterize the sedimentary stack are best seen in the Grand Canyon. In the Basin and Range, these sediments are broken up by the faulting and in many areas have been eroded away, but where they are present, they are important host rocks for younger ore deposits. No significant metallic ore deposits formed during the Paleozoic, although local deposits of quartz (such as those at **Diamond Point**), geodes, chert, and calcite were formed.

By the start of the Mesozoic, the sea had mostly retreated from Arizona, and North America was part of the Pangaea supercontinent. During the Triassic, the landmass that today is Arizona was situated near the equator and had a tropical climate. Northeastern Arizona was a low plain with mountains to its south and southwest and the Mesozoic sea to its west. Streams flowing across the plain from the mountains deposited sediment and organic matter, including trees, other plants, and animals, giving rise to Arizona's Petrified Forest, a spectacular deposit of fossilized wood near present day Holbrook. Some of the major ore deposits in Arizona that are not associated with igneous activity, most notably the uranium deposits and some copper deposits, such as at the **Grandview Mine** in the Colorado Plateau, also formed during the Triassic and Jurassic.

About 175 million years ago Pangaea broke apart, North America began moving west, and subduction started along the west coast. The location of the subduction zone controlled the geologic evolution of Arizona for most of the next 200 million years and was responsible for a majority of the ore deposits in the state. From about 200 to 150 million years ago, the subduction was at the southern edge of Arizona. Igneous activity associated with the subduction formed the renowned **Bisbee** ore deposit as well as a few smaller gold, copper, and silver deposits in extreme southern Arizona.

From 150 to 80 million years ago, the subduction and igneous activity was south and west of Arizona, and no metallic ore deposits formed in the state during that time. However, beginning 80 million years ago and continuing until about 50 million years ago, the subduction zone and igneous activity moved back to southern Arizona. In the western United States, this period is referred to as "Laramide" and is characterized by widespread intrusive and volcanic igneous activity and the deposition of economically important ore deposits. This activity was generally focused in the Basin and Range and to a lesser degree in the Transition Zone. The tectonic activity during the Laramide produced a large number of metal-bearing igneous intrusions that are known as porphyry deposits. These copper deposits are so named because they usually form in association with an igneous intrusive stock with a porphyritic texture. Porphyry deposits are extremely important both from an economic perspective and because they are a source of many of the most spectacular Arizona mineral specimens. The best known porphyry copper deposits—**Inspiration**, **Christmas**, **Ray**, **Morenci**, **Mineral Park**, **Bagdad**, and **New Cornelia** mines—are spread across the southern third of the state and have yielded more than 25 million tons of copper and considerable amounts of gold, silver, and molybdenum. Other Laramide deposits include mineralized veins as well as replacement deposits in which carbonate rocks, especially the Paleozoic limestones, breccias, or some types of volcanic rocks have been replaced by sulfides. The

A sunset view of a series of benches in the massive Morenci open pit

The Clifton-Morenci porphyry deposit formed during the Laramide, was discovered in the second half of the nineteenth century, and is producing ore, lapidary material and specimens today.

Phelps Dodge photo

79, **Vekol**, **Old Dominion**, **Magma**, and **Tombstone** area mines are examples of other Laramide deposits.

After the Laramide there was a period of tectonic quiescence in Arizona. The west coast subduction zone had flattened out, and the activity had moved east into Colorado and New Mexico. By the mid-Tertiary, 35 to 15 million years ago, subduction was slowing, and Arizona was experiencing widespread igneous activity and crustal extension. Based on the ratios of the metals present, three types of mid-Tertiary vein deposits are significant: gold-rich deposits such as the **Oatman** area mines; silver-rich deposits such as the **Red Cloud** and North and South **Geronimo** mines; and base metal (copper, lead, and zinc) deposits such as those found at the **Grand Reef** and **Iron Cap** mines. Some copper-gold mineralization is also associated with mid-Tertiary low-angle normal faults as seen at the **Rawhide Mine** and the **Mystery** mine group (also known as the **Heson Mine**).

By fifteen million years ago, the San Andreas Fault had formed, and all of the subduction activity in Arizona had stopped. Extension continued with widespread normal faulting forming the Basin and Range. Extensive basalt eruptions are associated with the Basin and Range activity. This basalt covers almost a quarter of the state and hosts Arizona's famous peridot as well as a few zeolite occurrences.

The final stage in the geologic history of the ore deposits—the oxidation and enrichment of the earlier sulfide ores—took place over the last 10 million years. The sulfide ore deposits were subjected to what is known as supergene enrichment. Meteoric waters reacted with the primary sulfide ore minerals and dissolved the metals. These solutions formed new minerals and concentrated or localized the ores. Over a few million years, the surface water dissolved pyrite and other sulfides and leached the metals from the earth's surface. As these solutions descended from the surface, some of the metals precipitated as oxides, carbonates, sulfates, and a variety of other minerals. Some of the descending solutions eventually reached the water table and formed supergene blankets of enriched sulfide ores.

In Arizona the most significant supergene enrichment is associated with porphyry copper deposits. The state's copper deposits are remarkable not only for their economic importance but also because the oxidized zones above the porphyries are the sources of the colorful minerals for which the state is known among collectors. **Bisbee**, **Morenci**, **Ray** and **Bagdad** all have igneous rocks that originally had modest grades of sulfide mineralization; however, because of supergene enrichment and interaction with surrounding sedimentary deposits, these porphyries gave rise to giant copper mines. The Precambrian and mid-Tertiary deposits have been subjected to the same processes; thus, azurite and other oxidized minerals are found at the Precambrian **United Verde** and the other massive sulfide deposits. Most of the famous wulfenite and vanadinite occurrences in the state are the result of the oxidation of the mid-Tertiary silver-rich veins.

The long list of minerals found in Arizona reflects the state's complex geologic history. The geological processes concentrated the metals and made the minerals; man extracts that wealth, and occasionally someone preserves a mineral specimen—a geological souvenir, two billion years in the making.

Upper: Pyrite, 4 cm tall, from the 6th level of the 79 Mine, a Laramide deposit near Dripping Springs in Gila County
Collected circa 1970 by Frank Valenzuela
John Callahan collection

Lower: Secondary smithsonite, 5.4 cm tall, also from the 79 Mine
Marty Zinn collection

Both photos are by Jeff Scovil

CHAPTER 2

Prospects of the Pimería Alta

Above: Double bird pendant, 2.4 cm tall Hohokam culture (circa 750–950 CE)

Private collection
Jeff Scovil photo

Left: Arizona's Canyon de Chelly National Monument protects Anasazi [from the Navajo word for "Ancient Enemies"] ruins which were built into natural alcoves in the canyon walls and date to the mid-1300s. The Mummy House ruins, pictured here, were named for two mummified bodies that were found on the site. Earlier inhabitants of Canyon de Chelly built pit houses into the floor of the canyon. President Herbert Hoover authorized the National Monument in 1931 to protect the canyon as an archeological and scenic treasure.

Dean Hueber photo, April 2006

Some 12,000 years ago, the first humans migrated into the region that would become Arizona. Much of the northern half of North America was under a retreating glacier, the Laurentide Ice Sheet. Arizona was much wetter than it is today; the valleys were lush and provided a hospitable environment for Arizona's first citizens. The land was home to megafauna including mastodons and mammoths, as well as bears, horses, giant sloths and beavers, camels, and a variety of other animals. Tools, weapons and ornaments made from local materials including basalt, jasper, argillite, and obsidian are found in archeological sites dating to the Paleo-Indians, from whom it is estimated that ninety percent of Arizona's modern Native Americans have descended.

Copper has always dominated the Arizona landscape, and numerous Arizona archeological sites report caches of copper minerals including azurite, malachite, and turquoise. The early inhabitants of Arizona particularly prized sky-blue turquoise; it was mined and traded through the southwest and Mexico. Blake (1898) wrote of an extensive pre-Columbian mining operation on **Turquoise Mountain**, located about 20 miles north of present-day Kingman. These ancient mines were a series of benches and terraces cut into quartz veins, which contained gem-quality turquoise. The ancient workings were quite extensive, the labor-intensive mining methods a testament to the value of the gem. The Ancients mined turquoise by fire and quenching: The rock was heated with wood fires and then suddenly quenched with large volumes of water. The sudden temperature changes cracked the rock, which was then removed with stone hammers. These ancient mines have been since removed by twentieth-century mining at the **Mineral Park** open-pit copper mine.

There also is substantial evidence that there was prehistoric mining in what would become the **Cerro Colorado** mining district (see page 24) located 30 miles south of Tucson. The valleys formed by the Santa Cruz, Sonora, and San Miguel rivers provided natural routes for the original tribe members to move through the area. The terrain was less rigorous and the rivers ensured a water supply. The area probably produced necessities such as flint (tools and weapons), clay for pottery, coloring (ocher) and turquoise, malachite, chrysocolla, and jasper for ornamentation and adornment.

The Spanish conquest of Mexico in the 1520s created a sense of endless treasure of gold and silver to be discovered in the new lands of North

Turquoise: Stone of the Sky

When the Spanish conquered Mexico in the sixteenth century, they noted that the Aztecs coveted a blue-green stone that they called chalchihuitl. This stone was considered magic and was said to come from a far off land. The Aztec term actually referred to a number of different green-blue minerals, but the most prized of these was turquoise. Surviving turquoise jewelry has been shown to contain stones that originated in ancient New Mexico and Arizona; it was probably the first material mined in Arizona. Prehistoric mining sites have been identified at localities including Mineral Park in the northwest part of the state, Turquoise Mountain near Bisbee, and along the Salt River Canyon. In 1943 a burial was excavated at Ridge Ruin near Flagstaff, dating from about 1150 CE. It contained over 600 artifacts, including elaborate turquoise mosaics, lignite mirrors, and ceremonial objects such as wood wands with wing-shape turquoise inlaid sections. From prehistoric times to the 1870s, Native Americans fashioned turquoise into beads, pendants, mosaics, ceremonial objects and carvings. In the late 1870s, the Navajos began setting turquoise in silver. At that point, the Navajo and other Native tribes became and artisans, and turquoise jewelry became an icon of the southwest.

Upper: Turquoise pendant
1.8 cm wide
Hohokam culture (700–1200 CE)
Heber, Arizona

Lower: Turquoise and argillite
2.7 cm in diameter
Sinagua culture (1100–1400 CE)
Black Canyon, Arizona

Private collection
Jeff Scovil photos

Left: Slate Frog carving
4.9 cm tall
Hohokam culture (500–900 CE)
Salt River Valley, Arizona

Right: Quartz projectile points to 4 cm tall
Hohokam culture (circa 750–950 CE)

Private collection
Jeff Scovil photos

America. A few short decades after conquest, the fabulously rich silver deposits at Taxco, Pachuca, Zacatecas, Fresnillo, and Guanajuato were discovered, exciting the conquistadors' appreciation for the indigenous fables of gold and silver—even greater treasure certainly remained to be found.

In 1528 a Spanish exploration party leaving from Mexico became shipwrecked off the coast of Florida. For the next eight years, the five survivors of the ill-fated expedition made their way back to Mexico. Along the way, they crossed Texas and traveled near the present-day Arizona-Mexico border. The lost Spaniards heard stories of seven cities of gold to the north, which became know as the "Seven Cities of Cibola." These stories electrified the Spaniards in Mexico, and in 1539 Francisco Vázquez de Coronado (1510-1554), Governor of Nueva Galicia, sent Friar Marcos de Niza and Estevanico, a survivor of the earlier shipwreck, to find evidence of Cibola. When Niza returned, he claimed to have seen a city made of gold although he had not actually entered it, as the inhabitants were hostile and had killed Estevanico. Niza's account prompted Coronado to assemble more than 330 soldiers and 700 supporters and lead an expedition to claim the

city for Spain. Coronado left Mexico in February 1540 and crossed into Arizona along the San Pedro River. Although Coronado spent two years looking, he never found cities of gold or even traces of rich mines; nevertheless, legends of mineral treasure in Arizona were firmly entrenched in the minds of the Spanish.

The first documented discovery of an Arizona ore deposit came in 1583, when Spaniard Antonio de Espejo led a small military force from Mexico to rescue two missionary priests who had reportedly been captured by Indians in northern New Mexico. It has long been speculated that Espejo had a less heroic rationale for leading this force: he was granted a pardon for various crimes to lead the rescue party, and he hoped that he would discover a great silver strike similar to Zacatecas or Guanajuato, or maybe stumble onto "Cibola." What the Espejo party found as they entered Pimería Alta (the area that would become the Arizona Territory) was a harsh land: untamed desert, rugged mountains and valleys, and an indigenous population that had adapted to the conditions and lived a life unfamiliar to Espejo.

When the expedition reached the Hopi pueblos in northern Arizona, they learned of Native American mines to the south, in the present day Verde Valley. Espejo found a shallow surface working on what, centuries later, would become the **United Verde** deposit (see page 65). Espejo proclaimed the mine rich with silver, but Diego Pérez de Luxán, the chronicler of the expedition, thought that the ore was only copper and was therefore poor and uninteresting (Young 1964).

Part of the process of claiming the mine for the Spanish crown was marking it with a wooden cross. In 1876 a group of American prospectors claimed to have found the old mine workings around **Jerome** and, much to their surprise, an ancient wooden cross. While it is hard to believe that a wooden cross would survive almost 300 years in the Arizona sun, the Espejo expedition and the tale of the found cross have contributed to the mystique of the lost mines of the Spanish cross that dominates Arizona mining lore.

The copper deposit discovered by Espejo was visited by Marcos Farfán de los Godos in 1598. Farfán was commission by Juan de Oñate, the new governor of New Mexico, to find the mines described by Espejo. Although Farfán wrote in glowing terms about the potential for mining in the area, there is no documentation that Spanish mining ever occurred.

The Jesuits entered the area in 1691 and began building a series of missions, extending from northern Sonora, Mexico into southern Arizona and as far north as present-day Tucson. Father Eusebio Francisco Kino (kinoite is named for him) was one of the two founding fathers of these missions but did not have much interest in mineral wealth, although at the turn of the eighteenth century, he made note of the numerous mines along the Santa Cruz Valley, west of the Santa Rita Mountains and south of Tucson. Kino wrote, "In these new nations and new lands there are many good veins and mineral lands bearing gold and silver; and in the neighborhood and even in the sight of the mission some very good mining camps of very rich silver ores are now being established." Beyond these glowing accounts, there is no evidence of Spanish mining in Kino's time. Kino died in 1711, and the next recorded discovery was about 20 years later.

Right: Two unidentified men stand in an archeological dig in Verde Valley. Photographer Edgar Mearns reported (1884) that "a great deal of pottery and many stone implements" were unearthed during the excavation of these aboriginal ruins.

Edgar Alexander Mearns photo, 1884
Library of Congress Prints and Photographs collection

Above: A circa 1300 CE Pinedale Polychrome olla, 30.5 cm wide (repaired) recovered from an excavation in east central Arizona

Ken Horn collection
Jeff Scovil photo

Facing page: Images of the lush, green enclaves of the Verde Valley are not generally the first that come to mind when picturing Arizona, but these oases had been providing for the native population for centuries. This "Fern Grotto" is today protected as part of the Tonto Natural Bridge State Park in Payson.

Dennis Edson photo, June 2004

In 1736 Yaqui prospector Antonio Siraumea discovered a placer silver deposit west of present-day Nogales. The site was called *Arizonac*—the root of the name for the state of Arizona—and the discovery became known as *Planchas de Plata*, meaning "slabs of silver." This was a surface deposit of native silver nuggets occurring as float. Shallow pits were dug in the alluvium to find and recover the silver lumps, one of which reportedly weighed more than 2,500 pounds! The rush to Arizonac was short lived, but it did bring in a group of prospectors and miners who conducted small-scale mining operations throughout the eighteenth century. As has been the case with American prospectors 100 years later, these areas were subject to various native inhabitants trying to keep these intruders out of their lands. As Spanish military garrisons were established near the modern-day border of Arizona and Sonora, more prospecting and mining occurred.

In 1746 the Jesuit missionary Jacobo Sedelmayr, wrote of living in Pimería Alta and mentioned the potential for mining, "I do not speak of mines of gold and silver, for this is not my province. Nor do the Indians of the Gila and Colorado have such things in mind either, for theirs is no such greed. This none the less does not preclude the possibility of such riches being found in the regions." Near the area where Sedelmayr lived, there are sketchy Colonial records of gold mining, mentioning in particular gold mining at **Guevavi** in the 1760s. The O'odham Indian settlement of Guevavi, meaning "big spring," was the site of one of Father Kino's missions. It is about 15 miles north of Nogales near Mount Benedict in the Patagonia Mountains. Although it is clear that the Spanish mined and recovered minerals from Arizona, no specimens are known to have survived.

In 1766 an official expedition from Mexico City that included a Captain of the Royal Spanish Engineers, Nicolás de LaFora, visited southern Arizona. In LaFora's diary he noted that there were silver mines in the Huachuca Mountains and very productive mines west of Tubac in the **Arivaca** region (see page 24). In particular, LaFora wrote about a mine or series of mines he called **La Longorena**. This mine is lost to history but has long been rumored to be the modern **Ajax Mine**, a small property three

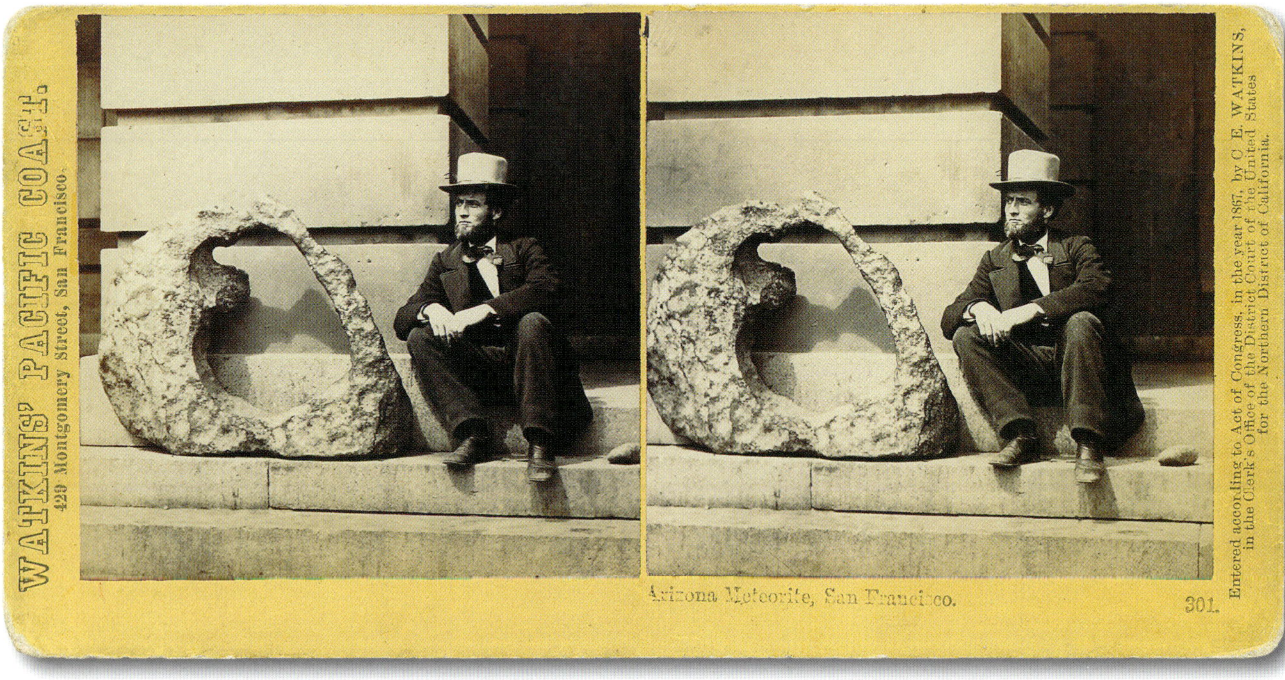

The Tucson meteorites have a long history: the first written record dates back to 1845. José Francisco Velasco, a Mexican government official, wrote from the state of Sonora, Mexico, that several virgin masses of iron were found in present-day Box Canyon in the Santa Rita Mountains near Tubac. Velasco also stated that a medium-size piece was selected and taken to the presidio at Tucson, where the blacksmith was using it as an anvil. He did not elaborate on the means whereby the 1,400-pound (688-kilogram) mass was transported the 25 or 30 miles to Tucson. Horses, or more likely burros, probably dragged the great iron across the desert over several grueling days.

The "Tucson meteorites" are two different masses. The first recovered and largest piece is known as the Ring for its distinctive oval ship. The second is an oblong mass that weighs 633 pounds (287 kilograms). It is commonly referred to as the Carleton Meteorite, named for the Civil War General who obtained it for the Smithsonian Institution. In the interum, the Carleton served as an anvil for a second Tucson blacksmith.

Both masses are chemically similar and are thus thought to be from the same fall. The meteorites are classified as an "ungrouped iron," which are chemically distinct from all other iron meteorites. Interestingly, these meteorites are 8 percent silica. They are the type "locality" for brezinaite (Cr_3S_4), which was first described as an inclusion in that silicate.

Both of the Tucson meteorites wound up at the Smithsonian Institution. The Ring was shipped to Washington, D.C., via San Francisco, where it was displayed on the steps of the Custom House before being freighted across the Isthmus of Panama to the east coast.

The above stereocard "Arizona Meteorite, San Francisco," was taken in San Francisco by acclaimed photographer Carleton Eugene Watkins (1829–1916) and is in the collection of Leonard Walle. When viewed through a stereoscope such as the handheld model below, the offset images on a stereocard provide the viewer with the illusion of depth of field.

–Mike Jensen

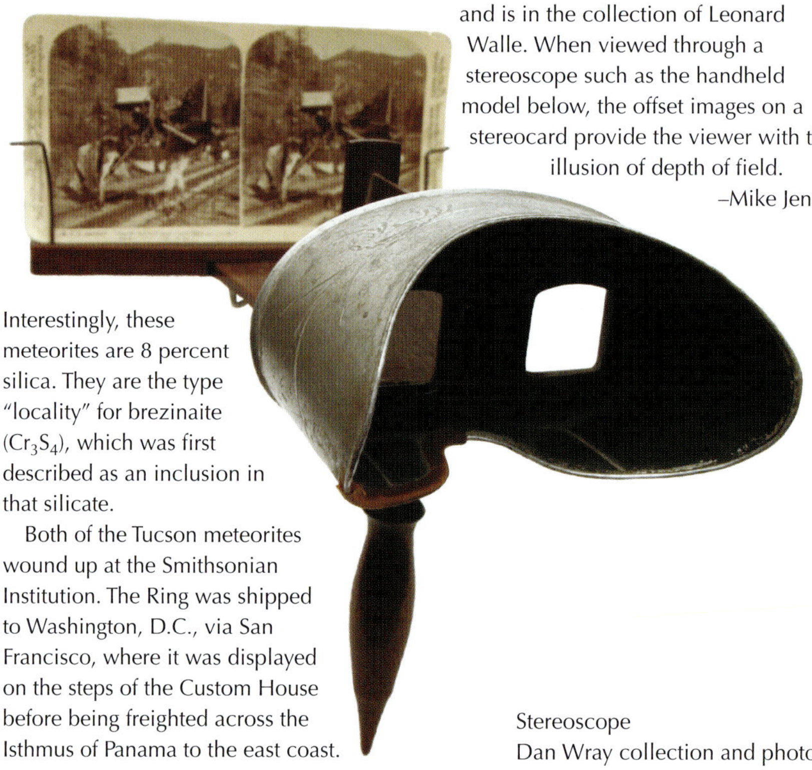

Stereoscope
Dan Wray collection and photo

miles south of the town of Arivaca.

In 1767, the King of Spain ordered the Jesuits out of the area, and one year later the Franciscans had moved in. The Apache continued attacking the various garrisons, which were continually undermanned, undersupplied, and underpaid. Tucson and the outposts to the south—Tumacácori, Guevavi, and Arivaca—were the only settlements that remained safe.

Arivaca must have missed the attention of the raiding Apaches because it was off the normal trails, and its rich silver mines made staying worth the risk to the miners. A 1777 report states,

> *There are many mines of very rich metals some twenty miles to the west in the vicinity of Arivaca. Three of these mines are especially productive. One yields eight ounces of pure silver to every twenty-five pounds of ore. A second yields forty-five ounces to every carga (100 pounds) of ore. The third mines yield a little less than this.*

Jesuit Priest Eusebio Francisco Kino established Tumacácori as a mission in January 1691. Established one day before nearby Guevavi, Tumacácori is the oldest mission site in Arizona and is today a National Historical Park. According to the National Park Service, the original mission was on the east side of the Santa Cruz River but was moved to the present site after the Pima Rebellion of 1751. The first church edifice was built on the new site. In 1800, the Franciscans began work on a structure to rival the Mission San Xavier del Bac, south of Tucson. But the project was slowed by lack of funds until 1821, when Father Juan Bautista Estelric sold 4,000 head of the mission's cattle to fund continued building. The church's bell tower was never completed.

Carlo Gentile photo, 1870
Library of Congress Prints and Photographs collection

There are other brief references to mining in Arizona in the late eighteenth and early nineteenth centuries. It is clear that mining was important in Colonial times, but there is scant physical evidence of the exact location of the mines. There are many fanciful claims of finding Spanish artifacts in mines including those at **Ajo** and in the **Tucson Mountains**, but these claims were probably more promotion than fact. Almost no records exist for the time period between when Mexico claimed its independence from Spain in 1821 until the Gadsden Purchase.

During the 1840s Arizona was caught in the American cry for "Manifest Destiny." With the end of Mexican-American War on February 2, 1948, and the signing of the Treaty of Hidalgo, the tract of land north of the Gila River that is modern-day Arizona became part of the United States. Mexico's ceding of northern Arizona was part of the much larger land transfer that included California, New Mexico, Colorado, Wyoming, and Utah. Although the transfer was extraordinary, the exact boundaries were left to surveyors to determine, and ultimately the United States and Mexico required a second purchase and treaty, the 1853 Gadsden Treaty, to define the modern boundaries of the state.

The massive new territory stretching from the Colorado River in the west to the 103rd meridian of longitude (west of Greenwich) in the east was called the New Mexico Territory; the western half was probably home to fewer than 200 "Americans" in 1855. There was tremendous skepticism as to whether the hot deserts of Arizona were a wise purchase. General William Tecumseh Sherman expressed the popular Washington, D.C. sentiment when he stated, "We had one war with Mexico to take Arizona, and we should have another to make her take it back." However, the California Gold Rush had captured the American psyche, bringing thousands of treasure seekers to California. Most of those would-be miners came by boat, but some came along a southern land route that took them through Tucson and followed the Gila River toward the Colorado River. Prospectors noted the potential for gold in Arizona but hurried toward the Mother Lode in the Sierras. Once the California goldfields began to be depleted, however, prospectors filtered back into Arizona.

CHAPTER 3:

The New Arizona Territory

Left: The Arivaca Ranch area, west-southwest of Tubac, is in the Basin and Range Province and was acquired from Mexico as part of the 1853 Gadsden Purchase. Herman Ehrenberg and Charles Debrille Poston discovered and mined rich silver deposits in this area.

Victor Tackett photo, June 2009

Below right: Gold nugget, 5.4 grams Collected June 1863 by E.H. Foster from La Paz Claim, La Paz County,

Smithsonian Institution collection (NMNH 55750, Catalogued Nov. 1884) Ken Larson photo

Below left: Gold nugget, 4.5 grams Collected July 1863 from Myer's claim, near Fort Yuma, Colorado River

Smithsonian Institution collection (NMNH 55749, Catalogued Nov. 1884) Ken Larson photo

As the California gold rush wound down, treasure hunters began to disperse across the west, some bound for Arizona. One member of the prospector diaspora was former Texas military man Jacob Snively (circa 1809–1871) who in 1858 discovered a gold placer in a bend in the Gila River about 20 miles east of Yuma. Snively's finds sparked a short-lived rush and the founding of Gila City, where nuggets to 22 ounces were reportedly found (Trimble 1989).

Well known trapper and guide Pauline Weaver (1797–1867) was responsible for a second major Arizona gold strike. Weaver first traveled west from Tennessee in the early 1830s. He roamed between California and New Mexico and in 1862 discovered gold about 50 miles north of Yuma on the Colorado River. Weaver followed up the discovery with trips into the surrounding mountains and made a significant find in a small wash, which he called El Arroyo de la Tenaja. Weaver filled a goose quill with gold flakes. He returned to Yuma with the gold and showed it to Don José Redondo, who raced to El Arollo and found several nuggets. A new rush ensued, and the mining town of La Paz was founded. During the next two years, miners recovered nearly 8 million dollars worth of gold around La Paz. The town's prominence was so great that it nearly became the territorial capital, but its fame was short-lived.

More than anything else, the search for mineral wealth defined Arizona in the period from 1855 to 1875. The 1860 census states that the population of Arizona Territory was 1,541 (Indians were not counted in the census), nearly all of them either directly or indirectly involved with mining. Integral with the founding of Arizona, men such as Charles Poston, Herman Ehrenberg, Frederick Brunckow, Samuel Heintzelman, and William Wrightson brought capital and a desire to exploit the riches that they knew were waiting in the new frontier.

But the Civil War (1861–1865) cast a pall over Arizona. When the war started and just as the discoveries of gold and silver were bringing in a rush of prospectors and investment, most of the federal troops were withdrawn from the territory. The indigenous tribes viewed the influx of whites as an invasion of their ancestral lands, but the presence of soldiers had kept them at bay and had kept the prospectors and supporting communities reasonably safe—no small feat given that there were only five forts in the entire territory: Fort Buchanan and Fort Breckinridge (also know as Fort or Camp

UTAH
NEVADA
CALIFORNIA
NEW MEXICO
MEXICO
SONORA
N
W
E
S
Colorado River
Colorado River
Fort Defiance
San Francisco Peaks
Flagstaff
Fort Mohave
Fort Whipple
Prescott
Walker mining district
Rich Hill
Weaver (Weaverville)
Bradshaw Mountains
LA PAZ
La Paz
Vulture Mine
Hassayampa River
Phoenix
Gila River
Castle Dome mining district
Hull Mine
Fort Yuma
Gila City
Yuma
SOUTHERN PACIFIC RAILROAD
GADSEN PURCHASE
Ajo
Fort Breckinridge
Fort Grant
Santa Cruz River
Tucson
Cerro Colorado mining district
Santa Rita Mountains
Fort Buchanan
Heintzelman Mine
Mowry (Patagonia) Mine
Tubac
Patagonia
Alto-Salero Mine
0 50 MILES
0 50 KILOMETERS

Fort Defiance, New Mexico (1851)
Seth Eastman, oil on canvas, 1873

The Army first used Fort Defiance as a base in 1851. By 1852 the fort boasted a number of pine-log and adobe structures built around a parade ground. For ten years the post housed elements of several regiments engaged in campaigns against the local Indians. The onset of the Civil War in 1861 caused the Army to abandon the installation, although it saw a flurry of activity when Kit Carson briefly occupied the fort in 1863.

The title of the painting places the fort in New Mexico, and during the fort's active service, it was in that territory. But the site became part of Arizona in 1863, when Arizona was given territorial status.

U.S. Senate Collection

Grant) in southern Arizona; Fort Mohave and Fort Yuma on the Colorado River; and Fort Defiance (pictured above) in the northeast. All were abandoned in 1861, and open warfare erupted between Anglos and the various tribes. The hostilities brought a halt to almost all of the mining activity in southern Arizona.

Both Indian and settler were guilty of atrocities, and in 1871 a group of vigilantes from Tucson murdered at least one hundred Apaches who had peacefully moved to a camp near Fort Grant on the San Pedro River. The massacre of Indian woman and children elicited outrage; still the non-Indian population demanded that the Apaches be moved to a reservation.

The bloodshed precipitated action in Washington, and President Grant ordered an immediate response with a two-pronged strategy: suing for a meaningful peace with all tribes, and should that fail, waging all-out war on individual tribes. In a few years, the Tonto Basin campaign led by General George Crook (1828–1890) had largely subjugated the Arizona tribes. Although there were numerous incidents after 1875, the stability and safety of the territory generally increased, allowing expansion of commerce and exploration of the last frontier in the contiguous United States.

Within one year after geologist Raphael Pumpelly was hired as the geologist for the Santa Rita Company mines, the Civil War broke out and the company, like others in the area, halted operations in southern Arizona. Pumpelly visited the Heintzelman Mine, some 40 miles west of the Santa Rita properties, to collect on a debt before leaving the area. He met with Charles Poston and accepted payment on the debt in the form of a load of ore worth $2000 per ton from the Heintzelman. Pumpelly left Arizona on April 24, 1861, with a wagon full of ore. In 1866 he became a Professor of Mining Sciences at Harvard University. He amassed a large collection of minerals that became part of the Harvard Mineral Museum when we left in 1875. The 3.2 cm wide silver on sphalerite pictured above was likely part of the Heintzelman load acquired in 1861 and is one of the oldest documented mineral specimens from Arizona.

Mark Hay collection
Ex Harvard University collection
Jeff Scovil photo

Cerro Colorado Mining District and the Heintzelman Mine

In 1854 Herman Ehrenberg (circa 1817–1866) and Charles Debrill Poston (1825–1902) traveled together through northern Sonora (Mexico) and the Gadsden Purchase to assess the silver-producing potential. Ehrenberg and Poston had met by chance in San Francisco in late 1853. While in southern Arizona, they collected samples, charted the region, and documented the mining potential of a number of prospects including the Ajo area and the old Spanish silver workings west of the Tumacácori mission in the Santa Cruz Valley near Tubac.

After returning from his journey, Poston headed east to secure capital for a mining operation in what would become the Cerro Colorado mining district, which he believed held promise. In 1856 in Cincinnati, Ohio, Poston, Ehrenberg, General Samuel Peter Heintzelman (1805–1880), firearms magnate Samuel Colt (1814–1862), and others organized the Sonora Exploring and Mining Company.

Heintzelman had led the fight to suppress the Yuma Indians and in the war's aftermath established Fort Yuma on the Colorado River. He met Ehrehberg and Poston when their expedition passed through Yuma. With Heintzelman at the company's helm, the Sonora Exploring and Mining Company purchased and began developing **Arivaca Ranch**, an expansive area west and south of Tubac. But southern Arizona was a dangerous place in the 1850s. The lack of security limited development of the mines until the United States built and manned Fort Buchanan in 1856 in the Santa Cruz Valley near present-day Sonoita.

Upper: Rare silver scrip, probably issued in about 1863 by the Arizona Mining Company (formerly the Sonora Exploring and Mining Company)

Holabird-Kagin Americana specimen

Lower: Charles Debrill Poston (1825–1902), "Father of Arizona" and first delegate from the Arizona Territory to the U.S. House of Representatives

U.S. War Department photo

On January 21, 1857, Sonora Exploring and Mining Company engineer Frederick Brunckow located rich silver ore near the Cerro Colorado district. He named the prospect the **Heintzelman Mine**. The silver vein was developed almost immediately, and by July 1857 a case of ore was shipped from the mine to the Cincinnati Merchant's Exchange. In September and October, more specimens were shipped to the Ohio Mechanics Institute's annual fair. Dana (1868) described the minerals stromeyerite and tetrahedrite from the Heintzelman. Although the written records of the minerals are clear, very few samples have been preserved from these early efforts at the Heintzelman Mine. Those described by Dana are in the collection of the Yale Peabody Museum, and Santa Rita Mining Company geologist Raphael Pumpelly saved at least one other (pictured facing page). The mines produced nearly $3,000 a day in silver until 1861 when the Civil War started and Fort Buchanan was abandoned.

The Illinois-born Charles Debrill Poston (1825–1902) has often been called the father of Arizona. In 1857 he became the *alcalde* (mayor) of Tubac and the supreme authority in the region. He printed his own money and presided over a company town of nearly 1,000 people. But almost immediately after troops left the region, constant Apache raids forced a curtailment in mining in the district that lasted for a decade. Poston left southern Arizona and joined General Heintzelman's staff in Washington, D.C. In an effort to stay afloat, Sonora reorganized in 1863 as the Arizona Mining Company, but was never again particularly successful.

While in Washington, Poston worked to gain territorial status for Arizona apart from that of New Mexico—he knew that Tubac was a distant concern of the territorial government in Santa Fe. To demonstrate the

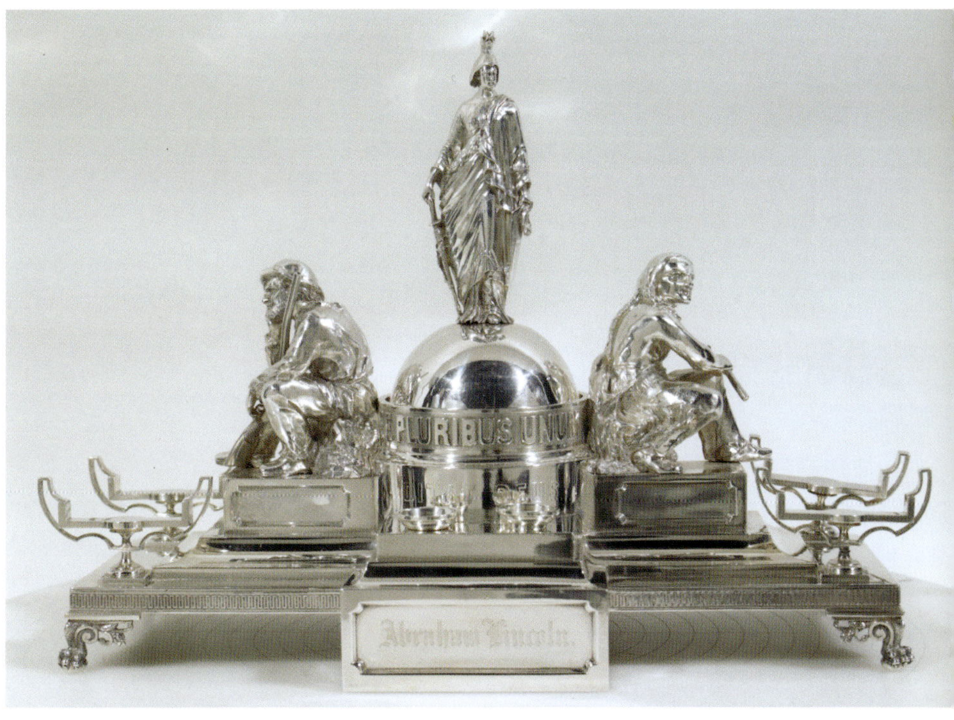

Right: Charles D. Poston presented this 32 cm wide inkwell, designed and cast by Tiffany & Co. of New York, to President Lincoln in early 1865. Made from more than 400 ounces of Arizona silver, the back is engraved with "From Charles D. Poston, Arizona 1863." A gift from Mary Lincoln Isham to the Library of Congress in 1937, the inkwell is today part of the Alfred Withal Stern Collection of Lincolniana in the Library of Congress Rare Book and Special Collections.

One of Arizona's early mining pioneers, Sylvester Mowry (1830–1871) was born in Rhode Island and served in the United States Army from 1852 to 1858. His service brought him to the southwest, where he gained an interest in prospecting. Mowry eventually purchased the Patagonia Silver Mine in southernmost Arizona in 1860, renaming the property the Mowry Mine. In 1862, he was arrested for selling lead to the Confederates. Mowry was ultimately cleared of the charges, but his mine languished during his imprisionment and was later destroyed.

Sylvester Mowry was the author of the 1863 book *The Geography and Resources of Arizona and Sonora*. This 1863 John Ross Browne drawing of Mowry appeared in Browne's book *Adventures in the Apache Country* (1863).

richness of Arizona's ore, Poston commissioned Tiffany & Company to craft a silver inkwell (pictured on page 25) from ore from his mines. He presented it to President Abraham Lincoln in 1863. Territorial status was granted on February 24, 1863, after the Union regained the region from the Confederacy, which had incorporated Arizona and New Mexico in 1861.

In the Santa Rita Mountains east of Tubac, the **Salero-Alto Mine** operated from 1857 until 1861. The Salero appears in colonial records in the 1700s, but it is impossible to gauge the amount of actual mining that was done. In 1857 the Santa Rita Company, organized by John Wrightson (for whom the highest peak in the Santa Rita Mountains is named), H.C. Grosvenor, and Gilbert Hopkins, bought the property from local ranchers and began mining. They hired Oswego, New York-born Raphael Pumpelly (1837–1923) to develop the property. The mine exploited quartz veins that had stringers of galena and minor silver sulfides. Unfortunately, the Salero-Alto mining camp was a favorite target of Apache raids, and with the exception of Pumpelly, who later became a well-known geologist and ultimately a Harvard professor, all the principals in the company had been killed by 1861. In later periods the Salero Mine and adjacent claims produced galena and sphalerite specimens that grace modern collections.

Mining activity in and around the Cerro Colorado district attracted others including Sylvester Mowry (1830–1871) to prospect in the area. Mowry moved to Tubac in 1859 after resigning from the Army. He bought the **Patagonia Mine** located southeast of Tubac in the Patagonia Mountains a few miles north of the Mexican border from Army officers whom he knew from Fort Buchanan. He renamed his mine the Mowry and built a smelter nearby—the slag heaps remain to this day.

The **Mowry Mine** was a silver-lead mine that exploited a series of limestone replacement deposits. In the late 1870s dozens of mines were operating near the Mowry. In later years lead from these mines was shipped to the Silver Belt (see page 31) and used in that district's silver smelters. In modern times, the mines around Patagonia have been a source for fine Japan-law twin quartz specimens (see page 219).

Mowry was a colorful character who had graduated from West Point and traveled throughout the west with the Army protecting the construction of the Pacific Railroad. While in the Army, he made an unfortunate enemy in the person of James Carleton. Although Mowry was from the north and was a Republican, Carleton, a ranking officer in the Army, arrested Mowry in 1862 as a southern sympathizer. Mowry stood accused of selling lead from his mine to Confederate militias for bullets. He was cleared in court, but the mine had been vandalized in his absence. He traveled to London to raise capital to restart mining, but died there in 1871. Although now a ghost town, Mowry, Arizona was repopulated and the mines revived for a few decade-long periods before and after the turn of the century.

The Castle Dome Mining District

The Castle Dome mining district, north of Yuma, is today part of the Kofa National Wildlife Refuge. The district is one of the older mining areas in the state, and in the 1860s the first miners found surface trenches and

Below: Major John Wesley Powell is a towering figure in American geologic history. Born in New York in 1834, Powell had a burning interest in natural history. In the late 1860s, he led a series of expeditions to the Rocky Mountains and decided that he could learn much about the region by running the length of the Green and Colorado Rivers. In 1869 Powell assembled a team of nine explorers and four boats and pushed off from a gentle landing on the Green River in Wyoming. Three months later six men and two boats exited the Grand Canyon at the mouth of the Virgin River (under water today at Lake Mead). The expedition was harrowing, with nearly 900 miles of rapids, and completely unknown terrain. What is more remarkable about this journey is that Powell had lost an arm in the Civil War Battle of Shiloh. However, he was the lead rower and navigator for the entire river trip. Major Powell repeated the journey in 1871/1872. In 1875 Powell published a book based on his trips entitled *The Exploration of the Colorado River and its Canyons*.

The left half of a John Karl Hillers 1872 stereographic card photo of John Wesley Powell's boat (with chair attached) on the banks of the Colorado River.

Library of Congress Prints and Photographs collection

shallow workings that had long predated their arrival. Prospectors Conner and Jacob Snively, who had earlier sparked Arizona's first gold rush, hauled some of the silver-colored ore to the Colorado River in 1863. The ore was not the hoped-for silver but was galena; however, the galena packed enough silver to hold mining interest; thus began the mining history of the district.

Ore was wrested from the ground under the southwestern desert's insane conditions—hot (well over one hundred degrees Fahrenheit for much of the year) and dry. Development of Castle Dome was aided when the Southern Pacific Railroad made it across Arizona in the 1880s, and the rudimentary smelter in Yuma was no longer needed.

In fact until the opening of the Southern Pacific, transportation difficulties limited mining throughout Arizona. Before then any goods, including ore, had to be packed to the Colorado River. Freight costs were clearly an issue, with cargo on the Colorado River costing $65 per ton and another $280 per ton from the mine to the river. Only the richest ores could be handled: it took three times as long to transport the ores across the southern Arizona desert as it took to mine, sort, and bag it.

The Castle Dome mining district consisted of many small mines, none of which was particularly large or significantly deep. Most were high, dry, and shallow—mineral collector's paradises. The **Rialto Mine**, later called the **Hull** was one of the largest mines. Most of the Rialto's ore was produced prior to 1900.

The Castle Dome district is primarily an alluvial area with north-northwest mineralized faults cutting Mesozoic-age shale and limestone. Shafts through the alluvium provide access to many of the mine workings. There are also dikes of diorite porphyry cutting Mesozoic sediments and volcanic rocks. The north-northwest-trending argentiferous galena-bearing veins, rich in fluorite and barite, dip steeply to vertical fault zones cutting Jurassic shales, limestones, and diorite porphyry dikes. The ore shoots are best developed where diorite porphyry forms one or both of the vein walls. Channels and vugs in the ore contain hydrozincite, smithsonite, anglesite, wulfenite, vanadinite, mimetite, quartz, calcite, and aragonite.

There does not appear to have been much in the way of specimen preservation during this period. It's no wonder—most of a miner's time would have been spent trying to preserve his hide, not his mineral specimens. Still, in 1866 William P. Blake described cerussite as occurring, "in heavy encrusting masses upon the galena of the Castle Dome District … fluorspar (fluorite) … in crystals and large cleavable masses of various tints—white, pink, and purple and green, like the specimens from Derbyshire, England, in the veins of galena and blende ..." And in 1873, as reported in the *Mineralogy of Arizona* (1995), a man named Brush described a mineral "associated with some specimens of galena from Castle Dome District, Arizona, there is found a compact banded mineral, very much resembling some of the varieties of wood-tin from Cornwall … compact anglesite."

It is thus documented that a few hearty individuals, possibly driven by money, visited the district even at the risk of dying in the wasteland. Even more collecting was noted between 1876 and 1900, as more and more attention was focused on the minerals in the Arizona Territory.

Ajo

Traveling in the region from 1691 to 1702, Father Kino was probably the first Spaniard to see the Ajo area, which was then northern Sonora, Mexico. He kept detailed logs of his travels, but he made no mention of encountering or initiating mining at Ajo; thus, mining likely began sometime after 1702.

Historians surmise that Ajo was the first copper district in Arizona to have been worked by Europeans. When the first American explorers arrived in the area in 1854, they found abandoned workings, rawhide ore buckets, and crude tools from the attempts to mine small veins of copper, cuprite, and chalcocite in the three green hills that stood where the open pit is today. Those first workings were most likely the efforts of Mexican miners from Sonora who were looking for silver ore (Thomas and Gibbs 1983).

Interest in the mineral potential of Ajo was probably generated from the reports of U.S. Army troops and surveying parties working in the southern part of Arizona after the Mexican War of 1848 and the subsequent Gadsden Purchase. It was certainly fueled by the rich masses of chalcocite (initially thought to be silver glance), cuprite, copper, and malachite that were carried to California. In 1854, organized mining American style had come to Ajo with Major Robert Allen as president and J. Downer Wilson as secretary and treasurer of the newly formed Arizona Mining and Trading Company.

No matter how rich the copper mineralization was, the expense of transporting it to smelters in Swansea, Wales, and the cost of returning supplies proved excessive. Lack of water was also a limiting factor. Although Ajo would become one of Arizona's great copper mines (see page 148), it would remain a footnote until the turn of the twentieth century.

The Bradshaw Mountains

Near Prescott, a belt of Precambrian rocks is exposed in a series of mountain blocks. These small ranges, with names such as the Weaver Mountains and the Sierra Prieta, are physiographically located in the Transition Zone. The primary rocks are granites and greenstones, the latter host small amounts of gold that were locally concentrated by hydrothermal activity during a period of intense intrusion of mafic dikes, approximately 1.1 billion years ago. Erosion of the rocks around the dikes created Arizona's richest placer gold deposits, targets of Arizona's greatest gold rush.

After his success at La Paz, mountain man Pauline Weaver set his sights on the gold prospects in the interior of Arizona. But moving into the territory held by Native Americans was problematic even for Weaver, who was half Cherokee. In 1863 Weaver organized talks with the leaders of a number of Arizona Indian tribes. After securing their confidence, Weaver met with a group of veteran California prospectors led by Abraham Peeples. Peeples outfitted an expedition of 31 men and headed toward the San Francisco Peaks (present-day Flagstaff). The expedition worked its way up the drainage of the Hassayampa River, which marks the southern boundary of the Bradshaw Mountains. Setting up camp at the base of a boulder-strewn hill, the prospectors found good signs of gold. In their excitement, however, they neglected their pack animals, which wandered away. The next morning

Above: "Juan Chivaria [sic] the most distinguished chief in Arizona" poses in 1870 for a Carlo Gentile portrait. In 1863 Maricopa Indian Chief Juan Chivari had volunteered to help Rich Hill discoverer Abraham Peebles avenge the theft of 29 horses and mules by a band of Apaches. Chief Chivari and some of his men joined the party in pursuit of the Apaches, offering guidance and provisions and ultimately leading a hasty retreat, as the Apaches out numbered the pursuing posse. For the duration of the battle, however, the Apaches were out gunned and suffered many casualties (McClintock 1916).

Library of Congress Prints and Photographs Division

Facing page: Our oldest images of the southwest are preserved in grayscale and sepia tones, which can potentially lull one into forgetting that the even the earliest explorers and settlers of Arizona were daily witnesses to the pandemonium of color and texture that is Arizona. This was the scene in Organ Pipe National Monument, near Ajo, in Feb 2005.

Dean Hueber photo

members of the party searched for them at the top of the hill and were stunned to find an extraordinary placer deposit. The prospectors reported nuggets the size of potatoes scattered about the outcrop that would become know as **Rich Hill**.

The Peeples party quickly collected all of the available surface gold and sent Weaver to acquire additional supplies. News of the discovery spread rapidly, and a major gold rush was on. Within a few months, a tent city of 1,500 people had sprung up at the base of Rich Hill. The city was called Weaverville (later shortened to Weaver) in honor of the guide.

Henry Wickenburg and Jacob Waltz were notable members of the Peeples party. Wickenburg eventually split off from the main group and discovered one of the most famous lode claims in Arizona, the **Vulture** gold mine. Waltz, a German immigrant who came to the United States in 1839, drifted to the gold camps of California, where he met Peeples and apparently acquired the nickname "Dutchman" (from *Deutsch* man or "German"). Waltz worked in the Bradshaw Mountains for a decade before moving to the Phoenix area. He continued to prospect, evidently with some success because he paid his debts in gold. However, the source of his gold was never revealed, giving birth to the legend of the **Lost Dutchman Mine**. Waltz's gold may well have been high-grade ore from various mines in the Bradshaw district.

Almost simultaneously with the Peeples party's discovery of Rich Hill, another prospecting party led by Joe Walker was exploring the headwaters of the Hassayampa River, about 30 miles north. Walker was a frontiersman not unlike Weaver, but Walker had a strong promoter streak. He convinced 18 Californians to pull up stakes essentially overnight and journey across the southwest. Legend has it that they were avoiding the Civil War. Whatever the reason, they left Keyesville, California, in 1861, crossed Death Valley, prospected in Colorado, and then traveled through New Mexico. Two years after leaving Keyesville, they discovered rich placers along the **Lynx**, **Big Bug**, and **Turkey** creeks (Coggin 1987) in the Bradshaw Mountains. The Walker mining district was officially established on May 10, 1863.

The goldfields caught the attention of the Union Army, which needed gold to help finance the Civil War. Fort Whipple was established to protect the miners and discourage Confederate adventurism. The town of Prescott grew up around the fort and became the first territorial capital of Arizona.

Henry Wickenburg discovered the first major gold lode. Wickenburg noticed a large outcrop of gold-bearing quartz along the Hassayampa River, and there he founded the **Vulture Mine**, named for the carrion feeders often circling overhead (Wilson, Cunningham and Butler 1934). After working the mine for three years, Wickenburg sold the property to the Vulture Company of New York, headed by Benjamin Phelps. Phelps paid Wickenburg $20,000 in cash and a promissory note for $65,000. Wickenburg tried to collect on the note for the rest of his life, but he never received a dime. The Vulture Company built a 40 stamp mill, and the mine operated profitably from 1867 to 1872. While the ore in the original mine began to decline in grade in the 1870s, nearby discoveries were consolidated into a single, large-scale operation.

CHAPTER 4

The Silver Bonanza

Above: Constructing the railroad bridge across the Colorado River at Yuma
Enoch Conklin photo, 1877

Library of Congress Prints and Photographs collection

Left: The Silver King Mine was one of Arizona's great bonanza silver mines and a source of fine silver specimens. Although the town of Silver King, in the Pioneer District, was short-lived and completely abandoned by 1900, fine crystallized specimens such as this one in the Les and Paula Presmyk collection speak to the richness of the ores. The specimen is 6.4 cm tall.

Jeff Scovil photo

With the resolution of the Civil War and General Crook's successful 1872 to 1873 campaign to round up errant bands of Indians, the nation was positioned to focus on discovering and taming the Wild West. Lack of transportation hindered development until 1881, when the Southern Pacific Railroad opened, and the interior of Arizona was connected to the transcontinental rail system. Communities built connections to the main railway; equipment and goods began flowing into Arizona; and minerals, cattle, and cotton began flowing out. The railroad changed the Arizona landscape, allowing exploration and mining to flourish. From the early 1870s until the mid-1880s nearly every important Arizona mineral deposit was discovered. By 1876 there were more than 11,000 mining claims in Arizona.

Gold still dominated the imaginations of the prospectors, and even as gold rushes on the Colorado River played out, a few new gold camps were springing up. The mines at Schultz (see page 48), later known as **Tiger** (see page 127), ultimately mined for copper, began as gold prospects.

Greaterville, located about 30 miles south of Tucson along the eastern flank of the Santa Rita Mountains, was another 1870s gold camp. Although Greaterville's total gold production was only 30,000 ounces, it is still a popular destination for modern gold-nugget hunters, and most of the stream drainage channels are under claim. The camp takes its name from the village of Greaterville, founded in 1873 by a settler named Greater (Harnack 2000). In 1874 placer gold was discovered in a dry wash near the village. The find started a rush, and by the early 1875, the town boasted a population of nearly 500. Greaterville was the quintessential boom-camp. "Several dance halls, saloon, and stores filled most of the miners' needs. ... The jail was a round hole dug in the ground, into which offenders were lowered by means of a rope" (Sherman and Sherman 1969).

Although gold was king in the 1860s and the early 1870s, the territory was realizing fortunes in silver by the late 1870s, and for a decade silver reigned in Arizona.

The Silver Belt

The Arizona Silver Belt, one of the state's most important silver mining regions, is in southern Gila and northern Pinal counties, around the town of Globe. The belt produced both specimens and millions of dollars of ore in the late 1870s and early 1880s. The **Silver King** is the most famous mine in

Right: Silver, 6.6 cm wide
Silver King Mine, Pinal County
Paul Harter collection
Jeff Scovil photo

Above: Silver, 8.7 cm tall
Silver King Mine, Pinal County
Dick Morris collection
Jeff Scovil photo

the belt. The mineralization associated with the mine was discovered in 1871 when Colonel George Stoneman, commander of the new military department in the Arizona Territory, began developing a system of roads to connect the Army's forts and thus facilitate the Indian wars. Although the soldiers constructing the road were not knowledgeable prospectors, all were aware that Arizona was a land of riches, and they were on the lookout for "mother lode," which would make their military service worth the hardship. While working near the base of an escarpment (later named Apache Leap), a soldier by the name of Sullivan noticed some unusual, dark, heavy rocks. He picked up the samples but said nothing of his discovery. After mustering out of the Army a few months later, Sullivan went to work at Charles Mason's ranch near the town of Florence. He showed his samples, later identified as native silver (Short et al. 1943), to Mason, but Sullivan only told Mason that he had found the samples near the foot of Stoneman Grade. Sullivan disappeared shortly thereafter and was presumably killed by Apaches.

In 1874 Charles Mason interested three other ranchers in forming a prospecting party, and the four headed east toward Stoneman Grade. As they neared the base of the grade, they discovered a rich outcrop, which they claimed, naming it **Silver Queen**. Continuing to prospect, they discovered several other deposits, including the **Globe Claim**, about 20 miles to the east. While meeting with some success, the foursome was not convinced that they had found Sullivan's lost silver outcrop.

On March 22 of the following year while returning from their Globe claim with a pack train loaded with samples, the team of Charles Mason, Benjamin W. Reagan, William H. Long, and Isaac Copeland camped near the base of the Stoneman Grade, where they found more of the black nuggets (Short et al. 1943). They followed the float up to a small conical hill and made the initial location and claim on the Silver King deposit. They began to develop the Silver King Mine while also beginning work on the **Irene** and **Hub** claims in the Silver Queen area. The Irene and Hub were later renamed the **Gem and Hastings Group**, which was incorporated into the **Magma Mine** (see page 201) after the turn of the century.

At the Silver King, the Mason party systematically removed the surface of the hill where the vein outcropped. The ranchers turned prospectors were far more enthusiastic than they were knowledgeable; they dutifully hand sorted the ore, keeping the galena and throwing away the silver. Their first shipment to San Francisco lost them $1,200. When four experienced Comstock miners happened into the area and saw what the owners had done, they negotiated a deal to work the dump for equal shares. After setting up a small concentrator and stamp mill at Pinal City near Queen Creek, one of the miners took the concentrate to San Francisco and brought back a check for $50,000!

By the end of 1875, the Silver King Mine consisted of a 42-foot deep shaft with a 12-foot drift at the bottom. The shaft was centered on a vein of high-grade ore, which included chlorargyrite, acanthite, and native silver. The sorted ore assayed at $2,000 per ton; this 54 feet of shaft and drift produced ore valued at $50,000. As the mine developed, a small smelter was erected in Florence to further refine the ore and thus reduce the cost of shipping the ore to San Francisco. Interestingly the lead used in the smelter was purchased from the **Mowry Mine** (see page 26) smelter (Mowry himself had died 4 years earlier), 150 miles to the south (Short et al. 1943).

Acanthite on calcite, 4.7 cm tall
Silver King Mine, Pinal County

Les and Paula Presmyk collection
Jeff Scovil photo

Convinced the mine was in its last gasp, two of the original four partners had sold their stakes to the other two by June of 1876. In January 1877, Mason sold out to a Yuma merchant. Finally in May 1877, San Francisco and Comstock interests purchased the mine, incorporating it as the Silver King Mining Company.

In 1879 Arthur Macy was appointed Superintendent, and by 1882 all of the free milling ore had been exhausted. The Silver King Mine had reached a depth of 830 feet on a body of rich silver ore by 1883, but the grade of the ore was declining (Ransome 1903). By 1884 the shaft was below 950 feet and most of the ore was being mined from the 830 and 950 levels. The mine's last year of profitable operation was 1887. Enough ore had been blocked out to allow the mine to run until January 1889, but only with an assessment requested of the stockholders. After operating intermittently from 1889 to 1896, the mine closed for good in 1896. The shaft had reached a depth of 1,150 feet. The mine's total production had exceeded $6.5 million, with silver accounting for more than $5.9 million (Short et al. 1943).

The Silver King Mine is situated in a Laramide-age quartz diorite porphyry with a stockwork of silver-bearing veins surrounding an irregular, mostly barren core of white quartz. The conical hill the Mason team originally staked is about 65 to 115 feet above the surrounding surface and is a stockwork in the porphyry. The quartz veins and veinlets contained galena, sphalerite, tetrahedrite, and chalcopyrite with native silver, acanthite, stromeyerite, and proustite. Chalcocite, covellite, cuprite, and bornite were present in smaller amounts. Pyrite was present but only sparingly. Sphalerite

Silver King Mining Company Stock Certificate

In 1882 mining geologist William Phipps Blake visited the Silver King Mine. In his report (unpublished) he literally gushed about the native silver:

> *In the filamentous form, coarse and fine, very white, and striated as made up of bundles of fine wire, and ranging in size from the size of a pin or knitting needle to masses half an inch through, but solid and much twisted and gnarled ... in long wires, and in threads as fine as a hair or silk filling cavities or branching from the coarse wires. The filaments are often subdivided toward the extremity, and ramify like the tendrils of a vine, and they frequently bear aloft on their ends detached and beautiful crystals of calcite, the silver wires penetrating the crystals as if they were beads.*

The stock certificate pictured here dates to the mine's heyday in 1878.

Dick Morris collection

was the most abundant sulfide, especially in the lower levels of the mine, but no known specimens exist. Galena was the next most abundant sulfide but interestingly, did not contain silver. Chalcopyrite and tetrahedrite were the most common copper minerals; the tetrahedrite contained some silver. Bornite was not as prevalent as the other copper minerals but nearly always occurred in association with chalcopyrite. Stromeyerite occurred mainly on the third and fourth levels of the mine. Azurite and malachite occurred in small pockets in quartz near the surface in the open pit. Native silver was especially prevalent in the upper levels of the mine (Short et al. 1943)

Crystallized silver from the Silver King Mine occurs in a variety of forms, ranging from arborescent clusters, to spinel-law twin crystals up to 10 centimeters long (Robert Peele specimen), to wires up to 90 centimeters long (used by the mine superintendent as a hatband), although the wires are most commonly less than 2.5 centimeters long. If present, the matrix is generally a white to tan calcite.

Although just 30 miles east and north of the Silver King Mine, the silver deposits in the Richmond Basin and around McMillenville were much more remote and difficult to operate, particularly prior to General Crook's campaign to contain Apache aggression. But relative calm and word of the Silver King discovery enticed prospectors into the region.

The rich, but short-lived **Stonewall Jackson Mine** exploited an epithermal vein system. The ore was native silver and silver chloride minerals. An extremely rich pay zone within the vein consisted of virtually pure native silver, 2 to 5 centimeters wide. The mine produced silver valued at two million dollars during its short life (Dunning 1959). The shaft was over 800 feet deep, with less than 900 feet of lateral workings.

While a number of the mineral specimens from Arizona's Territorial days are well documented, specific locality information has been kept with relatively few of Arizona's silver specimens. Collectors have visually matched features such as matrix and crystal habit to the few verifiable reference specimens available in order to ascertain the locality from which their specimens originated. However, because any deposit can produce a single species showing a surprising range of colors, crystal habits, and other physical characteristics, such attributions are imprecise at best. Isotope or trace element analysis can provide strong clues to a silver specimen's locality, but not without comparison with data from a verifiable reference sample. One of very few such reference samples from the Stonewall Jackson Mine in present-day Gila County is pictured **below**. The label affixed to the back of the specimen (**left**) reads, "Silver, Stonewall Jackson Mine, Maricopa Cty. A.T." The front of the same 4.7 cm tall specimen is pictured on the **right** with a label indicating that it was traded out of the Territorial Museum in 1907.

Dick Morris collection
Jeff Scovil photo

Legend has it that prospector Charlie McMillen discovered the deposit in 1874. He flipped the property for $120,000, and rumors were soon flying of buyers who would have paid as much $300,000 for the mine. In the winter of 1877, author Enoch Conklin asked the prospector about his decision to sell the property. McMillen's response was published in Conklin's 1878 book *Picturesque Arizona*:

> *... a thing in my estimation, has no real—no intrinsic value. It has only a comparative one, and is governed entirely by the relative value of the things surrounding, or immediately associated with it. ... I had been roving about this country and in these mountains ever since '55, when I struck this little affair up here that we are talking about. ... I had put my foot on some others I say, and better ones. But I thought this little one would do to raise some money on to work the rest. You see I was broke—dead broke. Couldn't get trusted for an onion or a slice of bacon ... I wanted money to develop and open up my other mines; and I would have taken ... half the amount I got, if I couldn't have got what I did ... We are just sending down fifty thousand dollars worth of machinery to commence on—my partner, Mr. Flournoy and myself. We are now at work on the 'Hannibal.'*

Indeed other veins in the immediate area include the **Hannibal**, **Washington**, **Centennial**, **Virginias**, **Robert E. Lee**, **Henry Clay**, **San Francisco**, all of which were extensions of the Stonewall Jackson vein. The **Democrat** and **Little Mack** were, according to an Arizona Department of Mines and Mineral Resources report, "on a spur which runs into the Stonewall vein ... both produced very rich native silver ore, the total yield being estimated at $85,000." According to that same report, northwest of the

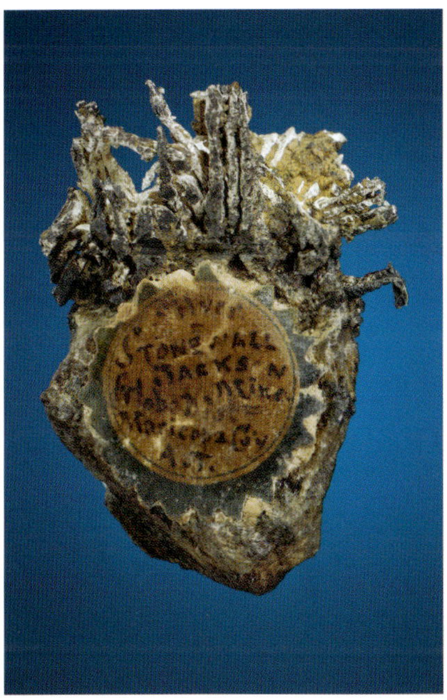

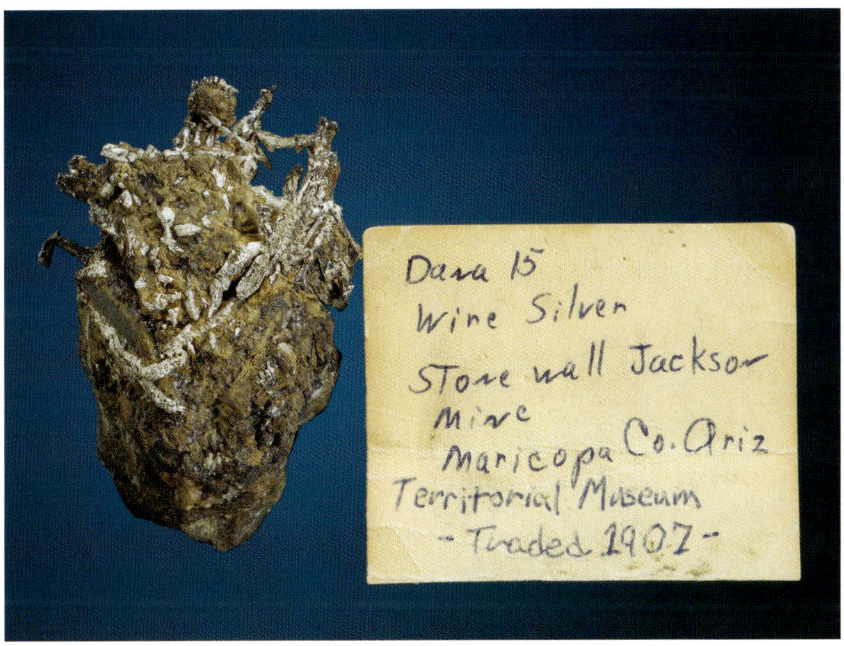

Silver with Minor Chlorargyrite
8 cm wide
Attributed to the Stonewall Jackson Mine, Gila County
Evan Jones collection
Jeff Scovil photo

Stonewall Jackson Mine, the North Star "shows a ledge from 3 to 8 feet wide, and has produced ore that has gone as high as $1,000 per ton. The mine is opened by a 120-foot tunnel and a 60-foot shaft."

In 1882 McMillenville was the site of a bloody skirmish between a band of Apaches aligned with Geronimo. The Apaches had escaped from the San Carlos Reservation and had raided numerous sites in the area before targeting McMillenville on July 7. The town had some warning that the Indians were on their way, and the women and children hid in the Stonewall Jackson adit. A pitched battle ensued, with the Apaches eventually leaving the area. The miners won the battle, but the ore had already played out, and McMillenville was soon abandoned.

The Stonewall Jackson Mine produced some of the Arizona Territory's finest silver specimens. Crystallized silver from the mine exhibits arborescent crystals, which show minor pitting. The crystals can be sharp and at least one specimen exhibits elongated spinel-law twins. Specimens are generally flat, consisting of masses of arborescent crystals. The largest known example was a 1,000-gram mass of arborescent crystals, 5 by 7 by 17 centimeters. The piece was divided into several specimens, the largest of which is 9 centimeters wide. Notably an 8-centimeter-wide specimen in Evan Jones's collection has a bit of associated chlorargyrite at the base of the crystals.

Richmond Basin is 6 miles west of McMillenville. The deposits are in Precambrian-age Pinal schist. The richest mineralization occurred in altered diorite and tight fine-grained, banded quartz associated with an east-west fault in the **Mack Morris** (aka **MacMorris**) and **LaPlata** mines, the former being the largest mine in the basin. The **Nugget Mine**, about a mile and a half southwest of Richmond Basin, was also well known at this time.

The Richmond Basin mineralization consists of native silver, argentite, acanthite, stromeyerite, chlorargyrite, and tennantite in masses and thin sheets through quartz. There is no calcite associated with the mineralization. Finely divided pyrite and chalcopyrite accompany the silver minerals.

A few silver specimens from the Richmond Basin are preserved in modern collections. These specimens are distinctive in that the silver is not crystallized; it has instead formed as nodules and masses. Hodge (1877) noted that masses of almost pure silver nuggets, weighing from several pounds to more than 500 pounds were found in the Richmond Flat area. This area is marked "Nugget Mesa" on the USGS topographic map. In *The Mineral Collector* (4:5), Maynard Bixby reported on an 1880s trip to the area,

> *At Richmond Basin, about 20 miles distant* [from Globe], *rare specimens of crystallized silver in rolled nuggets have been found loose in the soil. I obtained one weighing several pounds and a larger piece weighing about 30 lbs. was taken out several years ago. Placer silver is more abundant at this camp than at any place I have known, and in fact these nuggets are the largest and rarest of this metal in the United States. Fluorite occurs here in large purple crystals on the dump of one of the prospects.*

In his report to the governor in 1909, William Blake described a large, nearly pure silver nugget taken north of **Globe** in 1895. The 448-ounce lump now resides in the Smithsonian Institution; it came from an extension of the Richmond Basin. Blake went on to say that the nugget had lost all of

Above: The Apache warrior Geronimo is shown here in an1888 postcard. The photo was cropped from a larger image of four members of the band and was taken after their surrender to General George Crook in March, 1886. Geronimo escaped the following day.

Camillus Sydney "C.S." Fly photo

Right: Acanthite on quartz, 5.5 cm wide
Globe district, Gila County
Formerly in the Earl L. Calvert collection

Evan Jones collection
Jeff Scovil photo

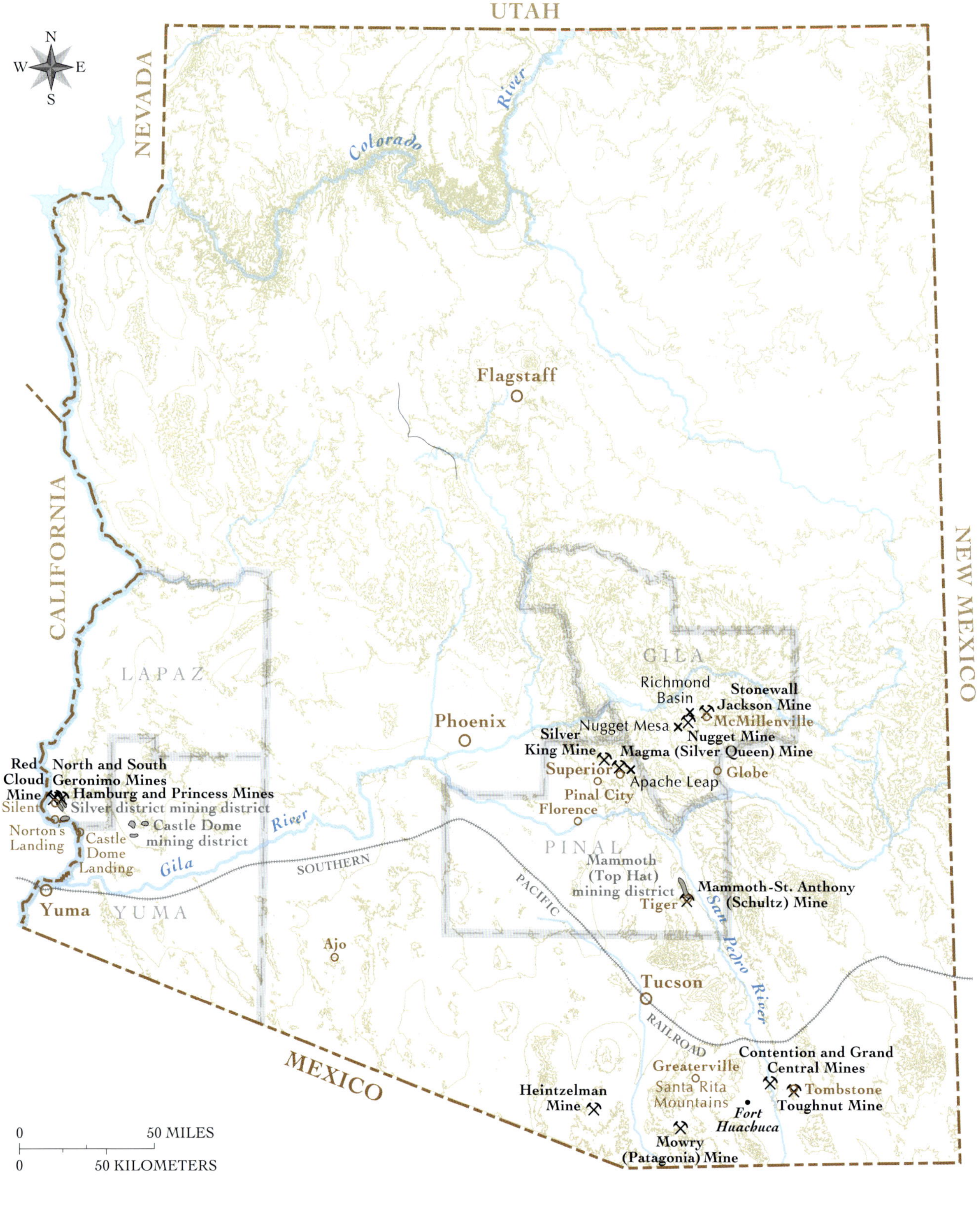
UTAH
NEVADA
CALIFORNIA
NEW MEXICO
MEXICO
N
S
E
W
Colorado River
Flagstaff
Phoenix
GILA
LAPAZ
PINAL
YUMA
Richmond Basin
Stonewall Jackson Mine
McMillenville
Nugget Mesa
Nugget Mine
Silver King Mine
Magma (Silver Queen) Mine
Superior
Apache Leap
Globe
Pinal City
Florence
Red Cloud Mine
North and South Geronimo Mines
Hamburg and Princess Mines
Silent
Silver district mining district
Castle Dome mining district
Norton's Landing
Castle Dome Landing
Gila River
SOUTHERN PACIFIC RAILROAD
Yuma
Mammoth (Top Hat) mining district
Tiger
Mammoth-St. Anthony (Schultz) Mine
San Pedro River
Ajo
Tucson
Contention and Grand Central Mines
Greaterville
Santa Rita Mountains
Tombstone
Toughnut Mine
Heintzelman Mine
Fort Huachuca
Mowry (Patagonia) Mine
0 50 MILES
0 50 KILOMETERS

its surface definition and was fully rounded by attrition.

Unfortunately, the mineralization in Richmond Basin was not extensive enough to be profitably worked for very long. Within a few years, the mines had played out. Several entities attempted to develop and mine the area over the years, but none were successful. The underground workings have long since caved in and are inaccessible. Area mine dumps occasionally yield pieces of tarnished native silver, but even the recent surge of metal detector-armed collectors has not produced many additional nuggets of silver.

Below left: Silver nugget, 13 cm tall

Les and Paula Presmyk collection
Jesse La Plante photo

Below right: An undated photo of the "31-pound" silver nugget found near Globe, which is today on display at the Smithsonian Institution in Washington, D.C. According to Paul Pohwat, mineral collection manager at the Smithsonian, the actual weight of the nugget is 448 ounces (28 pounds).

Sammy Jones Photograph Collection
Arizona State University Library collection

Left: Arizona's Major Silver Bonanza Mines and Districts

Tombstone

In 1877 in the rugged foothills east of San Pedro River, legendary scout and prospector Ed Schieffelin discovered Arizona's most infamous mining district. The soldiers stationed at nearby Fort Huachuca, still struggling with a 20-plus-year war with marauding Apaches, warned Schieffelin that the only thing he would find in the area was his own tombstone, so when Schieffelin discovered a vein system at a place locally known as Goose Flats, he dubbed his find "Tombstone." The town that grew up around the discovery came to symbolize the Wild West, playing host to the Earp

lawmen, Boot Hill cemetery, and the shootout at the OK Corral. Despite its wicked reputation, Tombstone was Arizona's most productive silver district, and for a few short years, it pumped out bullion and helped to fuel exploration across southern Arizona.

Schieffelin probably had some idea that there were valuable ores in the San Pedro Valley. Frederick Brunckow, the German-born engineer who worked for the Sonora Exploring and Mining Company and discovered the Heintzelman Mine, had also discovered mineralization in 1857 near the San Pedro River. The claim became known as the **Bronco** (possibly a play on Brunckow's name), but the collapse of security with the onset of the Civil War precluded working the property until the mid-1870s.

When Schieffelin arrived in 1877, he visited the Bronco claim. Observing the northeast-trending mineralized ledges, Schieffelin located rich silver float northeast of the Bronco claim. He staked the **Tombstone** and **Graveyard** claims in the western part of the district and recorded them in Tucson. As there was apparently no assay office in Tucson at that time, Schieffelin could not prove the richness of his discovery.

Portrait of Edward Schieffelin, founder of Tombstone

Camillus Sidney "C.S." Fly photo, 1880

C.S. Fly was one of a number of pioneer photographers who documented the exploration and settlement of the American West. Italian Carlo Gentile was among the first, taking "some views" in Arizona in 1868. Known for his Civil War images, Timothy O'Sullivan brought back photos from the 1866 to 1869 King Geological Survey and the 1871 to 1874 Wheeler Survey. German-born John Karl Hillers was one of John Wesley Powell's photographers for the historic 1871 to 1872 journey through the Grand Canyon.

Private collection

Schieffelin's brother Albert had been working at the Silver King Mine, so Ed headed north hoping to get his ore assayed. When he arrived in Superior, he found that his brother was working at the nearby McCracken Mine. Ed sought work in the mines at Globe with plans to raise a grubstake and return to work his Tombstone claims. He showed his ore to Richard Gird, assayer at the McCracken. Gird declared the ore rich in silver and convinced Schieffelin to take him on as a partner. A deal was struck, and Gird and the Schieffelin brothers headed to the San Pedro Valley to begin a systematic exploration of the claims. The trio was not immediately successful on their Tombstone and Graveyard claims, so Ed expanded the search and located the **Lucky Cuss** and **Toughnut** deposits on two consecutive days.

Other prospectors, who had taken Gird's rapid departure from the productive McCracken Mine as the sign of something big, followed Gird and the Schieffelins. Two of these "rushers," Oliver Boyer (for whom the mineral oboyerite was named) and Henry Williams, located the **Grand Central** ledge. The Schieffelin group disputed Boyer and Williams's claim, and a part of the Grand Central was relocated as the **Contention** claim. By 1878 serious work had started on all of the claims.

Word of the rich discovery spread rapidly, and former Territorial Governor Anson Safford visited the area in early 1879. A true politician, he offered to make things happen in exchange for a quarter interest in the Toughnut ledge. Safford brought in Frank Corbin, and they, along with Gird, formed the Tombstone Mill and Mining Company, erecting a stamp and amalgamating mill that was commissioned in June 1879.

The Schieffelins sold their interst in the mine in 1880 for $600,000, but Gird held out a little longer and sold his share for considerably more. He then divided the additional profit with the Schieffelins—a rare action in any time. Ed Schieffelin hung around the Tombstone area until 1882 and then left for Alaska.

At the first miners' meeting in 1878, the district was christened *Tombstone*, and by the end of 1879, it was the most famous mining district in

Right: Cerussite, 8.1 cm wide
Tombstone, Cochise County

Evan Jones collection
Jeff Scovil photo

Below: At least nine mineral species were first described from mines in the Tombstone district, Cochise County:

Dugganite, $Pb_3Zn_3TeO_6(AsO_4)(OH)$ green mineral **upper** photo
Emmonsite, $Fe_2Te_3O_9 \cdot 2H_2O$
Fairbankite, $PbTeO_3$
Girdite, $Pb_3H_2(TeO_3)(TeO_6)$
Khinite, $PbCu_3TeO_6(OH)_2$
Oboyerite, $Pb_6H_6(TeO_3)_3(TeO_6)_2 \cdot 2H_2O$
Parakhinite, $PbCu_3TeO_6(OH)_2$
Schieffelinite, $Pb(Te,S)O_4 \cdot H_2O$
Winstanleyite: $TiTe_3O_8$, **lower** photo with jarosite

Upper: Dugganite, field of view 5 mm
Dave Bunk Minerals specimen
Jesse La Plante photo

Lower: Winstanleyite, field of view 5 mm
Mesa Southwest Museum collection
Jeff Scovil photo

the West (Butler et al. 1938). The town boomed. By 1881 more than 650 Cornish and American miners were working the mines, and another 150 were working in the mills. The promise of wealth, whether wrestled from the ground or from the pockets of miners, attracted some unsavory individuals, which led to the wickedness for which Tombstone is famous. The blaze of wildness that culminated in the shootout at the OK Corral was actually quite short lived, and by 1882 Tombstone had tamed.

The bulk of the silver came in the form of silver halides, but more exotic species were also found (Anthony et al. 1995). Crystallized mineral specimens were present although they were rarely preserved. Perfect 1- to 1.5-centimeter pyrite pyritohedra, usually pseudomorphed to jarosite, are known from the Grand Central Mine. But it was not specimens, it was the very rich bonanza-grade silver halide and gold telluride ores that put Tombstone on the map. Both chlorargyrite and bromargyrite are commonly found in all of the ores (Anthony et al. 1995). The district also produced economically important concentrations of gold. While some native gold was found, it was usually bound up in the telluride ores.

The Tombstone district was productive for two decades, but by the end of 1881, only a few years after discovery, the mines began an ironic struggle against water. Although Tombstone is in the desert, the **Sulphuret Mine** encountered ground water in July 1881 at a depth of 520 feet. The severity of the problem was not recognized until the Grand Central shaft hit water in April 1882. The large flow of water immediately halted further development of the mine. To offset the flow, Cornish pumps were installed in the Contention and the Grand Central mines, and the Girard Mill was erected near Tombstone to take advantage of the water being pumped from the mines. In spite of the water problems, 1882 saw Tombstone's highest annual ore production. The next four years were also fairly good ones, as the district staved off pending union and water troubles. But in 1886 the hoist works of the Grand Central Mine burned, and the Contention Mine was left to

"Wulfenite—Tombstone, Arizona, with Cuprodescloizite in microscopic radiations" was how collector Lazard Cahn had labeled this 5.7 cm wide wulfenite and mimetite specimen when he owned it. Cahn (Cahnite) was a long-time collector and dealer, who passed away in 1940. When his collection and stock were sold a few years later, 2,400 specimens went to Northwestern University. The university sold a group of specimens, including this one, in 1978.

This specimen certainly came out with other wulfenites during the heyday of mining late in the 19th century. Subsequent underground exploration of the Tombstone mines indicates that it likely originated in the Empire Mine.

Les and Paula Presmyk collection
Jeff Scovil photo

handle the groundwater inflow alone. The pressure forced the mine to close for part of the year. Other mines continued to develop even as silver prices dropped. By 1892 most of the mines were being worked by lessees, and for the rest of the 1890s, the only operating mines in the Tombstone district were the **Tranquility**, **Silver Thread**, and **Telephone**. As each of the mines was worked below the 500-foot level, large inflows of groundwater were encountered. Cornish pumps and other strong water-moving systems were employed, but the water problems in combination with the 1893 silver panic curtailed activities. The great Tombstone Mill and Mining Company stopped operating in 1896, having produced 10.55 million dollars in silver, 1.15 million of which were paid in dividends prior to 1888 (Butler et al. 1938).

F. N. Guild (1910) described an anglesite specimen he acquired from an unspecified mine in Tombstone in the late 1800s. The specimen appeared to be a pseudomorph after galena. The sample weighed several pounds and originally consisted of an aggregate of cubic crystals. These in turn altered to the lead sulfate in successive stages, giving rise to an appearance that Guild described as an "alligator-skin" structure. The successive layers about the original galena granule varied in color from nearly white to dark gray. Crystallized smithsonite as well as azurite and hemimorphite was also noted from the Toughnut and **Westside** mines (Butler et al. 1938). Other good specimens were most certainly encountered during the early years at Tombstone, although both historic specimens and records are scarce.

Modern collectors have turned up nice wulfenite and mimetite combinations from the **Empire**-Toughnut area in the 1980s, but collector activity in the long abandoned mines is restricted to occurrences that are above the water table.

Castle Dome

Although the district was continually active, mineral collecting in the Castle Dome was an oddity prior to 1876. Life in the camps verged on impossible given the remote location and the high temperatures experienced for better than half of the year. Castle Dome was the first area in the newly minted Arizona Territory to be approved for a railroad. While lack of funds prevented the road from being constructed, mining, and mineral collecting continued just the same. The mines in the district exploited silver-bearing lead minerals, including galena and cerussite. While no native silver has been described, the district was quite prolific in terms of total silver production.

Castle Dome was one of the earliest districts in the Arizona Territory to have been visited by one of the most remarkable figures in Arizona mineralogy. William Phipps Blake was the first Territorial Geologist and later the first the Director of the Arizona School of Mines and still later the Chair of the Geology, Mining and Metallurgy Department at the newly founded University of Arizona (see page 106). Blake received his PhD from Yale in 1852 and came to the southwest as a geologist and mineralogist for the Pacific Railroad Exploration Expedition in 1853.

Twenty-five miles to the north, Castle Dome peak could be seen from in front of Dr. A.G. Hull's property at Castle Dome, Yuma County, on October 11, 1926. The 26-acre Castle Dome town site was privately purchased and the buildings restored to be part of a living museum.

E.F. Burchard photo
Courtesy of the USGS

Blake was a tireless explorer and author, writing more than 200 articles on geology, mostly of the mines in the southwest. In one of these articles, Blake (1881) described, "vanadinite in the lead-bearing veins of Castle Dome District, Arizona, associated with wulfenite, cerussite, galena, and fluor-spar." He had occasionally collected minerals in the district and at one point became president of the Castle Dome Mining and Smelting Company. One of the specimens in the University of Arizona collection, labeled, "anglesite with galena, cerussite and [fluorite], Castle Dome, Arizona," is attributed to Professor Blake (Domitrovic et al. 1998).

At the turn of the century, mineral collecting was becoming more common in the United States. In *The Mineral Collector* (6:12), Philadelphia mineral dealer A.E. Foote was advertising:

> *Pink fluorite. Rare color in a common mineral—seldom observed from localities other than Switzerland. A correspondent made a special trip and secured for us a limited supply, which yielded some excellent octahedral cleavages. A specimen, one to three inches from tip to tip, is particularly striking on account of the unusual color and its resemblance to the crystallized Pink fluorite from St. Gotthard—a mineral with appreciated value. From Yuma County, Arizona, at 50¢ to $3.00.*

Based on one of the author's (tlp) experience collecting in the mines around Castle Dome and Yuma, he believes that the specimens advertised by

Dr. Foote came from the mines in Castle Dome. While Castle Dome fluorite is a far cry from the Swiss classics, the locality did produce pale pink crystals. The green to gray-green fluorite more commonly found in the district alters to light pink or lilac when exposed to sunlight.

In a December 1919 article in *The American Mineralogist*, William F. Foshag reported on the mineral localities in Yuma County, mentioning bright, lemon-yellow wulfenite crystals up to one centimeter on edge on etched crystallized anglesite from the Castle Dome district. After Foshag's description, there was no reported specimen recovery in the Castle Dome district until the 1940s.

The Silver District

Arizona's Silver district is set in the harsh, dry southwest corner of the state but is mercifully located just a few miles from the Colorado River. The **Red Cloud Mine** was the district's top producer, both of ore and of specimens; the wulfenite from the mine is an American classic. In fact Red Cloud wulfenite has been recognized as unique and worth preserving since mining began in the 1870s.

Prospectors first visited the Red Cloud area in 1862, finding silver-rich lead ores. Interestingly the mine was not named for its vivid red wulfenite but for a nearby Spanish trail, which in turn was named for a local tribal chief. The claim was patented in 1877, and the nearby town of Silent was established. The most intensive period of mining took place from 1878 through 1885 (Wilson and Thompson 2008). As much as $1 million in silver was estimated to have been produced during this period, although others more conservatively estimate the production to have been half that amount (Wilson 1933). One thing is certain, given the mine's remote location, only the highest grade ores could be mined. The ore initially had to be shipped five miles by wagon to Norton's Landing on the Colorado River. There it was placed on ships destined for the Selby smelter in San Francisco. But the grade of the ore ultimately justified building a mill on the Colorado River.

A bald eagle stands on a sandbar in the lower Colorado River notfar where the Red Cloud's 1880s mill would have been. The photo was taken from the California side of the river, looking east, Arizona is in view. Today 30 miles of the lower Colorado as well as the desert area between the Red Cloud and the river are protected as the Imperial National Wildlife Refuge, established February 14, 1941, 29 years to the day after Arizona became the 48th U.S. state.

Randy January photo, Feb. 9, 2011

From the beginning, the original prospectors reported finding red wulfenite crystals scattered across the vein. Miners initially dug a 30-foot open pit along the vein. They followed the high-grade stringers for silver, also extracting the lead ores as they went. The underground workings ultimately included a decline that stopped at the 570 level and a vertical shaft to the 270 level. Virtually all of the underground mining and ore production had been concluded by 1890.

The Red Cloud mineralization is a low-temperature, epithermal vein, occupying faults and fractures (the Red Cloud Fault) and bearing calcite, quartz, barite, and fluorite. The vein is high angle and mineralization may be related to late-stage volcanic activity associated with Basin and Range style faulting. Persistent veins in the area can be traced on the surface for over a mile and a half. The mineralization is controlled by two major faults: the Red Cloud on the west side of the valley and the McNeal on the east side of the valley. The Red Cloud Fault dips to the east and the McNeal dips to the west, both have a northward strike. The two faults intersect just south of the mine. The hanging wall of the Red Cloud vein is andesitic volcanic

INCORPORATED UNDER THE LAWS OF THE STATE OF NEW YORK.

Number 24

5000 Shares

The Red Cloud Gold & Silver Mining Company

MINES IN ARIZONA.

CAPITAL $ 140,000 SHARES $ 1⁰⁰ EACH

OFFICE IN MIDDLETOWN, N.Y.

This Certifies, that Gurdon B. Horton -

is entitled to Five Thousand Shares in the

Capital Stock of THE RED CLOUD GOLD & SILVER MINING COMPANY,

transferable on the books of the Company, by the holder, in person or by Attorney, on the surrender of this Certificate.

In Witness Whereof the said Company has caused its Corporate Seal to be affixed hereto, and to be signed by its President and Treasurer.

Middletown, N.Y. Sept 28 1880

Treasurer. President.

Charles Horton

STOCK NON ASSESSABLE.

Opened as a silver mine in the 1870s, the Red Cloud Mine, La Paz County, produced what is probably the world's finest wulfenite. These exceptional specimens have been recovered and sold since the earliest days of mining. Dealers such as George L. English and A.E. Foote were offering pieces as early as the mid-1880s. The stock certificate above was issued on September 28, 1880. The silver mine was closed by 1890, but the locality took on new life in the twentieth century as a wulfenite specimen locality.

Mark Hay collection

rock; the footwall is composed of Jurassic granodiorite. The footwall portion of the vein consists of massive black calcite mixed with oxides and less common pods of altered galena and lead oxides, and has not been mined as heavily as the upper, hanging wall portion, where the silver values were apparently located. The vein is estimated to vary in thickness from 10 to 23 feet.

There are 33 primary and secondary minerals found in the veins (Parker 1966). The silver-lead-zinc minerals occur in limonitic quartz-barite-calcite-fluorite fissure fillings. The primary minerals are silver-bearing galena, lead oxides, argentite, sliver halides, cerussite, smithsonite, willemite, and anglesite. The Red Cloud vein is chiefly composed of limonite, hematite, quartz, fluorite, and calcite along with considerable amounts of fault gouge and brecciated wall rock fragments, more or less stained by pyrolusite. Cellular spongy masses and ropy limonite are good lead-ins for wulfenite and mimetite-bearing pockets.

Not far from the Red Cloud, the **Geronimo Mine** (later called the

Pure Potential Mine) is on the same fault structure. In February 1892 in *The Naturalist's Leisure Hour and Monthly Bulletin*, Dr. A.E. Foote noted that,

> *... the red skeleton crystals of vanadinite from Yuma County, similar to the pyromorphite from Ems* [Germany], *are the largest Vanadinite crystals ever seen, 10 cents to $1.00. The Yuma County minerals are certain to be more expensive in the future, as all mining operations have been suspended in the Silver District.*

The North and South Geronimo mines are the likely sources of Dr. Foote's specimens; however for many years collectors debated where the famous vanadinite had come from. Richard Bideaux and many others searched for the lost locality, finally finding that it was the Geronimo mines.

By the early 1890s, a number of colorful mineral species were drawing the attention of museums, dealers, and collectors to Arizona's Silver District. In 1881(a), William Blake described "beautiful crimson-red crystals of vanadinite from the Hamburg Mine and fine wulfenite, sometimes in octahedral crystals, from the Red Cloud Mine, the Oakland Boys claim, and other points in the Silver District, Arizona."

In 1881, Yale University mineralogist Benjamin Silliman noted that,

> *This hitherto rare species* [vanadinite] *promises now to be comparatively abundant. In the so-called "Silver District," in Yuma County, Arizona, about 50 miles north of Fort Yuma, is a large area traversed by veins of quartz carrying argentiferous galena, with salts of lead, but no gold, and rather extensively explored. The lead salts which I have seen from this region are wulfenite, of remarkable beauty, vanadinite, and massive anglesite with galenite. Vanadinite occurs in three mines, near together, the Hamburg, the Princess, and the Red Cloud. The crystals of vanadinite are extremely beautiful, alike for brilliancy of color, luster, and perfection of form. Wulfenite crystals of rare beauty are found in the Red Cloud mine already mentioned as furnishing the vanadinite. The specimens sent me are from a depth of about 300 feet. They show very solid tabular crystal of large size, brilliant luster, and rich orange-yellow to orange-red color.*

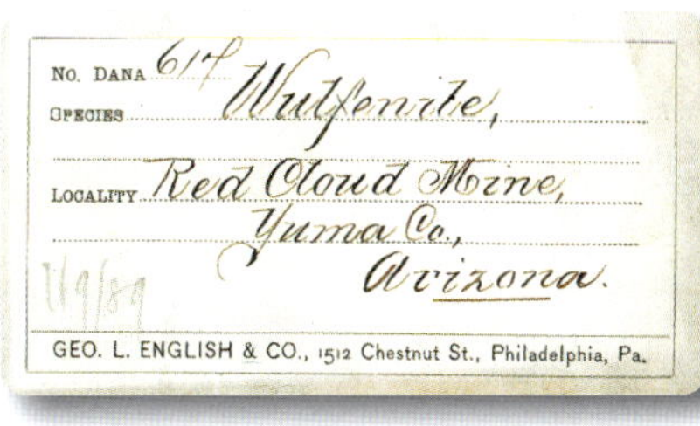
No. DANA 614
SPECIES Wulfenite,
LOCALITY Red Cloud Mine,
Yuma Co.,
1/9/89 Arizona.
GEO. L. ENGLISH & CO., 1512 Chestnut St., Philadelphia, Pa.

Noted Philadelphia mineral dealer George English traveled to the Red Cloud Mine in late 1888 or early 1889 and acquired some of the earliest preserved Red Cloud specimens. This 4.9 cm wide specimen, in the collection of Paul Harter, includes a George English label, which has been dated through the Mineralogical Record label archive to 1889. The largest crystal is 1.5 cm on edge.

Jeff Scovil photo

Wilson and Thompson (2008) revealed that Silliman's wulfenite crystals originated with F.F. Thomas, a furnace operator at the Red Cloud. Thomas had put together a suite of wulfenite crystals from the mine when the workings were at their height.

A.E. Foote was carrying Red Cloud wulfenite during the mine's most productive years: 1882-1886. George English was advertising similar specimens between 1888 and 1889. Both of these early dealers even traveled to the mine and bought specimens from the miners (Wilson and Thompson 2008). Specimens from the 1800s are in the collections of the Seaman Museum at Michigan Technological University, the American Museum of Natural History, in New York, and Harvard's Peabody Museum of Natural History in Cambridge.

By 1896, advertisements in *The Mineral Collector* were referring to the "old" specimens from the Red Cloud Mine—Arizona was on the map as a producer of world-class mineral specimens.

In the February 1899 issue of *The Mineral Collector* (V:12, p.v), Ward's Natural Science Establishment touted,

> *a fine assortment of orange-yellow, well crystallized (wulfenite) specimens, the last that can be obtained from the famous Mammoth Gold Mine, Pinal County, Arizona; and offer them at prices ranging from 25 cents to $12. The latter a magnificent specimen, 5 x 3½ inches, thickly strewn with beautiful crystals.*

This 8.8 cm wide specimen, labeled "Schultz Gold Mine," was likely collected in the late 1800s and may have been among those handled by Ward's right around the turn of the century, before the mine collapse in 1901.

Mark Hay collection
(ex Harvard collection)
Jeff Scovil photo

The Old Hat Mining District

Even during silver's reign, a few gold prospects were discovered and developed in the Arizona Territory. Frank Schultz staked the first gold claims—the **Hackney** and **Aaven**—on the Collins vein in 1879. He located "the **Mohawk** vein in 1881 and the **Mammoth** vein in 1882" (Bideaux 1980). The Old Hat mining district, about 50 miles north-northeast of Tucson, was thus established.

The Mammoth and Collins veins, precursors to the twentieth century Mammoth-St. Anthony Mine at Tiger, produced 350,000 tons of ore between 1881 and 1901 (Peterson 1938). Interestingly from 1881 until 1936, the Mammoth mining camp produced over $5 million, 83 percent of which was gold. During these early years, the ore was hauled 10 miles east to the town of Mammoth, which was located along the San Pedro River and was the closest assured water supply. About 1896 a Bleichert double-rope aerial tramway was erected to transport the ore more cheaply between the mines and Mammoth, where the mill had been built about a decade earlier. Also in 1896 a 200-ton per day cyanide plant was added to the old mill to treat the tailings in which 40 percent of the gold remained. In 1898 a British company acquired the Mammoth and Collins mines, and they were reorganized as the Mammoth-Collins Mining and Milling Company.

From very early on the mines were producing specimens, which can still be found on the market from time to time and are typically labeled "**Schultz**

Gold Mine." In 1897 George English & Co. advertised in *The Mineral Collector* (IV:3, p.xii):

> *A great strike at Mammoth, Arizona. Two pockets were struck by our collector containing brown and black, sparkling descloisite associated with brownish-red crystals of Vanadinite 10c to $2.00 for fine specimens are prices which ought to suit every collector. A few good Wulfenites from Mammoth associated with scarlet Descloisite 25c to $2.50. A few specimens of Leadhillite from this locality, 50c to $2.00. Massive Anglesite with Linarite and Chrysocolla, 10c to $1.00.*

One of the constant struggles at Mammoth-Collins was against the "heavy ground" (a miner's term for weak, fractured rock which is prone to collapse). This coupled with the mining company's inattention to proper timbering caused the mines to endure several major collapses, the most debilitating occurred on April 15, 1901. More than 750 vertical feet of ground and timbering collapsed, and the ensuing repairs and litigation closed both mines for more than a decade. Years later this incident was evident when the area was exposed through the open pit. The Mammoth Mine finally reopened in 1913 under a lease with Great Western Copper Company.

Upper: When collectors ventured into the long-closed Mammoth-St. Anthony Mine in April 1973, their efforts were repeatedly confounded by debris from the mine's epic cave-ins. Wayne Thompson is pictured here contemplating a collapsed drift between the Mohawk-New Year and Mammoth shafts.

Les Presmyk photo

Right: Vanadinite on mottramite
5.5 cm tall
Mammoth Mine, Schultz

This specimen was in the collection of E.O. Stratton, a rancher, miner, and entrepreneur in Arizona from the late 1880s through the early 1900s. His collection included specimens from Bisbee, the Magma Mine, and a few vanadinites from Schultz. At the time, this specimen was pretty good. Although Foote Mineral Co. offered leadhillite out of the mines on the Mammoth and Collins veins, the best specimens did not come out until the 1940s and early 50s.

Les and Paula Presmyk collection
Jeff Scovil photo

CHAPTER 5

The Rise of Copper

Copper and copper minerals were among the earliest materials mined in Arizona, but given the isolation of Arizona's mines, copper was generally only recovered as a byproduct of mining precious metals. Its importance changed dramatically in 1879 in the workshops of Thomas Edison. The invention of the electric light bulb created monumental demand for remote delivery of electricity and for the ideal conductor—copper. Not only is copper an excellent conductor of electricity, it is also soft and malleable enough to be drawn into wires. About the same time as Edison was inventing the incandescent bulb, Thomas B. Doolittle invented a process to draw individual wires that were miles long. Alexander Graham Bell's invention of the telephone further increased the need for copper, and its value skyrocketed.

In 1885 George Westinghouse developed alternating current (AC) generators, and in 1889 Nikola Tesla developed a more efficient AC motor and the means for transporting electricity across copper lines for hundreds of miles. By 1890 copper was arguably the most important commodity in the United States; it fueled the manufacturing revolution that made the country a superpower. In 1875 the total copper production in the United States (mostly Michigan) was a few tens of million pounds per year. By the start of the twentieth century, the United States was producing more than 600 million pounds of copper annually, three quarters of it coming from Arizona.

Arizonans had dabbled in mining copper before 1875; most notable were the modest efforts at Ajo. But by the early 1880s, the ore deposits that would become the great mines—**Clifton-Morenci**, **Old Dominion**, **Bisbee**, and **Jerome**—had been discovered and mining was becoming an important industry, producing tens of millions of pounds of copper annually and accounting for nearly half of the territory's economy. A tremendous legacy of mineral specimens is associated with these mines, which are also Arizona's defining collecting localities.

Above: Thomas Edison's electric lamp, patented January 27, 1880.

Left: Azurite on malachite, 4.5 cm wide
Copper Queen Mine
Bisbee, Cochise County

Paul Harter collection
(ex Jim and Dawn Minette)
Jeff Scovil photo

Old Dominion

The Old Dominion mines, near Globe, were some of the first large-scale copper-mining efforts in Arizona. Most of the Old Dominion production history is twentieth-century, but its story starts with the first prospectors who scoured the Silver Belt. The foursome of Mason, Reagan, Long, and

A turn of the century view of the "Old Dominion Concentrator" on a circa 1910 postcard

John Taylor collection

Copeland, the group that ultimately discovered the **Silver King** deposit (see page 31) had first staked several silver claims in the Globe area. Others followed and located claims to the north and to the east. The silver properties were rich but shallow, and by 1883 all were exhausted.

Copper claims had also been staked, and by 1879 the original Old Dominion claim had been filed and sold to an eastern capitalist, who resold it a year later for $1,000,000. The Old Dominion Copper Mining Company was thus organized, and additional capital was secured to develop its properties and build two smelters. Those properties turned out to be almost worthless but the purchase of the Old Globe Copper Company properties was more prudent. The **Old Dominion**, **United Globe**, **Commercial**, and **Iron Cap** became the most productive mines on the Old Dominion vein system. Other significant mines included the **Buffalo Mine** (a great cuprite producer) and the Great Eastern.

By the early 1890s the two major mining companies in this district were United Globe and Old Dominion; the latter owned by Phelps Dodge. In 1901 Phelps Dodge had purchased United Globe, consolidating the two operations. It took until 1898 for the railroad to reach the area; in the meantime, the high cost of freight precluded extracting all but the richest ores. Even the Silver King Mine and the rich deposits at McMillenville and Richmond Basin had a tough go of it until then. Finally as geographic and political obstacles were overcome, the Gila Valley Globe & Northern Railway was completed in December 1898. One month later, the Old

Dominion made its first shipment of 500 tons of copper bars in 20 cars.

The original Old Dominion discovery was along a large vein that extended to great depth, eventually descending 26 levels below the surface to nearly 2,200 feet. The veins are primarily quartz and carry pyrite, chalcopyrite, bornite, and specular hematite along with local concentrations of cuprite, sphalerite, and galena. In several places the vein system contacts a limestone unit, and there are "pods" of replacement ores. It is in these pods that the best mineral specimens were found.

Old Dominion mineral specimens were being offered for sale as early as 1897. George L. English & Company was advertising quartz on chrysocolla from Globe in *The Mineral Collector* (IV:3, p.xii), offering "A few exquisite specimens of this most beautiful of combinations, 50c. to $4.00. Order by return mail or you will be too late." In a report for the *The Mineral Collector* in the same year (IV:5, p.71), Maynard Bixby reported,

> *Of all the magnificent velvet malachite found in former years, not a single fine piece remains in the camp. Fine octahedral cuprite crystals on copper were formerly obtained, but now only from some of the collections.*
>
> *The beautiful quartz on chrysocolla is now found only in the collections*

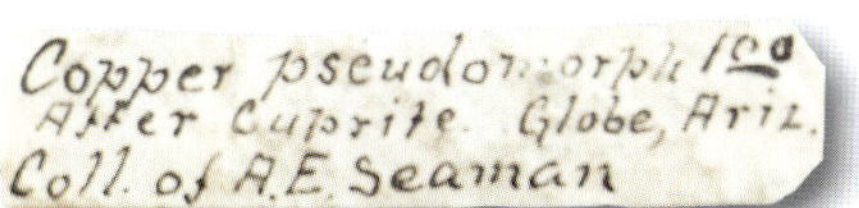

Above: Copper pseudomorph after cuprite, 5.2 cm tall
Old Dominion Mine, Gila County

Evan Jones collection (ex A.E. Seaman)
Jeff Scovil photo

Right: Calcite on quartz on chrysocolla 7.1 cm tall
Old Dominion Mine, Gila County

Paul Harter collection (ex Levi Smith)
Jeff Scovil photo

of the town. A late find was a very few octahedrons of copper in groups, and occasionally very perfect single crystals.

Chalcotrichite occurs rarely at the Old Dominion Mine, generally coating or enclosed in calcite, and probably less than a dozen specimens have been found. A few specimens of calcite have been collected very rich in native silver, usually crystallized in beautiful and delicate feathery forms.

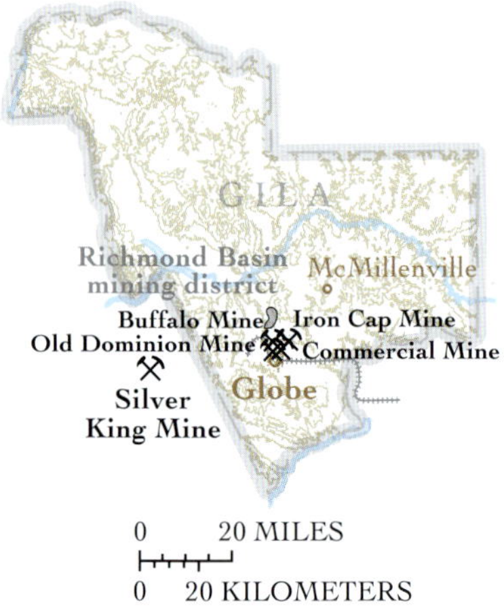

Old Dominion cuprite is legendary among collectors. Open watercourses contained crystals larger than 2.5 centimeters on an edge. The mine bosses ran scaling bars through the watercourses to destroy the crystals in an attempt to prevent the miners from collecting them, but some survived nonetheless. The cuprite takes two distinctive forms. The first consists of flattish clusters of lustrous dark red octahedra, which probably formed on and after dendritic copper. These clusters, which can be up to 10 centimeters long, were recovered from the Buffalo Mine. The second form consists of equant clusters of reddish octahedra up to 2 centimeters on an edge. The octahedra of this second form have a distinctive etched appearance, sometimes with the pits containing yellow iron oxides.

A few of the etched Old Dominion cuprite specimens have been preserved in public and private collections. Some of these distinctive specimens have been mislabeled "Bisbee." Others including one that passed through the collection of famed mineral dealer Art Montgomery and into the hands of mineralogist Dick Bideaux have reliable provenance. One notable, 5-centimeter knob of lustrous (although etched), crystals with a topknot of white, nail-head calcite crystals to 1 centimeter, formerly in the California Institute of Technology collection, is labeled "Old Dominion Mine." Interestingly, Ransome (1903) described gold wires on cuprite from the original Old Dominion Mine that was located a few miles north of the mine that ultimately carried that name. Unfortunately examples of the find have not been preserved.

Drusy quartz on chrysocolla is the second most collectible species from the mines. While the color of Old Dominion chrysocolla can be as intense a blue as that from any other mine, it commonly formed on an iron-rich, light

Below: This 6.2 cm wide knob of lustrous (although etched), cuprite crystals with a topknot of white, nail-head calcite crystals from the Old Dominion Mine, Gila County, was formerly in the Caltech collection.

Les and Paula Presmyk collection
Jeff Scovil photo

to dark brown, vuggy material, rather than on the massive chrysocolla matrix seen in specimens from other localities. Specimens larger than 10 centimeters are rare.

Worthy specimens of calcite, native copper, malachite and the occasional azurite crystal have all come from the Old Dominion Mine. They can be found in a number of old collections, mainly in the eastern United States. And surprisingly a number emerged in the latter half of the twentieth century, having been preserved by collectors in the Globe-Miami area.

For all its richness, the Old Dominion Mine has been closed since the 1930s; its only product is water, which it furnishes to the city of Globe as well as to active mining operations in the area.

Below: Malachite on quartz on chrysocolla, 8.2 cm wide
Old Dominion Mine, Gila County

Les and Paula Presmyk collection
Jeff Scovil photo

Left: A circa 1909 gelatin silver print of the Old Dominion Copper Company complex

Library of Congress Prints and Photographs collection

Upper: A 1901 Waldmar Lindgren photo of King Gulch, near Metcalf, looking down toward Chase Creek shows the terrain in the area. Copper porphyry deposits occupy the foreground and most of the ridges in the background.

USGS collection

Facing page: Azurite and malachite 12.9 cm tall

This polished slice showing concentric 'eyes' of royal blue azurite and variegated green malachite was number 2642 in the historic Washington Roebling collection. The label bears Roebling's handwriting, as well as the early name for the locality, "Clifton, Ariz."

Carl Acosta collection
ex Washington Roebling
Jeff Scovil photo

Clifton-Morenci

Usually referred to as the Clifton-Morenci Mines or simply Morenci, several immense open pits exploit the classic porphyry copper deposit, which is related to a series of early Tertiary magmatic intrusions. The uppermost part of these intrusions generated ascending hydrothermal fluids rich in copper sulfides. These hydrothermal fluids interacted with a thick overlying sequence of Paleozoic limestones, and later supergene enrichment created the copper carbonate mineralization—azurite and malachite—that make the district famous. Today Morenci is the largest copper mine in North America relying on ore that averages less than 0.6 percent copper.

In the 1860s the future home of the Morenci-Clifton district was remote and in the heart of hostile Apache territory. In 1864 a group of prospectors led by Henry Clifton left Silver City, New Mexico, to look for gold and silver in the upper Gila River drainage (Joralemon 1934). Clifton's group discovered a rich deposit of malachite in the limestone cliffs over looking a small river, which would be named Chase Creek. But in the 1860s, the value of copper could not justify mining in the remote and hostile region. The prospectors returned to their base, and the story of their discovery circulated piquing interest in the region.

In 1870 a group of ranchers visited the area ostensibly in the pursuit of horse thieves. The group was led by a Captain Chase and included brothers Robert and James Metcalf, Joe Yankie, and Charlie and Baylor Shannon (Coloquhoun 1924). Once the party reached the region described by Clifton, the Metcalf brothers and Yankie formed separate scouting parties. Joe Yankie traveled up Gold Gulch where he found gold in the sand; the Metcalfs found the copper outcrop that would later be the Longfellow Mine as well as green-stained rocks of the deposits that would later bear their name. They returned to Silver City with Captain Chase with the potential riches, particularly of a gold prospect, on their minds.

Two years later in June of 1872, Isaac Stevens and Robert Metcalf led a group back to Chase Creek after the first of many peace treaties with the Apaches was signed. They located a half of a dozen claims in what would become the Copper Mountain mining district and founded the town of Clifton. These original claims would become the Arizona Central, Copper Mountain, Yankie, and Longfellow mines. Bob Metcalf continued to prospect and staked the Shannon and Metcalf claims, 6½ miles north of the Copper Mountain district that would became the Metcalf district.

The former group founded the settlement of Clifton and began to extract the rich ores from the Longfellow Mine, but turning a profit in the isolated district was difficult. With no way to economically ship the material to be processed; the Longfellow was working at a loss. The Metcalf brothers finally had no choice but to sign over controlling interest in the operation to Las Cruces merchants and brothers Henry and Charles Lesinsky, who had been funding the Longfellow by advancing food and the supplies to the camp. Robert Metcalf was able to hold onto the Metcalf claims, and in 1875 he established the Metcalf mining camp.

Newly installed at the Longfellow, the Lesinskys along with their partner

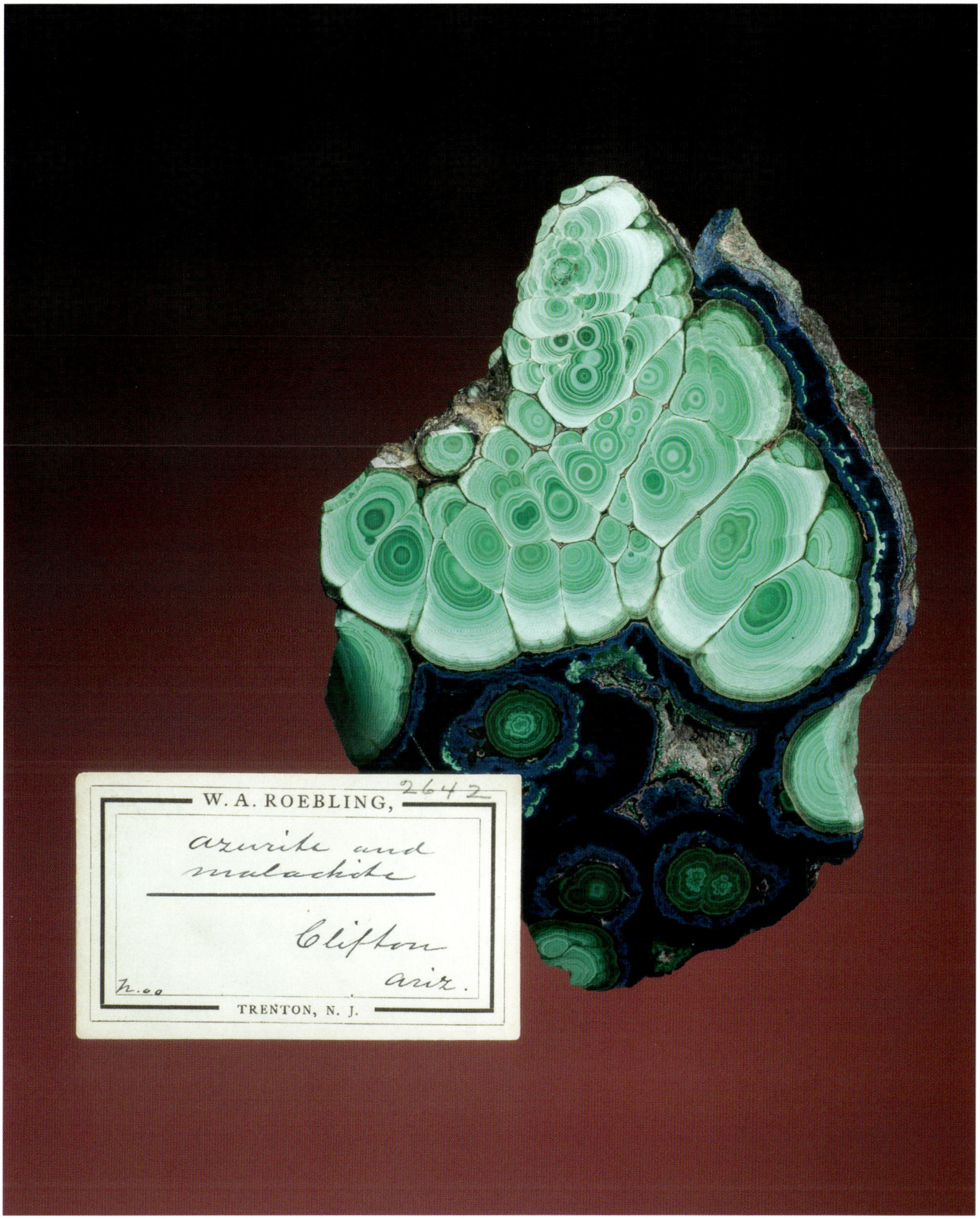
2642
W. A. ROEBLING,
azurite and
malachite
Clifton
ariz.
TRENTON, N. J.

and uncle Julius Freudenthal knew that the ore had to be smelted on site. So in 1873 they built a small furnace near the mine; its bellows were worked by hand. The following year, they had a copper-jacketed furnace in place along the San Francisco River in Clifton. The bellows on this one were water-powered.

Initially teams of oxen hauled the copper ore 1,200 miles east to Kansas City with the teams carrying supplies on the return trip. Mining was initially restricted to the surface, but as the rich copper ores were depleted, underground mines were developed. For the first ten years of operation, Longfellow ore averaged about 25 percent copper, all of which was in the form of either native copper or oxidized copper ores—azurite, malachite, and cuprite.

The Longfellow was

> *... famous throughout the country, not only for the richness of its ore, but for the ease with which the ore could be mined and smelted, producing black copper at one smelting. The ore occurred mostly in contact with a porphyry dyke which had acted as a dam, and permitted the copper liquors to eat out the limestone to a depth of 300 feet in places, and to widths varying from 10 to 200 feet, and for a length of at least 400 feet along the contact. Below the ore was the unaltered limestone. The copper had evidently been leached out of the porphyry, which perhaps at one time covered this formation* (Colquhoun 1924).

The Lesinskys further developed the mine by building a 4½-mile-long, 20-inch baby-gauge railway from the base of their Longfellow Incline to Clifton in 1879. Their aim was in part to help thwart Indian (which were attacking the teams) and in part to reduce transportation costs. At first mules provided "horsepower" to the cars. The driver and mules would ride on a gravity-powered platform (with good brakes) when taking loaded trains to

Long drawn to mining, Prussian-born Henry Lesinsky (pictured **above**, circa 1890), his brother Charles and uncle Julius Freudenthal bought the Metcalfs' interest in the Longfellow Mine in Clifton. They owned and developed the mine for a decade before selling to Scottish-owned Arizona Copper Company.

Right: Teamsters drove freight trains such as this one the treacherous 1,200 miles from Clifton to Kansas City

Photographer unknown, circa 1883
Photo courtesy of the National Archives

Azurite, 9.5 cm wide
Detroit Mine, Morenci, Greenlee County
Les and Paula Presmyk collection
Jeff Scovil photo

the smelter; they only had to exert themselves when pulling the empty cars back up the grade. A steam engine named "Emma" or "El Vapor" made its debut in 1880 and replaced the four-legged power.

The Metcalf mines were developed simultaneously with the Longfellow by the Arizona Copper Company. The Metcalf mines are on the southern slope of Shannon Mountain. Large-scale development had begun there by 1879; the production of the Metcalf mines is impossible to ascertain as the ore was comingled with the Longfellow ores and smelted at a single facility. Eventually the Metcalf overshadowed the Longfellow, and by 1905 the "Metcalf mines …produced, roughly, 5,000,000 pounds of copper per annum" (Lindgren 1905).

Meanwhile, Captain E. Ward, William Church, Joe Yankie, and others had organized the Detroit Copper Company in 1874 to exploit their claims west of Clifton. Slower to develop than Clifton, that community consisted of a small contingent of tents and was called Joys Camp, named for Miles Joy who had overseen work on the original claims. Church took over as General Manager around 1880 and jump-started the operation. He changed the name of the camp to Morenci and organized the haphazard mining effort.

The persuasive Church then began pursing eastern capital to develop Detroit's assets. In 1881 he met with officials of Phelps Dodge in New York City to secure a loan. Phelps Dodge was in the business of exporting cotton and other commodities as well as importing metals, mostly tin, copper, brass, and zinc. The company sent Dr. James Douglas to examine the Detroit

Copper Company properties and based on his favorable report sent $50,000 out west and became part owner in the Detroit Copper Company. Thus began Phelps Dodge's entry into mining and metal production in the United States, a legacy that spanned more than 120 years until Freeport-McMoRan Copper and Gold, Inc. purchased Phelps Dodge in 2007.

In spite of all the development, Morenci-Clifton remained isolated and dangerous. In 1881 marauding Apache killed a number of miners in Gold Gulch before killing some of Morenci's teamsters. Indians were such a threat that the Detroit Copper Company dedicated half of the workers on each shift to stand guard (Conger 1987). Rustlers the likes of Kid Louis ran wild, and lawlessness prevailed in the settlements. But work progressed, and by 1882 Detroit had built a smelter on the north side of the San Francisco River about 3 miles south of Clifton.

In the same year, Scotland-based Arizona Copper Company bought out the Lesinskys and Metcalfs. Arizona Copper quickly raised $4 million to fund an infrastructure overhaul, including new machinery, railroads, and inclines. The actual cost of the project was almost twice the original estimate, and the overrun, along with a cave-in at the Longfellow Mine (the ore-rich pillars were regularly robbed), almost bankrupted Arizona Copper Company just a few years after its entry into the Arizona copper scene.

In 1884 the leaner Detroit Copper Company was becoming the dominant player in the district. Church decided that it would be cheaper for Detroit Copper to move the smelter to Morenci than it would be to build a railroad, and a new smelter was soon operating in the town. While the Detroit Copper Company had William Church to provide vision and innovation, Arizona Copper hired young engineer James Colquhoun who brought an unexpected ingenuity to the company.

Church brought Arizona's first copper concentrating plant into production to process the district's low-grade oxidized ores, and Colquhoun quickly followed suit. For the next several years, both Arizona Copper and Detroit Copper made advances in ore mining and processing, but the depression of 1892-93 hit both companies hard. Detroit Copper Company was forced to close the mine as copper had dropped to 10 cents a pound in New York; it was worth even less in Clifton.

The downturn proved an opportunity for Colquhoun, who became General Manager of Arizona Copper. Facing shutdown and default, Colquhoun made a deal with the company's bank to forego payment for one year while he developed a leaching plant to reprocess the old mine tailings. The company kept its doors open through the depression, and with the success of Colquhoun's leaching plant, it turned a profit.

After the depression, both camps were looking for a way to concentrate the low-grade sulfide ores that their mines held in abundance. Church's first attempt failed, but Colquhoun's was a success. The technology provided a means to free the copper from Clifton's massive sulfide orebodies and paved the way to the district's future as a copper superpower.

The deposits continued to change hands. Phelps Dodge became the sole owner of the Detroit Copper Company in 1897. The Shannon Copper Company made a brief appearance on the Clifton-Morenci scene at the

Left: George Kunz purchased this 4.5 cm wide, polished azurite and malachite slice from the Clifton-Morenci while on a trip to Arizona in 1884. Kunz sold it through Tiffany's to J. Pierpont Morgan; it was included in Tiffany's gold-medal winning exhibition of American Gems at the 1889 Exposition Universelle in Paris.

Herb Obodda collection
Jeff Scovil photo

Right: Obtained by Dr. Édouard Cumenge who visited the Clifton-Morenci mines in 1884, this 2.5 cm lustrous azurite crystal is perched on a dark gray-black matrix of iron and manganese oxides. The overall specimen is 3.5 cm wide; it probably originated in the Longfellow Mine.

Tony Potucek collection
Jesse La Plante photo

Facing page, upper: James Colquhoun
Emery Walker photo, 1896

Facing page lower: The Lesinsky brothers had the jail in Clifton built about 1881. The contract for blasting the cells into the face of the cliff was awarded to Mexican Margarito Barela, who was also its first prisoner, incarcerated for involvement in a "shooting up" celebration (circa 1881)

Photo courtesy of the National Archives

turn of the century. In 1899 Shannon purchased property from Arizona Copper Company. Although they had some very good years, Shannon's mines were shallow and only operated through 1919. Devastated by long strikes in 1915 and 1917, Shannon sold their assets back to Arizona Copper, which in turn fell on hard times as a result of declining ore grades, low copper prices, and high taxes. Arizona Copper ended its run in the district in 1921 by selling out to Phelps Dodge, and making the legendary company the sole operator of the mines in the Clifton-Morenci district.

Collectors have been preserving mineral specimens, primarily malachite and azurite, from the Clifton-Morenci mines since at least the late 1800s. Careful labeling was not a great concern in the early days, as such many specimens from nineteenth century Arizona's are mislabeled. Fortunately, specimens from Morenci are fairly distinct. Reference specimens and ample descriptions are available and as such many mislabeled specimens can be attributed at least to the correct mining district with relative certainty.

Native copper (Esbenshade and Thompson 2008), pyrite, willemite (Guild 1910), dioptase (Guild 1910), sphalerite, cuprite, and chalcocite (Lindgren 1905) were all described from turn-of-the-century Morenci, but the district's mineral specimen legacy is dominated by azurite and malachite.

The Longfellow Mine azurite and malachite after azurite specimens in Washington Roebling's collection that were donated to the Smithsonian in the 1920s exhibit distinctive spear-shaped crystals with multiple faces that are similar in habit although not as lustrous as those from Bisbee. Crystals approaching 1.5 centimeters in length formed on what appears to be a manganese oxide, with overall specimen reaching 10 centimeters. The Longfellow also produced small (less than 5-millimeter), highly lustrous, chisel-shaped azurite crystals, again on a black manganese-rich matrix. Distinctive medium blue rosettes to 1.5 centimeters that formed on hard, botryoidal malachite plates came from the Detroit Mine.

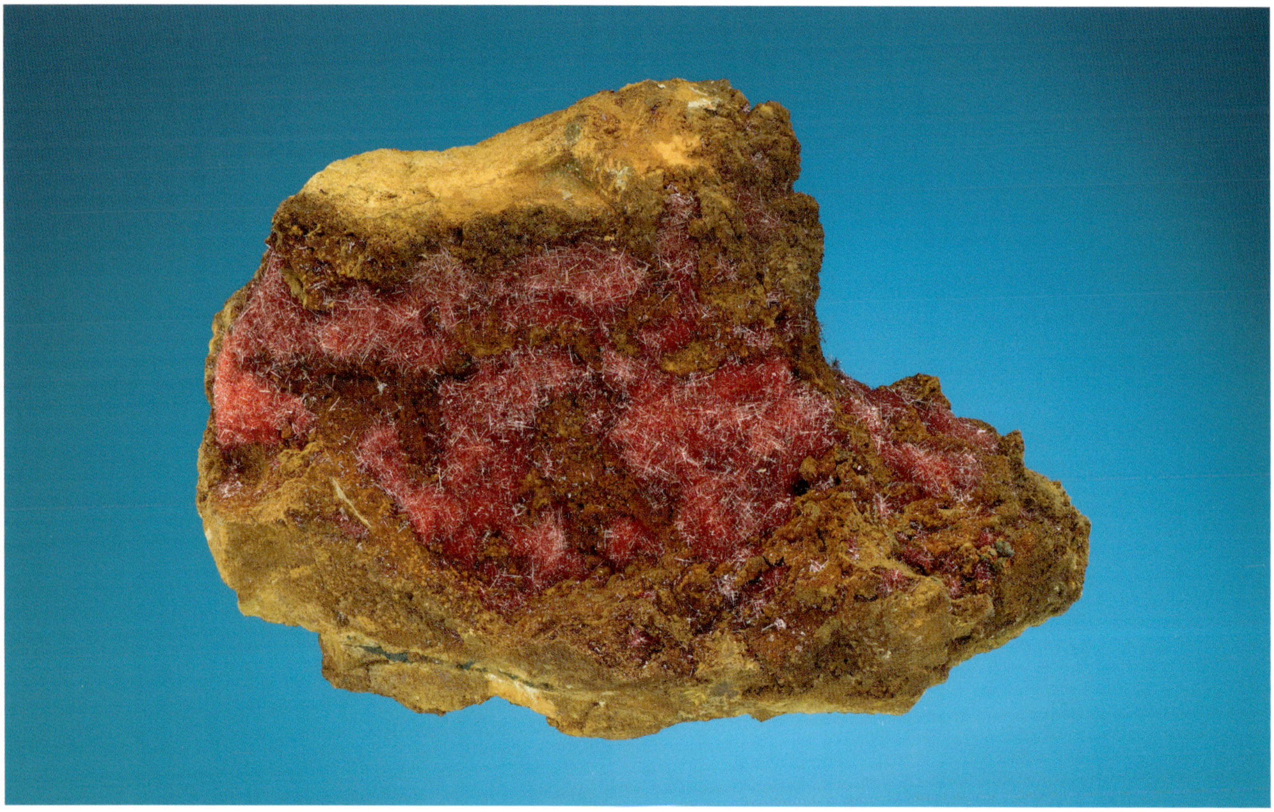

Fortunately for mineral collectors, U.S. Geological Survey geologist Waldemar Lindgren (1860–1939) was generous in his descriptions of minerals from all of the mines in which he worked. In describing the oxidation of the chalcocite zone at Morenci, Lindgren (1905) observed macroscopic minerals and noted that stalactitic masses of azurite alternating with crusts of malachite occurred at the Detroit Mine. He mentioned fine, radiating malachite crystals from the Standard Mine, which was one mile below Metcalf. He further noted an occurrence of azurite in some of the early stopes in the Copper Mountain Tunnel at Morenci and said that the partly contact-metamorphosed limestone had been converted to "a loose yellowish aggregate containing bunches of small azurite crystals and many small pale-green balls of radially fibrous malachite." Lindgren went on to describe chrysocolla from the Copper Queen Mine (at Morenci) that had formed on loose porphyry and was topped with fine crystals of azurite.

The matrix of Morenci specimens provides the most telling clues for distinguishing origins of the specimens. In general the matrix of Morenci azurite specimens is manganese-rich and does not contain malachite. The occasional bright, shiny Morenci azurite has been found, but most crystals are slightly dull to matte in terms of luster. Morenci was nowhere near as prolific a specimen-producer as Bisbee, but Morenci's early copper carbonate specimens, impressive enough to have been carefully preserved for more than one hundred years, hinted at the abundance that was to be recovered in the latter half of the twentieth century.

Right: Azurite and malachite, 5.5 cm wide
Morenci, Greenlee County

Tony Potucek collection
(Washington A. Roebling (Cat. No. R2675)
Jesse La Plante photo

Below: Photographers Lucas and Burlan captured this photo of the mines in Clifton in 1885, the only year that they worked together as a team and just a few years after Arizona Copper Company had bought out the Lesinsky brothers and begun its ambitious development project.

Beckstead collection

Facing page, upper: Cuprite variety chalcotrichite, 15.1 cm wide
Greenlee County

Dick Morris collection
Jeff Scovil photo

Facing page, lower: Underground at the Wilson Mine in Metcalf, circa 1890s

Robert Pearson collection

Lower: Some 50,000 years ago, a gigantic body blasted a hole almost 1 mile wide and 550 feet deep in the plains near present day Winslow, Arizona. The impact sprayed tens of thousands of nickel-iron meteorites across the surrounding plains.

Also known as Canyon Diablo, or the Barringer Crater, Meteor Crater is the world's best known and most studied meteor crater. And while the crater was long known and received many visitors (some of whom even collected meteorites) in the nineteenth century, no one is known to have taken a serious interest in it until 1891, when mineral dealer A.E. Foote as well as U.S.G.S. and Smithsonian geologists visited the crater and began surmising about the origin of the crater.

A patented claim was granted to Daniel M. Barringer's Standard Iron Company in 1903. Barringer believed that the crater was caused by a meteor and that a large iron body, potentially worth more than a billion dollars, was buried in its floor. The iron body was never recovered, and mathematician Forest Ray Moulton later demonstrated that a huge body would not bury itself but would instead vaporize on impact. In 1939 Harvey H. Nininger found tiny spherules of nickel-iron meteorites all over the surrounding plains, substantiating Moulton's theory.

Finally in 1960 USGS geologist Eugene Shoemaker recognized the two high-pressure minerals coesite and stishovite at the crater; both could only have formed as the result of a significant meteorite impact. The meteorite fragments themselves are the type or co-type locality for several other mineral species including haxonite, krinovite, lonsdaleite and moissanite.

While the iron mine never materialized, the "rock flour" silica generated by the impact of the meteorite was mined from the crater rim for a few years. And the crater has turned out to be a gold mine of sorts, the Barringer family owns the popular tourist attraction, charging an entry fee to visitors.

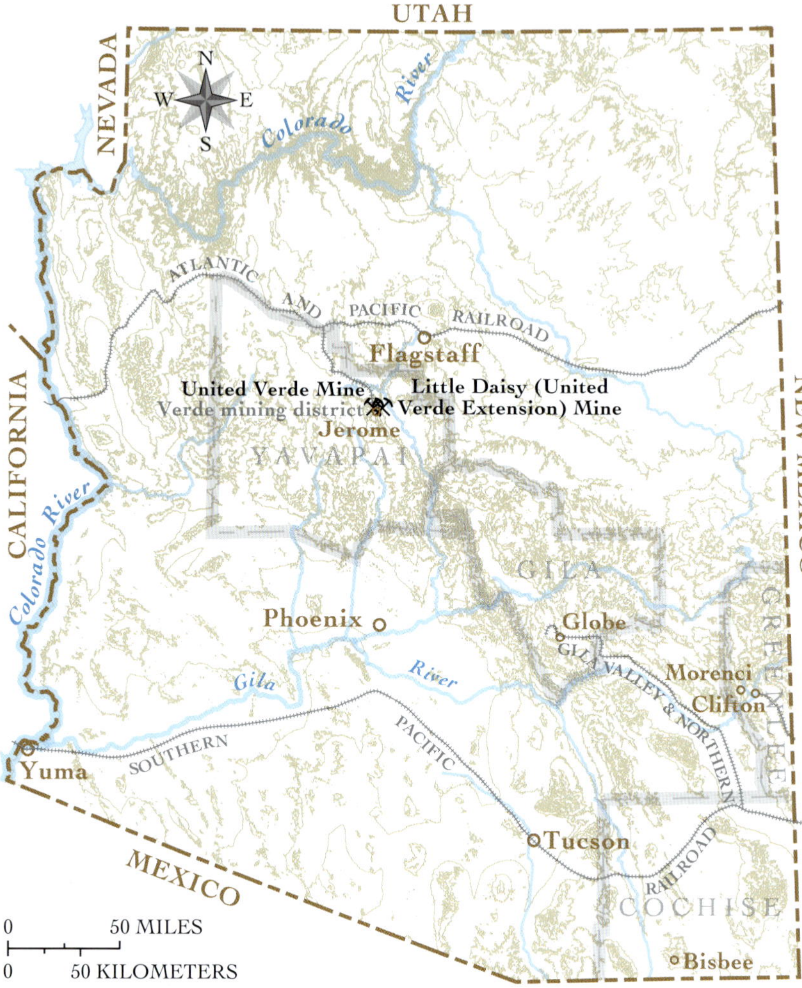

Meteor Crater in Arizona as viewed from the rim in 1891, G.K. Gilbert/USGS photo

Left: Principal Mines and Mining Camps During the Copper Period (~1885–1900)

Right: Circa 1910 postcard of the "Jerome Verde Mine" also known as the United Verde Mine

John Taylor collection

Below: Teacher, turned miner, turned merchant, turned banker, "Copper King" William Andrews Clark, purchased a 70 percent share of the United Verde Mine in 1887. It and other investments made Clark one of the 50 richest Americans of all time when he died in 1925. Clark is pictured here, circa 1917, with two of his seven children, daughters Andrée (left) and Huguette (right). Andrée died of menegitis at age 17. The reclusive Huguette died in 2011, just two weeks shy of her 105th birthday.

Montana Historical Society photo

Jerome

The mines in the Jerome, or Verde, mining district in Yavapai County were notably large and extremely rich in copper and other metals. They exploit Precambrian massive sulfide deposits that formed in the ocean from black smokers. While Jerome is not as well known for colorful minerals as Arizona's other great copper camps, the **United Verde Mine** is the type locality for twelve species. Four of which are hydrated sulfides produced by a fire in the 1890s in a level of massive sulfides (Anthony et al. 1995).

While farmers and prospectors had been through the area as early as the mid-1860s, soldiers scouting the area in 1875 were the first to recognize the mineralization near Jerome. In February 1876 John and Edward O'Dougherty and John D. Boyd located and filed claims on the outcrop that ultimately spawned the United Verde Copper Company. Others located claims in the vicinity, and although there was no railroad or port within easy reach, F.A. Tritle, the territorial governor, helped secure eastern financing for mines in the district. The United Verde Company was born in 1882, the same year the northern Arizona portion of the Atlantic and Pacific Railroad was completed (Anderson and Creasey 1958).

Bullion and copper matte production began in August 1883 from gold-, silver-, and copper-rich oxide ores mined on the surface. The United Verde Company immediately began turning a profit and paid a dividend of $60,000 in 1884. But in the same year, with the rich surface ores depleted and a precipitous drop in the price of copper, the mine closed.

Butte, Montana, copper magnate William A. Clark bought a seventy percent interest in the mine in 1887 when Phelps Dodge's James Douglas made a rare error and allowed his options on the mine to expire. After a visit to the mine in 1888, "Clark started buying the remainder of the widely scattered stock of the company and at the time of his death in March 1925, he and his family owned about 299,000 of the 300,000 shares issued" (Anderson and Creasey 1958). The United Verde made a rich man significantly richer.

Bisbee

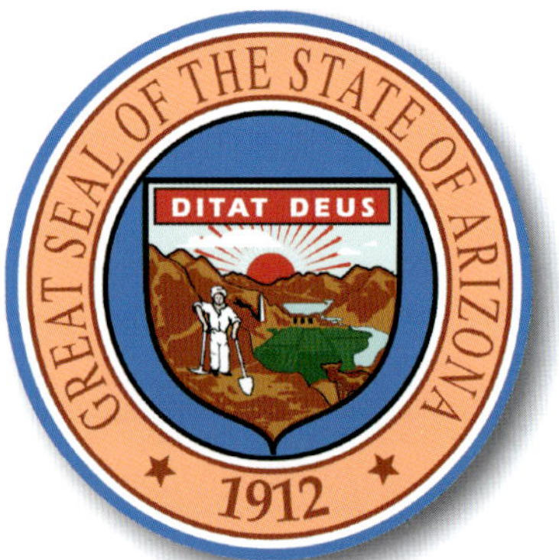

Upper: Bisbee founder George Warren posed for a formal portrait by Mary "Mollie" E. Goodrich Fly, wife and partner of C.S. Fly.

John Taylor collection

Lower: George Warren is pictured in the foreground of The Great Seal of the State of Arizona, the rendering of Warren is based on Mary Fly's above photo.

In 1876 the U.S. Government dissolved the Chiricahua Apache reservation in what would become Cochise County, and the Indians were forcibly removed to the desolate San Carlos reservation. Many refused to go. Hostilities in the area were renewed, and the Army began an effort to round up the "renegades," including Geronimo.

During the summer of 1877, a U.S. Cavalry troop from Fort Bowie under a Lieutenant John Rucker and with civilian scout Jack Dunn came into Mule Gulch searching for Apaches and hoping to find water. No Apaches were found, but a spring of bitter water flowed from the base of the rust-red hills in the wooded canyon. Dunn looked further up the gulch and found abundant, sweet water issuing from a limestone outcrop.

As Dunn returned to the waiting cavalry, he found cerussite float near a large, protruding mass of hematite. Together with Rucker, he staked the first claim in Mule Gulch and named it the "Rucker." Both men knew that more claims must be staked, but their duties forced them to return to Fort Bowie.

At Fort Bowie, Dunn and Rucker grubstaked hanger-on and sometime-prospector George Warren to stake more claims in Mule Gulch on their behalf. Warren went to the Mule Mountains all right, and he did stake a number of claims, but neither Dunn nor Rucker were named as co-owners.

Warren had cheated his sponsors, but justice was ultimately served. Before long George had either lost or been swindled out of his share of the claims he had staked. His ultimate punishment was to watch the claims he had once owned become some of the most valuable mines in the world while he scrubbed the dirt from the miner's boots off the saloon floors. His only reward was to be that the mining district was named for him; his image is also part of the Arizona Great Seal.

The 1877 discovery of minor lead mineralization at what would become Bisbee did little to foretell of the richness and beauty that lay below the surface of this remote desert canyon. Bisbee became Arizona's premier locality for mineral specimens; indeed it is arguably the most important specimen locality in the United States, both in terms of the quality and the number of specimens collected. Hundreds of thousands of specimens are preserved in collections worldwide. More than 320 distinct mineral species have been identified from Bisbee, making it one of the most mineralogical diverse localities in the world. No other U.S. locality has produced as many fine mineral specimens in the broad array of colors and forms over such a long period of time. During Bisbee's near-century of mining, there were few periods when good mineral specimens were not being found and preserved.

As with most Arizona copper deposits, the ores at Bisbee are related to a mineralized, porphyry intrusion. While the ores mined from the porphyry by the two open pits were important, it was the ores of many hundreds of scattered, limestone replacement deposits that yielded most of Bisbee's copper and all of its extraordinary mineral specimens. The Paleozoic limestones were locally mineralized with immense quantities of pyrite and lesser amounts of copper/iron or copper sulfides. Important amounts of lead and zinc sulfides as well as silver and gold were also emplaced during later

Right: "First copper smelted in Mule Gulch by George Warren," written in Ben Williams' handwriting, 18 cm wide

Bill Larson collection
ex Ben Williams
Mia Dixon photo

Below: Mining in the open cut of the Copper Queen Mine, 1881

Graeme-Larkin collection

pulses of mineralization. It is of course the supergene minerals for which Bisbee is noted—the copper carbonates, oxides, and native copper. These formed as uplift and erosion exposed the sulfides to an oxidizing, near-surface environment, something that occurred during two widely separate periods in its geologic history. It was the last of these uplift/oxidation events that brought the signs of mineralization to the surface as well as forming the best of Bisbee's supergene minerals.

Early efforts to mine and smelt the lead ended in failure, as did later attempts to exploit the several small and discontinuous copper outcrops nearby. The first hint of greatness came in August of 1880 with the serendipitous discovery of extraordinarily rich copper ores on the surface of the **Copper Queen** mining claim, where brush was being removed (Douglas 1881). Mining began in earnest soon after.

To be sure, these early years were, in hindsight, the classic era—the time when the mines at Bisbee became recognized as the source of extraordinary copper minerals, but it did not start that way. The original Copper Queen orebody was truly marvelous from a mine owner's perspective, as it was rich indeed, averaging 24 percent copper for a while and with ore right at the surface. Malachite was the principal ore mineral. It comprised fully one-half of all of the ore mined early on. Only minor amounts of azurite and cuprite were found mixed with the abundant malachite in this, the first orebody mined (Kuntz 1885). The same was largely true for the smaller, nearby **Copper Prince Mine**, which was exploiting an extension of the Copper Queen orebody on an adjacent property. Justifiably proud of their rich ores, both companies sent large examples of their ores to the Denver Mining Exposition of 1882, with the *Chicago Tribune* of August 17, 1882, noting "The largest specimen shown by this mine [Copper Queen] weighs 1,100 pounds and has a special case allotted to it by the Commissioner."

But there was more to the Copper Queen orebody than high-grade

Left: "Bullion Team at Bisbee"

A rare, original, circa 1880s, C.S. Fly albumen photograph of one of the mule teams hauling copper bullion to the railhead at Fairbank. The hand stamp on the back of mount reads, "Fly's Gallery/ Tombstone, Ariz. / C.S. Fly, Proprietor."

Becksted collection

Right: This 17.8 cm tall malachite specimen represents some of the very first ore mined from the Copper Queen Mine. Recovered in August 1880, it is one of the very few pieces to have been preserved from this historic orebody, which sent Bisbee on the road to becoming an immensely important mining district and, of course, an important specimen producing locality.

Freeport-McMoRan collection
Jeff Scovil photo

Above: Born in Québec, Canada in 1837, James Douglas was involved in the copper refining business in Phoenixville, PA, when asked by the partners of Phelps Dodge, a New York mercantile firm, to advise them on copper properties in Arizona—Morenci and Bisbee. He recommended both to the partners, who had the courage not only to invest in both but to become involved in the development of the mine at Bisbee. Douglas overcame all sorts of difficulties to make the Copper Queen Arizona's most important mine and in doing so he became the best known man in the world of copper. But to the people of Arizona and Bisbee in particular, Dr. Douglas was a caring, kind, and generous friend, one worked as hard to make their lives better as he did to make the mines successful. Importantly, this visionary man assured that the mineral treasures of the Copper Queen Mines were saved and freely given to scores of museums worldwide.

ores. A group of malachite specimens attracted the attention of George F. Kunz a few years later. In an 1885 paper, Kunz described several malachite specimens, one specimen

> *with a mammillary surface has a dozen stalactites measuring from 10 to 20 mm. in diameter and 5 cm. in height ... Malachite from the Copper Queen mine, forms thick crust and seams, geodes and cavities over one foot across, and entirely made up of fine acicular, velvety crystals—these often two to four mm. long—the whole surface presenting the rich sheen of green plush velvet.*

Kunz also noted

> *...wonderful pockets of transparent calcite and malachite, measuring 10 to 20 cm. in width, filled with tufts of brilliant green, translucent, acicular crystals of malachite from 5 to 10 mm. long, grouped with beautiful flat nail-head crystals of calcite, also transparent, measuring 35 mm. in width.*

Thus began the mineral specimen fame of Bisbee and its Copper Queen Mine, a fame that could have been short-lived, as by mid-1884 the Copper Queen orebody was nearly exhausted. Efforts by the Copper Queen Mining Company to locate additional ore had failed. Several years of diligent work by the Phelps Dodge-owned Atlanta Mining Company on the contiguous claim to the south, had been equally frustrating. Reluctantly, the owners of Phelps Dodge authorized Dr. James Douglas to make one last effort to find ore. A gloom of uncertainty hung over the young, ragged mining camp until fate intervened with the timely and near simultaneous discovery by the two companies of what became known as the **Southwest orebody**.

This large orebody extended into the property of both companies, bringing about the merger, or more accurately a marriage of convenience, of the two companies to preclude protracted litigation. Thus was formed the Copper Queen Consolidated Mining Company to mine the high-grade, specimen-rich Southwest orebody, which along with the flow of good ore, would yield a wealth of incredible mineral specimens. Dr. Douglas was president of the new company. Fortunately for mineral collectors this

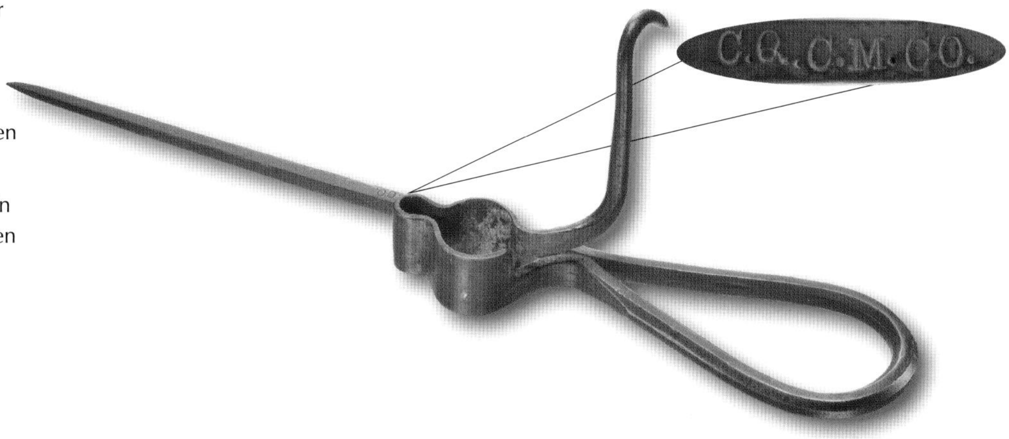

Right: Varney-style candleholder made in the blacksmith shop of the Copper Queen Consolidated Mining Company and marked "C.Q.C.M. CO." This is one of three known examples.

Becksted collection
Jeff Scovil photos

visionary man would ensure that innumerable specimens were saved, often with extraordinary effort. He was to lead the company to economic success during the most challenging of times, while providing the best wages and working conditions for the workers and investing in the community. Few men, before or since, have given so much.

Following the discovery of the Southwest orebody, but before the merger, the Copper Queen Mining Company began sinking the **Copper Czar** shaft on the recently acquired **General** mining claim (Williams 1884). This well-equipped, vertical shaft was much closer to the newly found ores and would be far more efficient for the exploitation of the new orebody than the now distant and somewhat small and inclined Copper Queen shaft.

Unfortunately it was not to be that simple: shortly before the August 1885 merger of the companies, the copper market began a precipitous slide. By year's end, copper had shed 60 percent of its value. With the transcontinental rail system 36 miles away from the fledgling district, the slide in copper prices made even the rich ores of this legendary orebody economically impossible to mine. The mines were closed for most of a year. Collectors would have to wait a bit longer for the jewels that lay just under **Queen Hill**, but a glimpse of what was to come was had with the limited discovery of impressive azurite crystals, some approaching five centimeters.

By early 1887 with improved market conditions, operations on the Southwest orebody restarted, and azurite crystals such as the world had never seen were soon emerging from both the Copper Czar and Copper

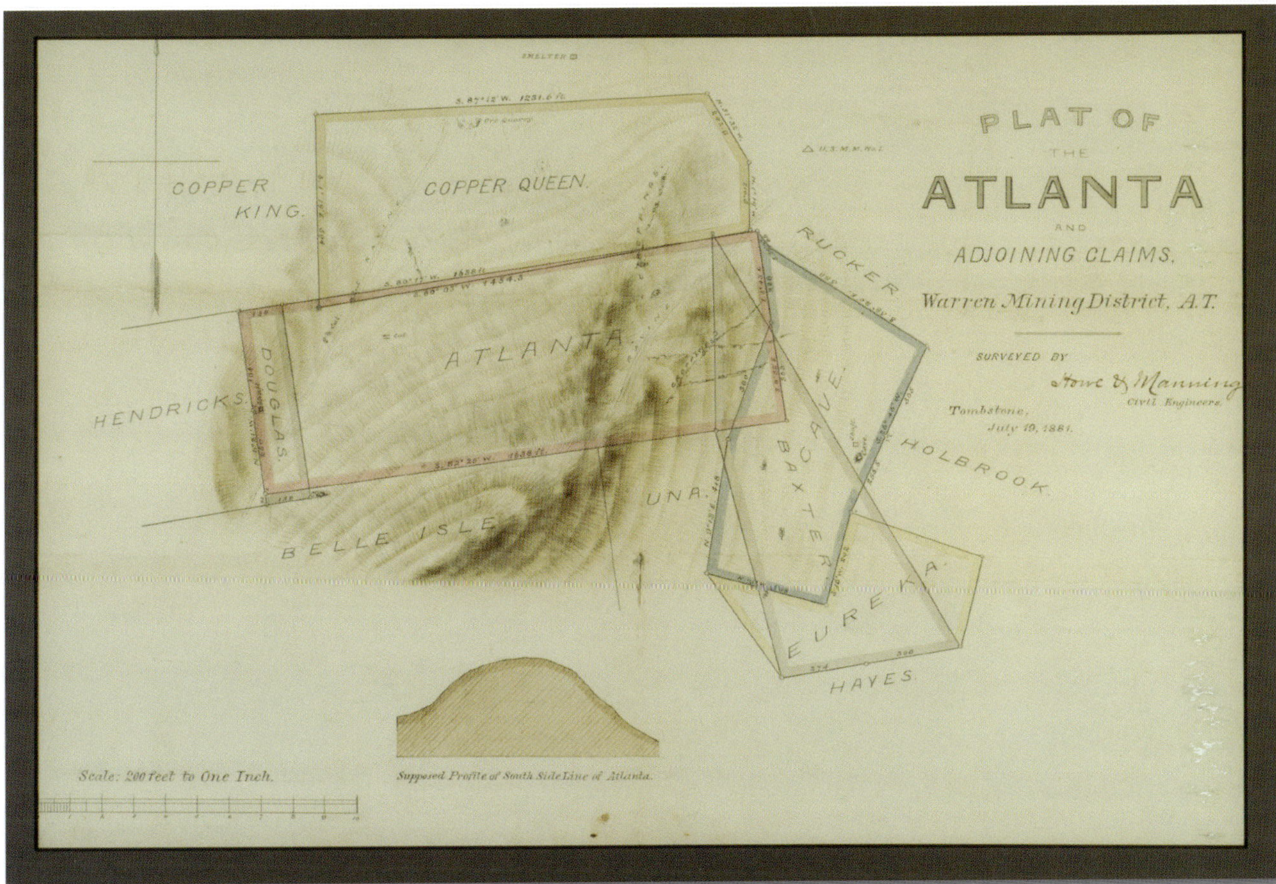

Above: A plat is a survey map drawn to scale that shows the horizontal projection of area occupied by a mining claim, with the side and end line lengths and bearings indicated. This plat of the Atlanta mining claim was drafted to illustrate the area which it occupied as well as to show it in relationship to adjacent claims, which is important as the other claims—Copper Queen, Hendricks, Baxter, and Cave—all overlap the Atlanta, as is common. The Atlanta had been located before all of the other claims except the Copper Queen; thus according to the General Mining Law of 1872 and the rules of the Warren mining district, the Atlanta had precedence to all of the overlap areas except the wedge-shaped sliver of overlap with the Copper Queen.

Plats were often used to demonstrate the claim location to a prospective investor as well as fix its position, as it was a common but unscrupulous practice to allow a claim to float, fixing the location only when the "best" ground had been determined. When a plat was used to help secure funding or to sell the property, the location of nearby ore and favorable rock outcrops were usually noted, as is the case with this plat. Indeed, this plat or one very similar to it would have been presented to Dr. Douglas and the partners of Phelps Dodge when they were deciding whether to purchase the Atlanta claim. Phelps Dodge ultimately purchased the unpatented claim for $40,000 in August of 1881. In February 1882, the Atlanta was resurveyed (*Tombstone Epitaph* 1882) and taken to patent, whereby full ownership and enjoyment of the surface and mineral rights passed from the U. S. Government to the Atlanta Mining Company.

Becksted collection
Jeff Scovil photo

Facing page, upper: Malachite after azurite, 5.4 cm tall

Bill Larson collection, Jeff Scovil photo

Facing page, lower: Malachite
38 cm wide

Freeport McMoRan Copper & Gold collection
Jeff Scovil photo

Azurite as a second-generation overgrowth on malachite pseudomorphs after azurite, 9.2 cm wide
Czar Mine, Bisbee, Cochise County
Les and Paula Presmyk collection
Jeff Scovil photo

Queen mines, and in volume. At the same time, other minerals were being recovered, including very fine malachite pseudomorphs after azurite (Friedrich 1888, Hills 1891) as well as fine botryoidal malachite masses.

The Southwest orebody was the source for all of the specimens, but for a while the ores were hoisted through both mineshafts as the company awaited the construction of a new, more efficient smelter adjacent to the Czar Mine as the Copper Czar was then being called. The shaft a specimen emerged from was often considered its locality; hence, the many near-identical Southwest orebody specimens are attributed to either the Czar and Copper Queen mines depending on which mine entrance was used to bring them above ground. When the smelter was completed in 1888, the original Copper Queen Mine was closed, but mining of the Southwest and other orebodies continued to produce ore and specimens through the Czar Mine.

Over the next 20 years, Copper Queen Consolidated developed and operated many mines in the Warren mining district. All of the newer mines were individually named, but they were still collectively referred to as the Copper Queen mines. Few collectors noted the specific mine that produced a particular mineral specimen, deferring instead to the long used and popular "Copper Queen Mine." Confusion as to the exact origin of many specimens persists to this day.

With the fine minerals emerging, it was not long before the noted mineral dealers of the time were in Bisbee. George English (1890) visited the district and wrote, "The finest azurites from any locality in the world (not even excepting Chessy), have been recently secured from the celebrated Copper Queen Mine at Bisbee. Both in size, perfection and brilliancy, they are unsurpassed." Most of these fine specimens were actually azurite overgrowths on malachite pseudomorphs after azurite (photo facing page); they are thus bluer than is typical for large azurite crystals. In many instances, the second generation of azurite covered only the tabular faces of the underlying malachite pseudomorphs after azurite, leaving the malachite edges exposed for even more attractive bicolor specimens.

Truly fine malachite in the form of large masses with velvet-like surfaces as well as stalactitic specimens was also found in this period. Aggregates of lustrous tufts of malachite crystals were offered by English (1890) as "...the most beautiful we have ever seen." He also raved, "Bisbee has so far produced the finest large cuprite crystals of any American locality, a few of them being fully as good as the best Cornish specimens."

During the latter part of the 1880s, the first of Bisbee's "type" species (those new to science) were recovered. The type specimens of spangolite (Penfield 1890, Palache & Merwin 1909) and paramelaconite (Koenig 1891) were probably found in the Czar Mine. While the character of the cuprite hosting both of these species was commonly seen in the original Copper Queen orebody, only modest quantities had been found. The other minerals associated with the type specimens strongly suggest a depth that at that time could only have been reached through the Czar Mine.

By 1888, Mr. James and Mr. Dodge, the principals of Phelps Dodge & Company, had acquired the **Holbrook** and **Cave** properties, located just south and east of the Czar shaft. Soon, rich carbonate ores were being mined and great mineral specimens were produced from the Holbrook Mine. Abundant geode-like pockets of satiny malachite with stalactites were frequently found in the soft, manganese-rich clay near the hosting limestone wall rock along the edges of orebodies. Botryoidal coatings of boxwork

Seven mineral species were first described from the mines at Bisbee. All are copper minerals; two of them—paramelaconite and spangolite—were described in the 1890s. An eighth type species, bisbeeite, $(Cu,Mg)SiO_3 \cdot nH_2O$, was described in 1915, but was later discredited based on work done not on the type material but on "bisbeeite" from Africa. Bisbee fans are lobbying for a review of the later work.

Chalcoalumite, $CuAl_4(SO_4)(OH)_{12} \cdot 3H_2O$
Graemite, $CuTeO_3 \cdot H_2O$
Henryite, $Cu_4Ag_3Te_4$
Kiddcreekite, Cu_6SnWS_8
Paramelaconite, Cu_4O_3
Shattuckite, $Cu_5(Si_2O_8)_2(OH)_2$
Spangolite, $Cu_6Al(SO_4)(OH)_{12}Cl \cdot 3H_2O$

A few of Bisbee's type species are among the most sought-after rare mineral specimens in the world.

Left: Spangolite, field of view 11 cm high
Czar Mine, Bisbee, Cochise County

Houston Museum of Natural Science collection, Jeff Scovil photo

Right: Paramelaconite, 1.9 cm tall
Czar Mine, Bisbee, Cochise County

Evan Jones collection
ex Jim and Dawn Minette collection
ex Bryn Mawr College
Jeff Scovil photo

Right: Azurite on iron-stained malachite
9.5 cm wide
200 level, Czar Mine, Bisbee

Scott Rudolph collection
Jeff Scovil photo

Above: Malachite on goethite stalactites
6.4 cm tall

Lance Hampel collection
Jeff Scovil photo

azurite and drusy azurite pockets were found in the ores as the mine deepened. The drusy pockets commonly contained stalactitic azurite as well. But as is always the case at Bisbee, azurite stalactites are azurite overgrowths of pre-existing stalactites of other minerals such as malachite or goethite.

Bright azurite roses, the best that have ever been found anywhere, were unearthed in profusion along the **Dividend fault**, on the 200 and 300 levels between the Holbrook and Czar mines. The roses occurred in the immense mass of plastic clays and along the hanging wall of this massive and treacherous fault. The first of these trademark specimens (see photo page 50) were mined in 1888, but the near impossible mining conditions created by the crushing power of the almost fluid clays prevented miners from recovering much ore and specimens from this area. For the next 40 years, efforts to thoroughly mine this 650-foot-long area were continually frustrated, but each attempt produced more azurite roses.

Brilliant radial clusters of azurite on botryoidal malachite came from elsewhere on the 200 level of the Czar Mine. In some cases, the malachite was lightly iron-stained (photo above). Few such specimens were produced, but they remain among the most striking azurite-malachite combinations ever found.

By 1890, mining on the 200 level of the Czar had advanced both to the east, connecting with the Holbrook Mine and to the south into the **Belle Isle** mining claim. The Belle Isle was a treasure house where great quantities of wondrously beautiful masses of botryoidal azurite, commonly with malachite, were found. The Czar exploited a number of scattered, rich orebodies that produced a wide variety of botryoidal forms and hues of azurite, often enhanced by stunning bands or splashes of malachite. A 3.5-ton brick of massive azurite and malachite was exhibited at the World's Fair in 1893 and is now the centerpiece of the mineral display at the American Museum of Natural History in New York.

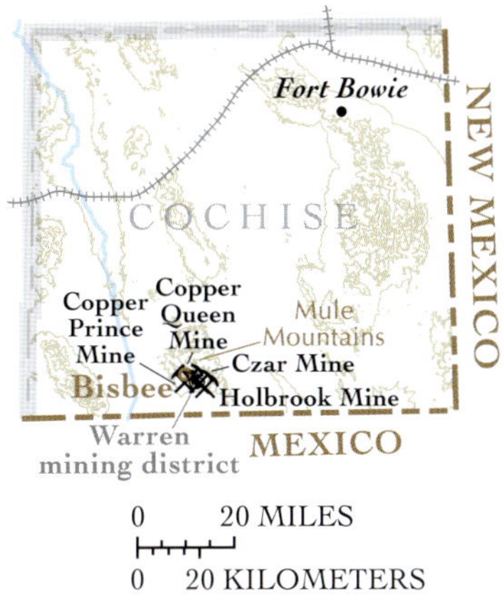

Copper Queen Consolidated made a surprisingly concerted effort to assemble a substantial collection of fine, quite large specimens for presentation at the 1893 World's Columbian Exhibition in Chicago. In addition to the large block noted above, scores of other fine specimens were on display. It may well have been the best public display of Bisbee minerals ever put together. Specimens from the Holbrook Mine were included in the Columbian Exhibition, as the Holbrook had by then been folded into the Copper Queen Consolidated mine group. *The Chicago Tribune* noted

> *... two malachite carbonate specimens from the Holbrook Mine of a green hue and covered with specks of silver... In the Holbrook Mine is a cave 1,200 feet long in which has been found remarkable formations of stalactite and stalagmite, several specimens of which give variety to the collection of copper minerals in the various showcases.*

The display of mineral wealth must have been impressive; indeed, the object of the display was to showcase the potential wealth that these specimens represented. That they were beautiful was a bonus. Arizona was struggling to achieve statehood independently from the largely agrarian territory of New Mexico. Many in Washington argued that if the two were admitted separately, Arizona would become an economic burden on the country. The thinking was that it thus must be combined with New Mexico as such a large state would have ample resources to ensure self-sufficiency. The battle against this perception was not to be won at the 1893 World's Fair, but many marveled at the richness and the beauty of the samples on display there (and at subsequent such events) from the Copper Queen's mines in distant Arizona.

The lapidary potential of the striking material mined from the Belle Isle

The Copper Queen Consolidated Mining Company collected this and a number of other specimens from the Holbrook Mine in 1891 for the sole purpose of displaying them at the World's Colombian Exposition of 1893 in support of Arizona's protracted bid for statehood. An important, private effort which eventually brought the goal to fruition. This piece is both the largest and finest of its type, bar none. 49.2 cm wide

Freeport-McMoRan Copper & Gold collection, Jeff Scovil photo

Right: Azurite, 5.9 cm tall

Steve and Carol Maslansky collection
Jeff Scovil photo

Lower: Azurite and malachite
8.4 cm wide

Paul Harter collection
Jeff Scovil photo

Facing page: This unique, 19.2 cm tall azurite with malachite was in the personal collection of Ben Williams, beloved and respected first superintendent of the Copper Queen Mine. Williams oversaw mining operations from 1880 until 1899. Bill Larson purchased Williams' collection at auction in the mid-1990s.

Bill Larson collection
Jeff Scovil photo

Claim was recognized as well. Valiant (1897) noted that in 1891 more than 2,000 pounds of "choice Bisbee specimens" were cut and polished in Denver, Colorado, to be used as cabinet specimens and ornaments.

One peculiarity of the oxidized copper ores at Bisbee is their close association with caves. Caves of stunning beauty had formed directly over a many of the orebodies as a result of the volume reduction of the ores that was caused by their oxidation. Many were large and some were downright huge; all were well decorated with typical cave calcite and aragonite speleothems. And in many instances the calcite stalactites were tinted green by copper substituting for calcium in the crystal structure and/or malachite inclusions. Aragonite stalactites and other forms acquired a striking bluish hue caused by the copper substituting for calcium. The array of copper-colored forms coupled with white formations and those streaked with the reds and browns caused by iron made these caves a visual feast. As the caves faced certain and necessary destruction through the course of mining, collectors recovered as many specimens as possible. In 1890 George English asked Walter Niven to collect one magnificent cave from which 29 boxes of specimens were recovered and sold by English (*American Journal of Science* 1891). While many caves were found over the next 40-plus years, surprisingly few of the fascinating specimens recovered have been preserved.

As mining expanded into deeper areas, cuprite became more common, as did some of the more rare and unusual minerals associated with cuprite such as connellite and spangolite. Native copper also became more common with depth. Native copper was typically found as isolated clumps and masses encased in the abundant supergene-derived clays. A simple washing removed the clay to reveal handsome clusters of intergrown, elongated crystals or copper masses with crystalline surfaces. Very large masses of pure copper were often left behind with the waste, as it was not worth the substantial effort to cut them into manageable pieces. One of the authors (RG) observed a mass of copper several meters in each dimension that had been encountered on the 400 level of the Holbrook Mine. It had been left in place, and the miners had worked around it.

The Holbrook Mine was producing quantities of botryoidal azurite that showed dramatic variations of hue as a result of minor zinc inclusions. Light green malachite was frequently associated with this azurite and made for striking specimens. At and above the 400 level, the Holbrook Mine produced innumerable fine specimens of azurite and malachite, both of which had unique velvety surfaces.

In 1895 the world's greatest azurite find was made in a small cave lined with crystalline azurite. A witness to the find described,

> *... a room, not too big, perhaps 50 feet in curved length and 20 feet high and 15 or so feet wide. The walls were all manner of irregular lumps of black azurite dotted with malachite. From the back* [ceiling] *hung limonite stalactites with azurite crystals here and there on them. The floor was mostly a thin crust of blue* [azurite] *on malachite.* (Graeme 1981)

The cave, on the edge of the Southwest orebody, occurred in massive, impure goethite. The opening likely formed along a marginal, oxidation

Above: Calcite, 22.4 cm wide
Southwest Mine, Bisbee, Cochise County
Freeport McMoRan Copper & Gold collection
Jeff Scovil photo

Facing page: Calcite and aragonite 22.1 cm tall
Czar Mine, Bisbee Cochise County
Graeme collection
Jeff Scovil photo

subsidence related crack.

Multiple periods of supergene activity are evidenced by the specimens. Azurite was deposited first. It was partially altered to malachite at much the same time as malachite was also being deposited. Minor goethite was deposited next, followed by another deposition of azurite, and finally a minor deposition of cuprite; much of the latter at least partly altered to malachite. Much of the malachite has a light coating of goethite and reflects minor corrosion by the iron-rich solutions. Broken faces of this malachite occurrence show a unique degree of luster and chatoyance.

Several forms of azurite were found in this exceptional cave. Much of it formed as spherical clusters of tabular crystals in grouping reminiscent of bunches of grapes. The clusters formed on goethite and/or goethite-stained malachite and are sprinkled with tiny, malachite coated cuprite octahedra. Hundreds if not thousands of fine, often large, azurite specimens were

This fine 21.3 cm wide specimen was collected from the Czar Mine in 1895. The miner who found it presented it to a waitress in the Family Restaurant at the mouth of Brewery Gulch (the proper end; the other, or upper, end was the red light district). The miner was trying to win the affections of the young woman, who kept the specimen but remained unmarried for some time. In 1902 she gifted the specimen to her brother Oliver Brown when he moved to Bisbee to work in the mines. When author Richard Graeme purchased Mr. Brown's collection in 1972, Brown relayed the story to Richard with great passion.

Graeme collection, Kevin Dixon photo

collected from this cave. Numerous specimens exceeding 50 centimeters across are in collections worldwide. Fine smaller pieces are relatively abundant and are treasured by their owners.

In some cases the azurite in the cave had formed as a bright, second-generation overgrowth on tabular malachite pseudomorphs after azurite and was on a goethite matrix. Crystals of this type in excess of five centimeters were not uncommon. Bates (1896) reported:

> ... *in 1895* [the Copper Queen] *yielded a large quantity of extraordinary azurites, malachites, cuprites and calcite-aragonite stalactites. The azurites occur on a musty brown matrix in masses of sharp crystals, or in isolated well defined crystals (some of which are doubly terminated), and on a hardened malachite in drusy crystals. Some quite large crystals are coated with, or have altered to malachite. Indeed, it is a common experience to discover azurites from this locality partially changed to malachite after being in the cabinet for years. Some specimens*

Right: This 7.8 cm wide azurite and malachite originated in a massive, crystal-lined cave on the edge Southwest orebody that was encountered in 1895. Hundreds of specimens were recovered from this occurrence.

Evan Jones collection
ex American Museum of Natural History
Jeff Scovil photo

Below: Holbrook Mine (left) Czar Mine and Copper Queen Smelter, 1896

Graeme-Larkin collection

are sprinkled over with minute green crystals of cuprite; others show the most delicate tufts of malachite, and some show all of these minerals in harmonious combination. The stalactites were found in fantastic fringes, in pipe-stems, in bunches of coraloidal forms and gnarled masses, from pigmies to giants in size. Some are snow-white; some are a delicate green. Others are white sprinkled with green dots, and some are of a silken sheeny luster.

Specimens of both crystalline azurite and botryoidal malachite continued to be collected from the Southwest orebody until it was exhausted, having been mined for many years.

During the latter half of the 1890s, other, smaller orebodies mined through the Czar shaft sent forth azurite and/or malachite in a variety of forms and hues. These added to the impressive array of minerals flowing from the nearby, prolific Holbrook Mine as well as the recently opened **Silver Spray Mine**. In all, more than a million tons of ore were mined from this single orebody, and countless specimens were collected.

Unfortunately, the ore deposits in the newer mines were at deeper levels where oxidation was less complete; thus, they produced fewer stunning specimens. But the flow of great Bisbee minerals was far from over. Older areas would be re-mined, and new and important oxidized ores were yet to be found. As the new century was about to break, Bisbee was the world's premier mineral specimen producer, unmatched in the variety, quantity, and quality of its mineral specimens. And, though barely known outside of Arizona, the 400-foot-deep Czar Mine had established itself as the source of the best of Bisbee.

Upper: Souvenir copper ingot from the Copper Queen smelter at Bisbee, circa1890

Roger Becksted collection
Jeff Scovil photo

Lower: Azurite on malachite
10.1 cm wide

Ron Gladnick collection
ex Ben Williams
Jeff Scovil photo

The Ben and Lewis Williams Loving Cups

For their first 19 years in production, the Copper Queen mines were operated under the close and capable management of Ben Williams, their first superintendent. Ben and his brother Lewis came to Bisbee in 1880 from the San Francisco firm of Bisbee, Williams & Co., to oversee an investment the firm had brokered in a copper mine located in Mule Gulch. Mining began in earnest with their arrival. The tiny mining camp came to be called Bisbee for D.W.F. Bisbee of the same firm.

Ben was responsible for the whole of the operations, while Lewis built and then oversaw the smelter. The success of the Copper Queen from the outset, greatly aided by the high–grade ores and ease with which they could be smelted, was credited to the able leadership of Ben, who proved himself an exceptional manager during the most difficult times. Both Ben and Lewis were liked and respected as men of impeccable integrity and as caring managers.

In November 1899, the town received the news of their resignations (*Tombstone Epitaph* 1899), and the mine workers and townspeople joined with the Copper Queen to give an appropriate token of their collective affection and esteem to the Williams brothers.

The people of Bisbee commissioned Shreve and Company of San Francisco to fashion two large, sterling silver (with gold decoration), three-handled loving cups. The cups were presented to each of the Williams brothers by a grateful populous on December 20, 1899. The cup that was given to Ben is pictured above. Ben and Lewis left for California the following day (Williams 1899), but they frequently returned to Bisbee.

Becksted collection
Jeff Scovil photos

CHAPTER 6

Coming of Age

Above: The Western Federation of Miners was a militant miners' union organized in 1893 in Butte, Montana. This Globe ribbon is one of the few surviving examples from Arizona mining camps.

Dave Bunk collection
Jesse La Plante photo

Left: Cuprite and copper, 3.5 cm wide
Irish Mag Mine, Bisbee, Cochise County

Mark Hay collection
ex Clarence Bement collection
Jeff Scovil photo

By 1900 copper defined Arizona; mining was the territory's largest industry, and the state was growing rapidly. At the turn of the century, the census listed Arizona's population as 123,000; a decade later it was 200,000. The four giant copper areas—Bisbee, Morenci, Globe, and Jerome—accounted for more than one third of the territory's population, and whole industries sprung up to support the production of 300 million pounds of copper annually. In 1910 the flag that would eventually fly over the Arizona Statehouse was designed with a large copper star at its center.

Copper dominated nearly every aspect of the lives of Arizona's citizens: politics, infrastructure, labor, and the fight for statehood. The copper giants faced many of the same issues; but the solutions had to be tailored to the unique needs of the individual camps. All had to build the complex infrastructure required to process hundreds of thousands of tons of ore annually. Each invested in their communities to attract and keep skilled workers. The mining companies innovated as they dug deeper and developed new means to process the changing and commonly lower grade ores that they were encountering.

Bisbee and Phelps Dodge were the quintessential story of Arizona in the first quarter of the twentieth century. By 1900 the Copper Queen Consolidated mines at Bisbee were the richest copper mines in the world. There were literally hundreds of miles of underground workings and the mines employed thousands of workers.

Phelps Dodge Corporation largely controlled the town, at least until the entry of other mining companies. The company owned the largest mercantile store. It also built schools and hospitals, created playgrounds, and stocked libraries. The company provided services to the community that, then and now, can be viewed as progressive, paternalistic, and/or profiteering. But regardless of their ancillary motivations, the corporations had to build infrastructure, including railroads, power lines and plants, water works, and telephone systems in order to viably mine the deposits and attract a skilled workforce to the remote desert town. In building that infrastructure, Phelps Dodge's influence was extended far beyond the confines of the towns in which it was operating copper mines.

Two transcontinental railroads, the Atchison, Topeka and the Santa Fe (usually just called the Santa Fe Railway) and the Southern Pacific Railroad opened up the state in the late nineteenth century. Between 1880 and 1920,

According to state statute, "The flag represents the copper star of Arizona rising from a blue field in the face of a setting sun." Charles Wilfred Harris designed Arizona's flag in 1910. Blue and gold, the state colors, were the colors carried by Coronado's 1540 expedition. The shades of blue and red are identical to those used in the United States flag, the 13 rays represent the original 13 colonies; the star its largest industry. The state legislature adopted the flag in 1917.

Mantags, or "brass," are kept on a large board in the timekeeper's office of an underground mine. Miners report to work and collect their tags before going underground. A look at the board instantly tells the shift boss whether all his men have reported for work. At shift's end, miners return their tags to the board. And again, the shift boss can see at a glance whether all of his men are out of the mine. A miner keeps his brass on his person at all times, and it serves as a form of identification in the event of a disaster.

Richard Graeme collection and photo

copper companies built 46 rail lines to connect to these arteries (Sheridan 1995) in order to make the remote, marginal operations more profitable. When Arizona became a state in 1912, there were more than 1,700 miles of track. The copper companies and the transcontinental railways played hardball with one another for a time before deciding that it was in the best interest of both to work together; the copper companies bringing the railroads needed business and the railroads offering favorable rates. The relationship was ultimately such that Walter Douglas, Dr. James Douglas's son, became president of Phelps Dodge, built El Paso and Southwestern Railroad, and sat on the board of the Southern Pacific Railroad.

As the largest employers in Arizona, the copper corporations dictated the conditions of labor, albeit offering better wages and working conditions in the state than those seen in other mines throughout the west. Arizona's mines, Bisbee in particular, required a stable and skilled workforce, and Phelps Dodge knew it had to strike a balance. But the worldwide labor movement came into open conflict with mining and manufacturing all over the country for a quarter of a century.

The first major mining clash in Arizona occurred at the Old Dominion Mine in 1896 when management imported Mexican labor to replace parts of the Anglo work force and dropped the prevailing wage by 30 percent. The Anglo miners responded by marching on the mine supervisor's home, kidnapping him, and forcing him to parade on the streets with a hangman's noose over his neck. Old Dominion relented, and the miners, flush with victory, voted to join the Western Federation of Miners (WFM), a radical union, which proclaimed, "complete revolution of social and economic conditions was required for the salvation of the worker." The mine operators stuck back by shutting down the mine and locking out the union members. The operators won the support of the territorial governor who declared martial law to protect the mine owners' property. Thus began the calculus of politics of Arizona as it made its journey toward statehood: populous revolution versus corporate domination.

The highly skilled miners were being replaced by increased mechanization at the mines. This trend began with the introduction of pneumatic drills, which when combined with trolley rail-haulage underground and concentrating plants on the surface, allowed for enormous quantities of rock to be mined and processed. Modernization permitted the mining of much lower grade ores, mostly porphyry, by block caving methods at both Morenci and Ray by 1910.

The early drills in porphyry were a deadly combination as clouds of silica-laden dust filled the air and the lungs of the drilling miner. Many succumbed to "miner's consumption" (silicosis) before a technical solution to controlling the drill dust was found. The dust problem and the trend toward more mechanized mining methods (i.e., miners being replaced by machines) were quickly made manifest as labor problems.

By 1900 the majority of Arizonians wanted statehood for the territory. The dramatic battle for statehood played out on a national stage. The debate was strongly along party lines—Democrat versus Republican—with control of the United States Senate in the balance. Republicans were concerned that if Arizona were to become a state, it would send two Democratic senators to Washington and upset the national order. Arizona was hardly alone in this partisan battle—the political balance was at issue with every state that was admitted at the end of the nineteenth century. In the early part of the twentieth century, a Republican senate proposed admitting Arizona and

At the Democratic and Republican Territorial Conventions, held in Bisbee on September 6, 1906, both parties voted overwhelmingly to reject the idea of joint statehood with New Mexico.

Above: Graham County delegate's ribbon
Becksted collection
Jeff Scovil photo

Right: A 7.6 cm wide copper cigar holder
Becksted collection
Jeff Scovil photo

Below: The Holbrook stony meteorite was seen falling to earth early on the evening of July 19, 1912, just five months after Arizona became a state. It fell just east of the town of Holbrook in Navajo County along the Santa Fe Railroad tracks. Witnesses to the event picked up many pieces, the largest being 6.6 kilograms. Shortly after the event, more than 14,000 individual meteorites totaling an estimated 218 kilograms were acquired by W.M. Foote of the Foote Mineral Co. and were distributed to collectors and museums around the world.

Peter Marmet collection and photo

Flag (1960–1966)
Jasper Johns (b. 1930)

The present 50-star flag of the United States of America is the only version of the flag that has been in use longer than the 48-star version used from July 4, 1912, following admission earlier in the year of New Mexico and Arizona as the 47th and 48th states. The 48-star flag was flown until Alaska was admitted to the union in 1959. The current 50-star flag has been in use since Hawaii's admission in 1960.

Artist Jasper Johns began working on his signature *Flag* series of paintings, prints, collages, drawings, and sculptures in 1954, immortalizing the now retired, 48-star American flag. Once in the collection of author Michael Crichton, the *Flag* pictured here is a collage of printed paper on canvas.

Private collection

New Mexico as a single large state; New Mexico had a larger population than Arizona, and it usually voted Republican. The idea enraged the Arizona populous and united business interests and the progressives. The debate got ugly as some Arizona residents repeatedly pointed out that the majority of New Mexico's population was Hispanic. Some of the unions echoed this concern, hoping to bar the mines from using Mexican labor. In 1906 the two territories voted on whether to join the Union as a single state: New Mexico overwhelmingly approved the resolution, while Arizona voted 16,265 to 3,141 to reject it. The issue of joint statehood died with that vote; the inevitability of Arizona statehood was clear.

In 1910 Congress passed an enabling act that would admit Arizona and New Mexico as states once they had approved constitutions. In Arizona, drafting of the constitution fell to a convention, which was a portrait of Arizona's politics. Three groups of delegates tried to hammer out an agreement: Republicans, Democrats, and Labor along with anti-corporation progressives. The Democrats eventually worked with Labor to build a powerful coalition—one of the most successful alliances of progressive groups in American history. The convention passed a liberal constitution, which included voting rights for women eight years before suffrage and the ratification of the nineteenth amendment. However, Labor's focus on prohibiting Mexican workers ultimately alienated the businessmen and significantly weakened the party's hold on the state. Arizona became the forty-eighth state on February 14, 1912.

After statehood Labor moved rapidly to consolidate the gains that it perceived it had won in the constitution. By 1915 most of the mines had endured major strikes, and the mine operators were forced to make considerable concessions. It looked like Labor was on a roll. The party's next target was the tax structure. It had been working before statehood to increase the assessed values of the mines, and this effort was being pushed energetically pushed forward. The move provoked the ideological argument

The Independence Day parade in Bisbee in 1912, just a few months following Arizona's acceptance as the 48th state in the United States of America

John Taylor collection

that raising taxes would price Arizona's ores out of the market. The copper companies began to target individual legislators and bought controlling interest in nearly every newspaper in the state. The outbreak of World War I brought a dramatic rise in the price in copper. The war also carried with it increased nationalism and anti-Socialism sentiments thus enabling the Republicans to capture the statehouse in 1916.

In early 1917 Arizona became a battleground that was mirrored across the globe. Small, militant unions focused their attention on American industry and appealed for a "worker's revolution." The movement was rooted in Marxism and targeted traditional unions and corporations alike. The Industrial Workers of the World (IWW or the Wobblies) spearheaded this movement by taking action against agriculture in the midwest, meatpacking in Chicago, and manufacturing in the east. In Arizona the union forced showdowns at the mines.

The group staged daily walkouts, and some 20 strikes at mines across the state were ordered. When the strikes began, the copper companies went on the offensive and branded the IWW as seditious. The companies used their newspapers to foment mistrust of the "Wobblies" and raised the specter of treason. They also promoted unions more sympathetic to the companies. Working with law enforcement to deputize nonunion miners and businessmen, the companies began taking action against the Wobblies. In

Above: A 1917 I.W.W. booklet *Sabotage*, written by Walker C. Smith, espoused the extreme organization's ideology and was ironically used in court to support the mine owner's position that the I.W.W. was conspiring to do unlawful damage to company property and that deportation was warranted. While the organization later denounced sabotage as doctrine, it continues to use the black cat theme, which came to symbolize the philosophy of sabotage.

Upper right: Company Guards defend the Old Dominion Mine during a 1917 miner's strike at Globe.

Arizona Historical Society collection

I.W.W. members are marched to Warren (**middle**) where they are loaded onto boxcars for deporation (**lower**).

John Taylor collection

Jerome a group of anti-Wobblie vigilantes herded more than 100 "troublemakers" together and held a mock trial. More than half of the captives were found guilty, placed in cattle cars, and deported 170 miles to Kingman. Successful in Jerome, the tactic became Phelps Dodge's policy in Bisbee. Walter Douglas, by then president of Phelps Dodge and the El Paso and Southwestern Railroad, gave a public speech that summed up the battle:

> *There will be no compromise because you cannot compromise with a rattlesnake … I believe that the government will be able to show that there is German influence behind this movement … it is up to individual communities to drive these agitators as has been done in other communities in the past* (Sheridan, 1995).

Early on the morning of July 12, two thousand deputized citizens rounded up strikers and sympathizers in Bisbee and the surrounding communities of Lowell, South Bisbee, and Warren. They were herded to a fenced ball field in nearby Warren. C & A mine manager John Greenway told the strikers that they were free to go if they renounced the strike, and about 800 men did. The remaining strikers—more than 1,100 men—were loaded into 23 boxcars, taken to southern New Mexico, and abandoned in the desert near the town of Columbus. The mass kidnapping appalled many, but despite legal actions over the next several years, no one was found guilty of any crime associated with the deportation.

Following the deportation, companies such as Phelps Dodge created their own employee associations to address employee concerns, and the corporations were soon firmly in control of their destinies. The labor struggles that marked the first two decades of twentieth century Arizona settled into stable and prosperous communities that supported and were sustained by big copper.

Bisbee

The turn of the century was a time of explosive growth for mining at Bisbee. Several new players entered the high-stakes copper game, most risking huge sums on the gamble that the ores, which the Copper Queen Consolidated Mining Company had mined for so long and so profitably, could be found in places where the company had not yet looked. A few won this gamble and with those successes came many wonderful and different mineral specimens.

First among the winners was the Calumet and Arizona Mining Company at the **Irish Mag Mine**. The C&A, as it was locally known, had burned through a good deal of money before finding ore of any real consequence in the "Mag." Here the ores were deeper, much deeper than the company had hoped. While scattered bunches of ore had been found on the 750 and 850 levels between 1900 and early 1902, a significant orebody was not discovered until reaching the 1050 level. This 1902 find made the C&A a viable company. For the mineral collecting world, the find produced the marvelous cuprite and copper specimens for which the mine became famous. But here too, confusion about the real source of the specimens would arise, just as it had at the Copper Queen.

Upper: Cuprite and copper, 4.4 cm tall
Irish Mag Mine, Bisbee

Evan Jones collection
ex American Museum of Natural History collection, Jeff Scovil photo

Lower: Polished slab of copper after cuprite, crystals to 2 cm wide, on (black) delafossite, field of view 5 cm wide
Irish Mag Mine, Bisbee

Tom Hughes collection
Jesse La Plante photo

Right: Brochantite, 6 cm wide
Shattuck Mine, Bisbee

Les and Paula Presmyk collection
Jeff Scovil photo

Lower left: Azurite on malachite
11.8 cm tall
770 level, Junction Mine, Bisbee

Graeme collection
Jeff Scovil photo

Lower right: Mining underground at Bisbee, circa 1910

John Taylor collection

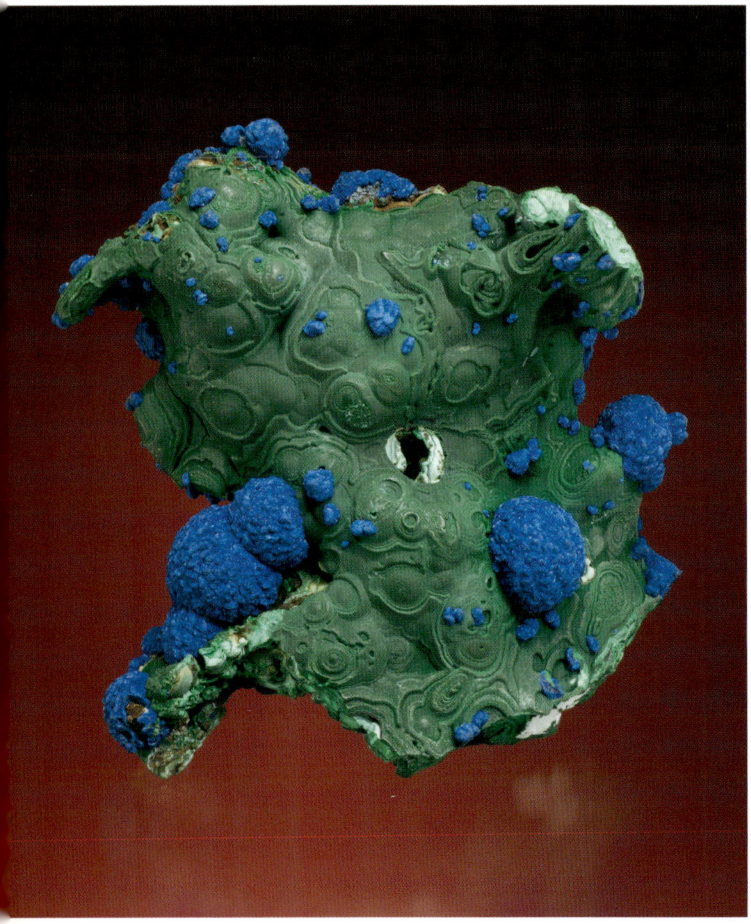

While there was never a mine at Bisbee named the C&A or Calumet and Arizona Mine, many specimens carry period labels with such mine designations. The early specimens are clearly from the Irish Mag Mine, although later specimens could be from any of the other several great mines developed by the C&A as the collective term of the "C&A Mines" was applied to all of these mines.

The Irish Mag cuprite specimens were the best that Bisbee had produced to this point. Most consisted of octahedra on copper or goethite with some exceptional crystals exceeding three centimeters on an edge. Some of the best specimens collected consist of black goethite densely covered with sharp, centimeter-size cuprite octahedral.

Upper: Dioptase on malachite
field of view 1.5 cm wide
Shattuck Mine, Bisbee

Graeme collection
Jeff Scovil photo

Lower: Delafossite
field of view 1.8 cm tall
Irish Mag Mine, Bisbee

Marcus Origlieri collection
Jeff Scovil photo

Uncommonly fine pseudomorphs of copper after cuprite also came from this mine. Faithful replacements of modified cubic crystals of up to 2 centimeters were found, although crystals of this size were rare. Rogers (1910) discusses these great specimens. Connellite, that uncommon blue, hydrous copper chloro-sulfate, which was frequently found in small amounts with cuprite at Bisbee, seems to have been more abundant in the Irish Mag ores than those of the older mines. Palache and Merwin (1909) studied connellite from the mine and noted the presence of chalcophyllite. Although a rich and profitable mine, the Irish Mag produced little in the way of azurite and malachite specimens. Other C&A mines, notably the **Junction** and **Cole**, did produce important numbers of specimens of both of these copper carbonates.

In March of 1905, while the Junction Shaft was being sunk, a zone rich with azurite and malachite was hit at a depth of 700 feet. This orebody was in a soft manganese oxide, which assayed at 22 percent copper (Calumet & Arizona 1905); thus implying that nearly half of the rock mined was one or both of these copper carbonates. Scattered, isolated azurite and malachite masses as large as 40 centimeters were found in the soft, black mud. Most of the azurite occurred as bright, light-colored overgrowths of spherical crystal clusters to 3 centimeters, or as pisolitic forms, all on irregular shaped malachite. The combination made for very attractive specimens. Thin crusts of malachite on box works of manganese oxides and scattered with light blue azurite specks also came from this area. Some of the azurite from this rich zone formed small, light blue, dull, rosettes, which commonly contain fossil fragments.

Rogers (1913) recognized the uncommon copper iron oxide delafossite in C&A's nearby **Hoatson Mine**. While the original occurrence was ordinary, later discoveries in the Irish Mag yielded the finest specimens of the species ever found. Calcite in vugs with copper was also found as fine specimens in this small but economically important mine (Tovote 1911, Tupper 1915).

Another winner in the copper mining gamble was the Shattuck and Arizona Mining Company with its **Shattuck Mine**. The mine was situated at the northern edge of a cluster of eight mining claims and about two-thirds of a mile south of the original Copper Queen Mine. The shaft was started in 1904, and the first ores shipped in 1906 (Weed 1918). Often called "the richest little mine in Arizona," the Shattuck produced a variety of

Right: A circa 1910 printed postcard of the "Shattuck-Arizona Mine"

John Taylor collection

Below: Malachite, 6.5 cm tall
Czar Mine, Bisbee

Evan Jones collection (ex George Feist)
Jeff Scovil photo

Left: Cuprite on graemite, 2.7 cm tall
Shattuck Mine, Bisbee

Graemite is named for author Richard Graeme who discovered the type specimen in 1959 in the Cole Mine. The species was described and named by Sid Williams and Phil Matter.

The specimen pictured here is the only other graemite specimen extant. It was recovered from the Shattuck Mine in the first quarter of the twentieth century, but the mineral was misidentified until Graeme's late wife Monica recognized it in 1992 in the drawers of the Arizona-Sonora Desert Museum when she was photographing specimens there.

Arizona-Sonora Desert Museum collection
Jeff Scovil photo

exceptional mineral specimens, most notably fine botryoidal malachite and azurite. One stope on the 200 level produced malachite stalactites to seven centimeters long that are adorned with clusters of brilliant 3-millimeter-long azurite crystals.

The Shattuck also yielded the very best brochantite specimens ever found anywhere, and in abundance. Splendent masses of jackstraw crystals, some 7 millimeters across and 5 centimeters in length, came out. Antlerite crystals to several centimeters long and in clusters of brilliant 5 millimeter-long crystals came from the mine; these too are the best of the species from anywhere. Palache (1939a and 1939b) described the tandem occurrence of these paramorphs using material from the Shattuck Mine.

Shattuckite and bisbeeite were both described as new species from material collected on the 200 level (Schaller 1915) of the Shattuck Mine. The validity of bisbeeite as a species is in doubt (Oosterwyck-Gastuche 1967), additional work on the type material is needed to fully discredit it. Interestingly, in addition to these copper silicates, a small number of dioptase specimens were recovered nearby. A small cave filled with stalactitic and botryoidal mottramite (Wells 1913) was found on the 600 level. The cave produced a wealth of interesting samples at a time when vanadium minerals were considered rather uncommon. On this same level, a huge body of hematite containing lustrous, botryoidal specimens reminiscent of the classic English material was encountered. Even some of the massive gold specimens from this mine are impressive for their richness; the contrast with the hematite matrix also makes them attractive specimens. By 1925 the Shattuck had largely been mined out and taken over by lessees.

In the early twentieth century, all of the mines of the Copper Queen Consolidated Mining Company were still producing a nearly constant, though reduced, stream of important specimens. That group of mines at this point included the new **Gardner**, **Sacramento**, and **Lowell** operations. While the Czar and Holbrook mines were still yielding fine examples of both

azurite and malachite, most were massive or botryoidal, many with drusy surfaces. The production continued for some time as small, new orebodies were discovered.

In time, however, the ubiquitous, mineral-rich, supergene clays around the **Holbrook** forced the great mine to close. The clays seemed in constant motion as mining progressed. By 1906 it had become impossible to keep the shaft aligned to allow for the safe passage of the cage used to hoist and lower men, materials, and ore. A new Holbrook shaft was sunk a few hundred feet away in hard limestone. It accessed the same ores and was likewise a source of great specimens. It was during this time that a small number of exceptional calcite with inclusions of cuprite were found in the **Czar Mine**. More than a few of these rare specimens are also associated with copper.

In 1925 Larsen and Vassar determined that some of the "chrysocolla" from the Holbrook Mine was indeed a species new to science. Bisbee's fifth type species, chalcoalumite was so named for its copper-aluminum composition. Along the way to the presently acknowledged seven type species, there were several missteps. Ford and Bradley (1915) published that "footeite," first described by Koening in 1891, was indeed connellite. Holden (1922) made a similar error when he described large connellite crystals from the **Lowell Mine** as the new species "ceruleofibrite." He corrected this own error soon after (Holden, 1924). Palache and Shannon (1920) mistook exceptionally large and fine crystals of vanadium-rich conichalcite from the manganese ores above the **Higgins Mine** for a new species, which they named "higginsite."

Below: Azurite pseudomorph after malachite, 11.5 cm wide
Gardner Mine, Bisbee

Graeme collection
Jeff Scovil photo

Left: Calcite with cuprite inclusions
11 cm tall
Czar Mine, Bisbee

Evan Jones collection
ex George Feist
Jeff Scovil photo

Above: The "Voirin" Azurite

Carl Mazzoni's father-in-law worked at the Czar Mine prior to World War I. He collected this 9.7 cm wide azurite on malachite specimen and perhaps others from the same pocket in the Czar Mine. The specimen eventually made its way to Carl, whose last position in Bisbee, was as the night foreman at the Dallas Mine. Al Voirin bought the piece from a couple of kids in the mid-1970s at the Bisbee show. The kids had the specimen in a gray box. They tried to show it to a number of collectors at the show; none but Voirin bothered to look in the box.

Voirin consigned his collection, including this specimen, to Wayne Leicht, who sold the specimen to Marty Zinn. Paul Harter purchased the specimen when Zinn's collection was sold in 2005.

Paul Harter collection
Jeff Scovil photo

Just before World War I, the newer mines of the Copper Queen Consolidated Mining Company were producing some of the largest and best azurite crystals ever found at Bisbee. Crystals to 7 centimeters long on matrix were found in the **Sacramento Mine**. These were secondary overgrowths of azurite on large malachite pseudomorphs after azurite. Mitchell (1920) notes similar though smaller crystals from the adjacent Junction Mine.

Many sharp malachite pseudomorphs after azurite to eight centimeters were found as singles and crystal clusters on the 1100 and 1200 levels of the Sacramento Mine. These were among the best of this form that Bisbee produced. Many of these fine pseudomorphs exhibited several periods of azurite deposition on pre-existing pseudomorphs after azurite, which had also altered to malachite, reflecting the fluctuations in the depositional environments commonly seen throughout the district.

Bonanza type ores reminiscent of those first mined at Bisbee were found on the upper levels of the Junction Mine in 1921 (Engineering & Mining Journal, 1921; Calumet & Arizona, 1922). Masses of malachite up to 10 centimeters thick with partial overgrowths of bright, light-blue azurite occurred in this orebody. Many of these specimens were quite large with some reaching lengths of 3 feet and weighing more than 100 pounds. A number of handsome specimens of bright blue botryoidal azurite with a compact radiating, fibrous internal pattern to 10 centimeters thick came from this area. Externally these specimens were typically composed of 2- to 4-millimeter-long, tabular crystals with scattered sheaf-like clusters of centimeter-size crystals parallel to the surface. The base of these azurites is a friable wad and/or pale yellow, dry-bone type smithsonite. Minor hydrozincite is common as white specks or blotches on the surface of the azurite. The slice of an azurite stalactite on the US postage stamp issued in 1992 is from this occurrence.

A second look at the limestones in **Queen Hill** above the original discovery site brought important new oxide ore finds and the development of the **Southwest Mine**. Of the many interesting calcite forms that came from this mine, the most impressive are the groups of scalenohedra with included malachite—true Bisbee classics—reminiscent of, but far superior to, similar specimens found 30 years prior in the workings of the **Copper Queen Mine**.

Bureau of Engraving and Printing artist Leonard Buckley designed the 1992-issued, 29-cent U.S. postage stamp, basing the work on a 5.8 cm tall, polished slice of azurite from Bisbee's Junction Mine (**pictured right**) that resides in the Smithsonian Institution collection.

Chip Clark photo

"Real photo postcards" are printed photographic paper (usually Kodak) and as such are usually, but not always, one-of-a-kind postcards. Some photographers made maybe a dozen copies of a single view, while others have made hundreds. Real photo postcards were in their heyday from 1905-1925. They are distinguished by the "silvering" of the photo that results from oxidation of the silver used in the coating of the paper, and where printed cards show "pixels," they are absent in real photo postcards. Real photo postcards gained collectible status in the last 25 years and can command hundreds or even thousands of dollars.

This real photo postcard of the Tom Reed Mill in Oatman was taken in 1917 or so.

John Taylor collection

The Mohave Gold Fields

The Black Mountains, a linear north-south 75-mile long mountain range paralleling the Colorado River, are situated in the northwest corner of Arizona. In the early part of the twentieth century, the area around the Black Mountains became one of Arizona's most productive gold districts, producing nearly 2.5 million ounces. There are four mining districts in the area: Gold Basin, San Francisco, Wallapai, and Weaver. The San Francisco, or Oatman, district is by far the largest, and is home to the town of Oatman, today an extraordinary ghost town.

The area was first visited in 1860s when soldiers from nearby Camp Mohave were prospecting in the region. In 1863 or 1864, John Moss found a high grade pocket of gold and took out $240,000 worth of ore from the **Moss vein**, which was named in his honor. Discoveries of other significant veins soon followed. A mill was started at Hardyville, and the settlement of Silver City was established. But in 1866 Hualpai Indians began raiding the camps, and the area fell dormant (Wilson et al. 1967). The region was reborn in 1900 when the **Gold Road vein** was discovered and began producing rich ore. The following year, a mining company sank the **Tom Reed** and

Ben Harrison shafts to a depth of 100 feet. Soon ore from both the **Hardy** and the **Moss** veins was being milled at the Leonora mill in Hardyville. Other mines, including the **Tom Reed Gold Mine** developed. The community (more accurately a wide spot in the road) of Vivian popped up in 1909. It grew into a town and was renamed Oatman in 1912.

A blind orebody discovered in the Oatman district's **United Eastern Mine** in 1915-1916 produced $6 million in gold. The find spawned wildcat speculation as every promoter in the west descended on the region, trying to scam investors on their "blind orebodies."

The ore had played out by 1930, but numerous gold specimens from the Mohave Gold Fields have been preserved. Very early on, minerals including fluorite (fluorspar) were noted from the Oatman district. Yale's Benjamin Silliman (1866) recorded

> *The rich specimens of free gold in quartz, of which over a ton weight were taken at one time in 1864...fluorspar, free gold, horn silver sometimes in distinct docecahedral crystals, and iron gossan ... The precious metal is sometimes embedded in a compact red jaspery quartz, presenting, when cut and polished, beautiful graphic goldstone.*

Silliman went on to say that the **Skinner lode**, south of Silver Creek, produced fluorite "as beautiful octahedral crystals of green, white, and purple ... in a quartzose and feldspathic gangue with occasional gray spots of minutely diffused sulphide of silver."

Jerome and Mine Fire Minerals

In 1896 a 25-year-long fire started in the massive sulfide orebody in the **United Verde Mine**. The ore spontaneously ignited when unstable sulfide minerals oxidized by exposure to air. In general a fire can only start when the ratio of surface area to volume of the sulfides is very high; dust, for example, is far more likely to spontaneously combust than is a tunnel face. A fire typically causes the orebody to crack, which provides pathways through which the fire can continue to burn.

The United Verde Company injected water, carbon dioxide and pressurized steam into the mine in attempts to extinguish the burning ore. Eventually, it sealed the burning areas, starving the ores of oxygen. Surface stripping the ore above the burned areas exposed fumaroles, which contained a whole suite of never before seen minerals. Lausen (1928) studied the fumaroles and found 11 minerals, 5 of which were new species. Although significant quantities of the mine minerals were collected in the first quarter of the twentieth century and dispersed to many collections, few examples remain today.

At one time thought to be an extension of the United Verde Mine, the **United Verde Extension** or **Little Daisy Mine** was discovered in 1914 by Jimmie "Rawhide" Douglas, son of Dr. James Douglas. The price of copper took a nosedive in the 1930s and the United Verde Mine was closed from 1932 to 1935, when Phelps Dodge purchased the operation. The United Verde and United Verde Extension were the only two significant mines in the 21-square-mile Jerome district, and both had closed by the mid-1950s.

The United Verde Mine is the source of six type species, four of which are hydrous sulfates that formed as the result of a fire in a sulfide level of the mine. Among these, ransomite was named for Frederick Leslie Ransome (1868–1935), Professor of Mining Geology at the California Institute of Technology. The type specimen ransomite is pictured **below**.

University of Arizona specimen
Patrick Mooney/University of Arizona
RRUFF project photo

United Verde's six type species are:
Butlerite, $Fe(SO_4)(OH){\cdot}2H_2O$
Gerhardtite, $Cu_2(NO_3)(OH)_3$
Guildite, $CuFe(SO_4)_2(OH){\cdot}4H_2O$
Lausenite, $Fe_2(SO_4)_3{\cdot}6H_2O$
Ransomite, $CuFe_2(SO_4)_4{\cdot}6H_2O$
Yavapaiite: $KFe(SO_4)_2$

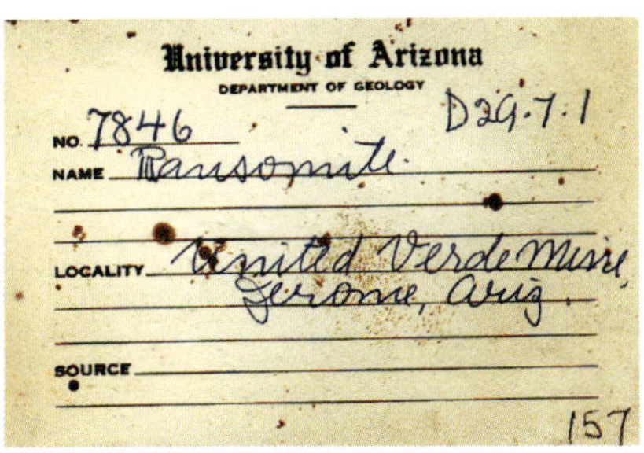
University of Arizona
DEPARTMENT OF GEOLOGY
NO. 7846 D29.7.1
NAME Ransomite
LOCALITY United Verde Mine
Jerome, Ariz.
SOURCE
157

Clifton-Morenci

By the start of the twentieth century, Phelps Dodge was the sole owner of the Detroit Copper Company, and the Arizona Copper Company had sold several prospects on the eastern side Chase Creek to Shannon Copper. In 1902 the Detroit properties were producing 18 million pounds of copper, Arizona Copper Company on the order of 30 million pounds, and Shannon a few million pounds. The three companies dramatically expanded the mining operations in the area during the first decade of the twentieth century. The towns of Clifton, Metcalf, and Morenci were all thriving, and their collective population had risen to nearly 25,000. The Clifton-Morenci district was the first of Arizona's mega-mines to exploit low-grade porphyry ores; the mining future of the area looked great in 1910.

But the district was hit with devastating strikes in 1915 and 1917, and when World War II ended, the collapse of the price of copper created further stress. Phelps Dodge eventually acquired both the Arizona Copper Company and the Shannon Company, and by 1925 was in complete control of the district. The consolidation was necessity as the district was mining extraordinary amounts of low-grade ore.

Malachite and azurite, 12 cm wide
Collected circa 1900 from an unspecified mine in the Clifton-Morenci area

Mark Hay collection
Jeff Scovil photo

Ben Harrison shafts to a depth of 100 feet. Soon ore from both the **Hardy** and the **Moss** veins was being milled at the Leonora mill in Hardyville. Other mines, including the **Tom Reed Gold Mine** developed. The community (more accurately a wide spot in the road) of Vivian popped up in 1909. It grew into a town and was renamed Oatman in 1912.

A blind orebody discovered in the Oatman district's **United Eastern Mine** in 1915-1916 produced $6 million in gold. The find spawned wildcat speculation as every promoter in the west descended on the region, trying to scam investors on their "blind orebodies."

The ore had played out by 1930, but numerous gold specimens from the Mohave Gold Fields have been preserved. Very early on, minerals including fluorite (fluorspar) were noted from the Oatman district. Yale's Benjamin Silliman (1866) recorded

> *The rich specimens of free gold in quartz, of which over a ton weight were taken at one time in 1864...fluorspar, free gold, horn silver sometimes in distinct docecahedral crystals, and iron gossan ... The precious metal is sometimes embedded in a compact red jaspery quartz, presenting, when cut and polished, beautiful graphic goldstone.*

Silliman went on to say that the **Skinner lode**, south of Silver Creek, produced fluorite "as beautiful octahedral crystals of green, white, and purple ... in a quartzose and feldspathic gangue with occasional gray spots of minutely diffused sulphide of silver."

Jerome and Mine Fire Minerals

In 1896 a 25-year-long fire started in the massive sulfide orebody in the **United Verde Mine**. The ore spontaneously ignited when unstable sulfide minerals oxidized by exposure to air. In general a fire can only start when the ratio of surface area to volume of the sulfides is very high; dust, for example, is far more likely to spontaneously combust than is a tunnel face. A fire typically causes the orebody to crack, which provides pathways through which the fire can continue to burn.

The United Verde Company injected water, carbon dioxide and pressurized steam into the mine in attempts to extinguish the burning ore. Eventually, it sealed the burning areas, starving the ores of oxygen. Surface stripping the ore above the burned areas exposed fumaroles, which contained a whole suite of never before seen minerals. Lausen (1928) studied the fumaroles and found 11 minerals, 5 of which were new species. Although significant quantities of the mine minerals were collected in the first quarter of the twentieth century and dispersed to many collections, few examples remain today.

At one time thought to be an extension of the United Verde Mine, the **United Verde Extension** or **Little Daisy Mine** was discovered in 1914 by Jimmie "Rawhide" Douglas, son of Dr. James Douglas. The price of copper took a nosedive in the 1930s and the United Verde Mine was closed from 1932 to 1935, when Phelps Dodge purchased the operation. The United Verde and United Verde Extension were the only two significant mines in the 21-square-mile Jerome district, and both had closed by the mid-1950s.

The United Verde Mine is the source of six type species, four of which are hydrous sulfates that formed as the result of a fire in a sulfide level of the mine. Among these, ransomite was named for Frederick Leslie Ransome (1868–1935), Professor of Mining Geology at the California Institute of Technology. The type specimen ransomite is pictured **below**.

University of Arizona specimen
Patrick Mooney/University of Arizona
RRUFF project photo

United Verde's six type species are:
Butlerite, $Fe(SO_4)(OH)\cdot 2H_2O$
Gerhardtite, $Cu_2(NO_3)(OH)_3$
Guildite, $CuFe(SO_4)_2(OH)\cdot 4H_2O$
Lausenite, $Fe_2(SO_4)_3\cdot 6H_2O$
Ransomite, $CuFe_2(SO_4)_4\cdot 6H_2O$
Yavapaiite: $KFe(SO_4)_2$

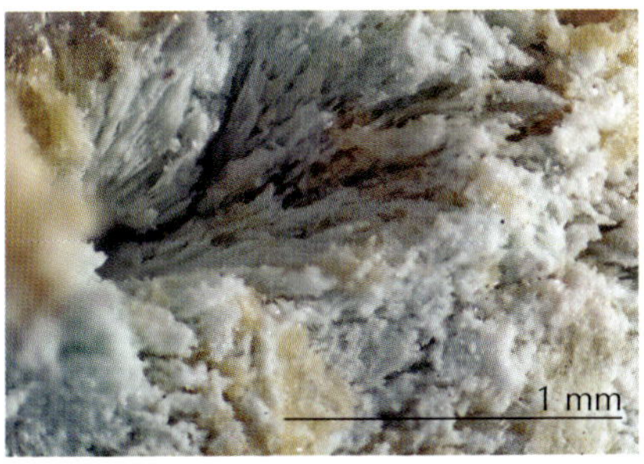

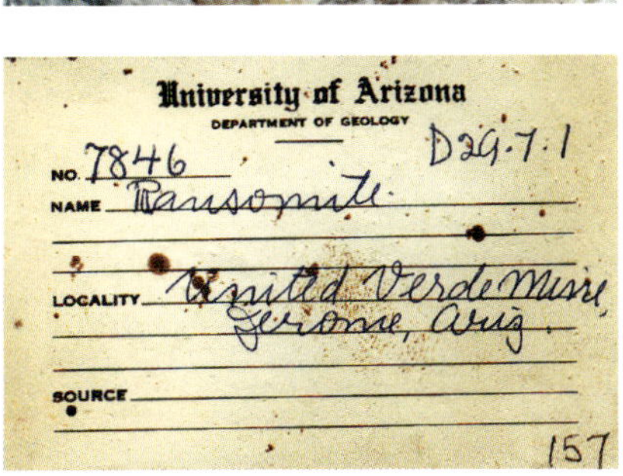
University of Arizona
DEPARTMENT OF GEOLOGY
NO. 7846 D29.7.1
NAME Ransomite
LOCALITY United Verde Mine, Jerome, Ariz.
SOURCE
157

Clifton-Morenci

By the start of the twentieth century, Phelps Dodge was the sole owner of the Detroit Copper Company, and the Arizona Copper Company had sold several prospects on the eastern side Chase Creek to Shannon Copper. In 1902 the Detroit properties were producing 18 million pounds of copper, Arizona Copper Company on the order of 30 million pounds, and Shannon a few million pounds. The three companies dramatically expanded the mining operations in the area during the first decade of the twentieth century. The towns of Clifton, Metcalf, and Morenci were all thriving, and their collective population had risen to nearly 25,000. The Clifton-Morenci district was the first of Arizona's mega-mines to exploit low-grade porphyry ores; the mining future of the area looked great in 1910.

But the district was hit with devastating strikes in 1915 and 1917, and when World War II ended, the collapse of the price of copper created further stress. Phelps Dodge eventually acquired both the Arizona Copper Company and the Shannon Company, and by 1925 was in complete control of the district. The consolidation was necessity as the district was mining extraordinary amounts of low-grade ore.

Malachite and azurite, 12 cm wide
Collected circa 1900 from an unspecified mine in the Clifton-Morenci area

Mark Hay collection
Jeff Scovil photo

Left: Grand Canyon National Park Santa Fe Railroad travel poster design printed from 1910 to 1959 Boston Public Library Print collection

Upper right: The Grand Canyon Railway steam engine circa 1905 Grand Canyon Railway collection

Lower right: This circa 1913 photo shows a miner on Humming Bird Trail, 400 feet above a mine on Mount Sheep. Numerous mineral deposits in the Grand Canyon were privately developed prior to its gaining national park status in 1919.. Library of Congress Prints and Photographs collection

Overleaf: "Earths Shadow at Yavapai Point," Grand Canyon National Park Dean Hueber photo, March 28, 2009

Grand Canyon National Park

The early twentieth century was indeed a time of transformation for the territory turned state of Arizona. The pioneering miners, missionaries, cowboys, and businessmen had tamed the frontier, provided the opening for tourists, scholars, and others to visit and emigrate to the state.

The dry, sunny climate attracted a population looking for respiratory cures, particularly for tuberculosis. The opening of the Santa Fe Railroad in the northern part of the state provided the perfect means for visiting one of the nation's greatest natural wonders—the Grand Canyon. In 1901 the Santa Fe Railroad built a spur from the main Santa Fe line in Williams to the South Rim of the canyon. The Grand Canyon Railway Company brought a rush not only of tourists but also of entrepreneurs who wanted to develop the South Rim at Bright Angel and eventually the canyon itself. Ralph Cameron built Bright Angle Trail, a toll road to the bottom of the canyon. He also proposed building hydroelectric plants in the canyon and camps along his toll trail.

To protect the scenic and geologic wonder, the Grand Canyon National

Park was established in 1919. By 1923 more than 100,000 tourists were visiting the canyon annually (Sheridan 1995), and Arizona had become known as the Grand Canyon State.

The University of Arizona Mineral Museum

The first quarter of the twentieth century also saw the establishment of the state's first mineral museum. In 1885 the Arizona Territorial Legislature had met in Prescott and passed an enabling act to establish the University of Arizona. The legislature set up the university under the Morrill or Land-Grant Act. The act promised about 46,000 acres of federal land to be turned over to the territory if the territory set up a college that taught "branches of learning as related to agriculture and the mechanic arts." Over the next six years the university was organized into three parts: the School of Mines, the College of Agriculture, and the Agricultural Experiment Station. The University opened its doors in 1891. The Board of Regents stressed that the university must be practical—and must have friends in the community—so it stressed the School of Mines. The philosophy of the School of Mines was that miners should be able to ascertain if their properties had valuable ore, and therefore, the School of Mines was to collect rocks, ores, and minerals and develop a teaching resource. The regents charged the first director of the School of Mines, Dr. Theo. B. Comstock, to conduct a geological and mineralogical survey of the territory and to "collect samples for study and preservation." This was the birth of one of the most important mineral museums west of the Mississippi.

The mineral specimens from the earliest history of the University of Arizona Mineral Museum are lost to antiquity. Although annual regents reports indicate that the School of Mines collection grew rapidly, the specimens were housed in the Arizona Territorial Museum and did not have a separate catalogue. One can only wonder what was coming out of the

William P. Blake arrived at the University of Arizona in 1895. He became Director of the School of Mines and assumed the responsibility of curator of the university's collections which were scattered throughout "Old Main," the university's main building. In 1905, the museum moved into the newly constructed Library-Museum Building (today the Douglas Building), pictured **above**. Part of the mineral display space in that new building is pictured **left**.

In 1918, the minerals were separated from the rest of the Museum and moved, along with the Mineralogy Department, into the second floor of the Mines and Engineering building (today the Old Engineering building), where it remained until 1958.

University of Arizona collection

Light blue chalcoalumite, $CuAl_4(SO_4)(OH)_{12}\cdot 3H_2O$, from the Holbrook Mine had been mistaken for "chrysocolla" until 1925, when Larsen and Vassar recognized it as a new species, naming it for its copper-aluminum composition.

Chalcoalumite and azurite, 20.5 cm wide
Holbrook Mine, Bisbee

University of Arizona Mineral Museum collection
Jeff Scovil photo

ground and preserved at the time!

In 1919 the Mineral Museum became an independent and formal entity and the collections of rocks, ores and minerals in the History Museum were transferred to the Mineral Museum under the first curator, Gordon Montague Butler, who was also the Dean of the College of Mines and Engineering and professor of mineralogy and petrology (Anthony 1980). Butler produced the first catalogue; the first specimens numbered include wulfenite and cerussite from Tiger; smithsonite, connellite, and chlorargyrite from Tombstone; and chalcopyrite from Jerome. Most of these specimens are study-grade by today's standards, but they document the glory days of mining in Arizona.

During Butler's tenure as curator, the Mineral Museum collection grew mainly through contributions from the faculty and other geologists that worked in Arizona. The list of contributors reads like an economic geology hall of fame: Frank Nelson Guild (the mineral guildite, is a copper-iron sulfate from Jerome; the type specimen is in the museum), Fredrick Leslie Ransome (ransomite, another copper-iron sulfate from Jerome), Maxwell Naylor Short (shortite), and Carl Lausen (lausenite, yet another iron sulfate from Jerome!). Butler donated his personal collection to the museum, including his butlerite type material (butlerite is an iron sulfate from Jerome) and many fine gemstones.

CHAPTER 7

Depression, War, Recovery

The early to mid-twentieth century was an extraordinary time for Arizona's copper mines. In early 1929 nearly 16,000 miners were producing more than $150 million in copper annually and supplying fully half of the nation's demand. But that changed on October 24, 1929 when the New York stock market crashed, dragging down the global economy and triggering the Great Depression. The demand for copper evaporated, and the metal's price dropped from 18 cents a pound to about 5 cents a pound. By 1933 Arizona's mines were only producing about $10 million in copper, and the mining workforce had shrunk to 3,500. The mines at Ajo, Miami, Morenci, and Ray closed; Bisbee cut its workforce by half.

The Great Depression hurt more than just the copper industry, the other pillars of Arizona's economy—cattle and cotton—were hit even harder. Making matters worse, the state endured several years of drought. Unemployment in Arizona reached 25 percent, and in the early 1930s the population of the state shrank dramatically as families fled the closing mines and struggling farms and ranches. Property values in the state, and consequently property taxes, plummeted. The assessed value of Arizona property in 1929 was more than $700 million, and by mid-decade it was less than $50 million. The tremendous stress was somewhat relieved by the massive programs of the New Deal. The Civilian Conservation Corps (CCC) built 28 camps in the state and between 1933 and 1939 spent $44 million on projects in national forests and monuments as well as on public lands. The Public Works Administration provided funding to build infrastructure including the Hoover Dam on the Colorado River.

But it was ultimately the massive investment in armaments for World War II and the national imperative to support millions of soldiers and win the war that lifted Arizona out of the depression. By 1943 Arizona was producing more minerals (primarily copper) than it had been before the 1929 collapse of the stock market. In addition Arizona became the home to a large number of military bases, including aviation training facilities such as Williams Field near Mesa, Luke Air Field west of Phoenix, Davis-Monthan in Tucson, and Marana Air Field north of Tucson. And a number of defense plants were built along the Salt River. In one short decade, Arizona grew from a subsistence economy to a booming and manufacturing service economy. The population of Arizona in 1940 was 500,000; by 1950 it was 750,000; and by 1960, 1.4 million people were living in the state.

Above: Three construction workers paint a slanted wall of riveted-steel plates on the Hoover Dam spillway in the 1940s. Planning for the famous dam began in the 1920s. It was constructed during the Great Depression and was dedicated in 1935. Power generated at the dam supply customers in Arizona, California, and Nevada.

U.S. Bureau of Reclamation photo

Left: Wulfenite on calcite, 7.6 cm tall
Hilltop Mine, Cochise County

Dick Morris collection
Jeff Scovil photo

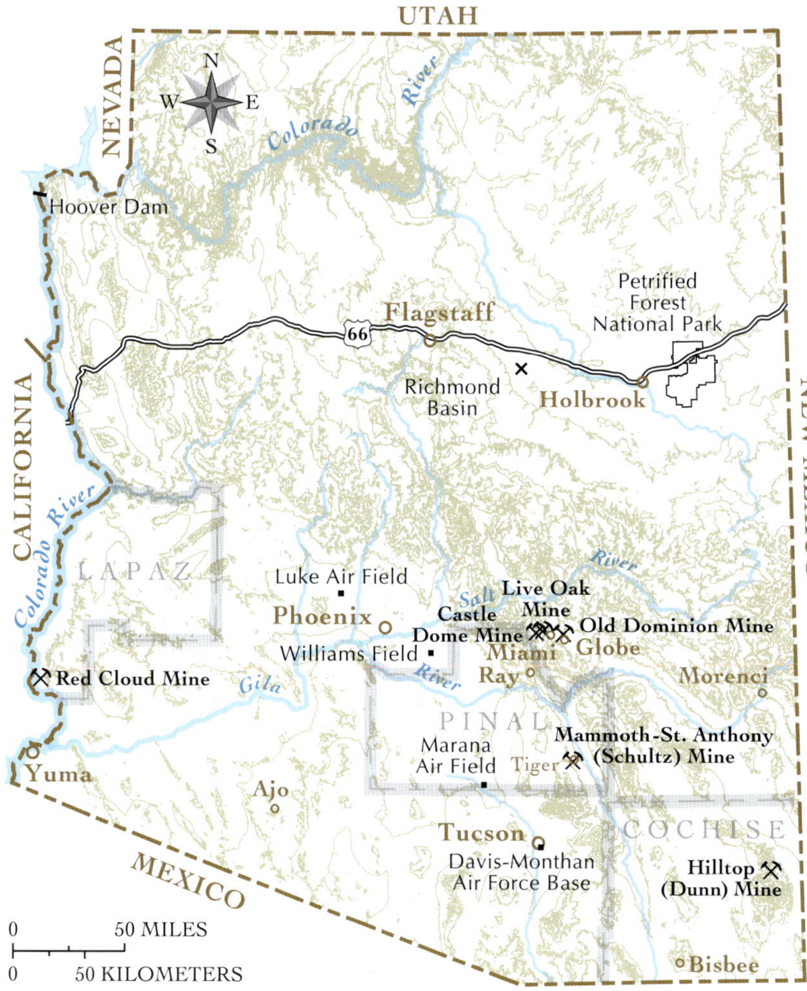

Right: Principal Specimen-Producing Localities During Depression, War, and Recovery Era Arizona (~1925 to 1950)

Below: Following World War II, most of the roads in Arizona were still unpaved, and the ones that were paved dated to the 1930s. Traveling to almost any locality was an weekend-long expedition. In contrast, many localties are today within a 2-hour drive from Phoenix. Dedicated rockhound Hafer Thompson, (Phoenix area dealer Wayne Thompson's grandfather), had to set up camp at his collecting localities in the late 1940s.

Thompson Family photo

The idling of the mines had an unexpected impact on mineral collecting in the 1930s. Contract mineral collectors scoured the mines of the west and discovered some of the spectacular specimens that grace many of today's museums and collections. The most famous of these collectors was Ed Over, a man who may have more fabulous mineral discoveries to his credit than anyone else in history. Ed moved to Colorado Springs as a teenager and later attended the Colorado School of Mines. In Colorado Springs, Over met mineral collector and scholar Lazard Cahn. He became enthralled with minerals and started prospecting for specimens. In the 1930s he teamed up with Arthur Montgomery, who often marketed Over's finds. Ed Over spent a considerable amount of time in Arizona, and with Montgomery's backing made spectacular discoveries at the Red Cloud and Hilltop Mines.

The rise of rock shops helped to stimulate mineral collecting. In 1938 U.S. Route 66, which traversed northern Arizona enroute from Chicago to Los Angeles, became the first completely paved highway in the United States. The "Mother Road" served as a major artery for the Dust Bowl

Right: A circa 1950 "real photo" postcard depicts a "Corner in American Meteorite Museum, Opposite Meteor Crater on Highway 66 in Arizona." The Meteor, or Barringer, Crater and museum is a privately-owned, mile-wide impact crater near Winslow, Arizona (see page 64) not far from Route 66.

John Taylor Collection

Above: Arthur L. Flagg, circa 1958
Photo courtesy of Richard Flagg

Lower right: Albert Einstein with his second wife (and first cousin) Elsa on a car trip in 1931 to the Petrified Forest National Monument, near Holbrook. Established in 1906, the Petrified Forest National Monument was expanded to include the Painted Desert and other cultural, scenic and geologic features in 1932. It was further expanded and designated a national park in 1962. Today the park preserves 218,533 acres including a part of historic Route 66.

Incidentally, "when [Einstein] was about 6, his grandfather gave him a small collection of minerals, and he soon became an ardent collector." (Parker 1986).

U.S. National Park Service collection

migrants during the depression and the military during the war. It also opened Arizona's national parks and monuments to burgeoning numbers of private automobiles, and a healthy tourist trade, including numerous rock shops, prospered along the 2,448 mile-long route.

During this period "rockhounding" became a popular pastime, and one of the most famous figures in Arizona mineralogy published the first guidebook to collecting in the state. Arthur L. Flagg had come to Arizona to work on a geologic map of the central Bradshaw Mountains and fell in love with the state's geology, mines, and minerals. He worked at several Arizona mines and permanently moved to the state in 1918. In 1935 Flagg founded the Mineralogical Society of Arizona. He later cofounded and served as president of the Rocky Mountain Federation of Mineral Societies. In 1944 Flagg published *RockHounds and Arizona Minerals*, writing "because Arizona, especially the mountainous parts, is literally peppered with mines and prospect holes, the area over which you may find mineral specimens is almost limitless."

Right: Collector Wally Mann bought this 4-cm wide, single wulfenite crystal, part of Ed Over's 1938 Red Cloud Mine pocket, from the late Betty Roberts in 1981. Betty had acquired the specimen from Rick Rollator who had purchased it from dealer Cal Graeber, who in turn had purchased it from Charlie Key.

Jeff Scovil photo

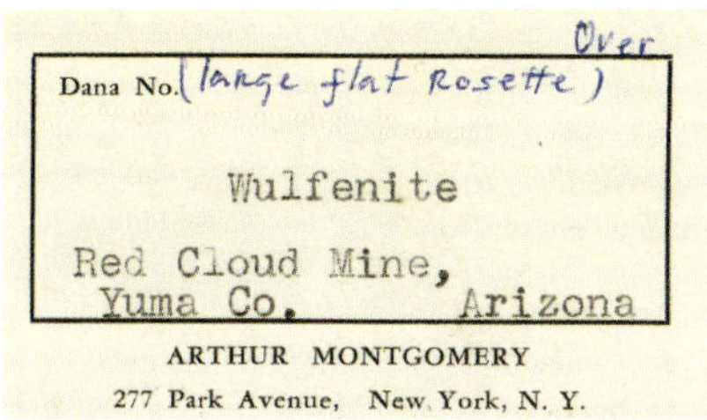

Over
Dana No. (large flat Rosette)

Wulfenite

Red Cloud Mine,
Yuma Co. Arizona

ARTHUR MONTGOMERY
277 Park Avenue, New York, N. Y.

Above and left: This 5 cm wide, single wulfenite crystal was also part of the famous 1938 Red Cloud Mine "Over pocket." It was in the collection of Over's partner Arthur Montgomery, whose label for the specimen is above. Montgomery later gave the specimen to Dick Bideaux.

Wayne and Dona Leicht collection
Van Pelt photo

The Red Cloud Mine

Noted collector Terry Szenics researched Ed Over's rediscovery of the historic Red Cloud wulfenite specimens. The story was published in *The Mineralogical Record* in 1975. A summary follows.

In the mid-1930s, mineral dealer Dr. Arthur Montgomery visited several east coast museums where he saw specimens from the Red Cloud Mine. By this time his partnership with Ed Over was well established, so Art wrote to Ed and suggested that he look into the Red Cloud Mine.

Today the trip from Yuma to the mine takes about 2 hours, and while La Paz County maintains the road, it is not unusual to get stuck in the sandy wash that a large portion of the road follows. In the 1930s Ed Over's mode of transportation was a Chevrolet coupe, and it took him 4 hours to travel the 50 plus miles from Yuma to the mine. On the road out of the mine after his first trip, Over got stuck twice on the way out and spent 10 hours digging himself out of the sand. He at that point decided to plan for weeklong trips to the Red Cloud.

Ed sought and received permission to look for crystals from the miners working the Red Cloud for high-grade stringers of ore left by the miners in 1880s. He started exploring the mine and found some areas with small crystals and color. Several days into his next visit, he was working an interesting seam, which required drilling and blasting. Ed drilled by "hand-steeling": the time-honored practice of hitting a series of progressively longer, hand-held drill steels with single-jack sledgehammer.

Ed followed the zone for about three feet and a pocket opened with a final blast. The vug contained a number of single crystals and clusters, and

Right: The largest crystal on this Red Cloud Mine wulfenite spray from Ed Over's 1938 pocket is an impressive 4.6 cm wide. At one time in the collection of Harvard University (purchased from Art Montgomery). Today the specimen is in the collection of Keith Proctor.

Wendell Wilson photo

Above: Legendary collector Edwin Over (1905–1963), pictured above in Fairfield, Utah, in 1939, had in February 1938, made an astonishing find of large single crystals and clusters of orange-red wulfenite at the Red Cloud Mine that elevated the locality to world-class.

Ed quickly realized that, while he had permission to collect, the miners might recognize the value of what he had found and establish a new democratic order (there were five of them). So he wrapped the specimens and put them into wooden powder boxes, placing the lesser specimens on top and showing those lesser specimens to the miners when he left the mine. Ed finished collecting the pocket and then headed back to Yuma.

He wrote to Dr. Montgomery,

> *On the 9th* [of February, 1938] *I tried another brecciated zone seam that has pocketed three times: have raised about 7 feet so far, last pocket is not quite finished, each pocket has been bigger and better, have an estimated 12 to 15 feet that I can raise before being cut off by old workings. Am shipping 8 boxes, 6-½ from this place. I can't compare this stuff with anything I've seen before, in fact the only Red Cloud specimens I can recall are the ones Pohndorf had. But I will say that these single crystals, clusters and groups are the finest wulfenites I've ever seen, and the most beautiful stuff I've ever taken out.*
>
> *Am sorry there are not more groups but without going into details, will say the pocket bottoms are covered with loose crystals, and groups are just pure luck. I think this seam is one in a million, when I get it worked out I'm calling it quits. Conditions for working are terrible, and there are various factors that make me think I should finish here about the first or soon after. My time is too valuable for chores right now.*
>
> *P.S. Note large rose-shaped clusters of crystals especially. Regards, Ed.*

Ed mailed the letter and then shipped the wulfenite crystals separately by "Model A Sedan Delivery." The boxes arrived before the letter, so when Art got through the endless wrapping, he was amazed to find specimens every bit as good as those that he had seen in the museums.

Throughout the 1940s, other dealers obtained and offered Red Cloud wulfenite to the public, but the 1938 Over pocket is the standard against which all Red Cloud wulfenite is held. Unfortunately more specimens are probably attributed to this pocket than actually came out of it. There is a big difference between a specimen out of the "Ed Over pocket" and a wulfenite collected by Ed Over during his time at the Red Cloud Mine (recall those lesser specimens shown to the miners). The great specimens from *the* Ed Over pocket are documented to have found their way into the collections of Harvard and the Smithsonian. Those kept by Ed Over and Art Montgomery went into the collections of Bill Larson and Dick Bideaux (the latter ultimately into the collection of Wayne and Dona Leicht).

Arthur Montgomery's hand lens (**below**), typewriter, and some of his letters, papers, and photographs have been carefully preserved in Terry "Skip" Szenics's "historical archive."

Jeff Scovil photo

The Hilltop Mine

Although wulfenite colors span the rainbow, the lemon-yellow is one of the most coveted. Only a few localities in the world produce this striking wulfenite, among them Toussit, Morocco; San Corallitos, Mexico; and Tsumeb, Namibia. But one Arizona locality rivals the others: the Hilltop Mine. Located in the California mining district in Cochise County, the mine is a cool respite compared to most southern Arizona mineral localities. It is high in the Chiricahua Mountains, nestled on about 2 square miles on the north fork of Pinery Canyon and the northwestern flank of Shaw Peak.

The Hilltop is a lead-zinc-silver mine, which exploits an irregular mesothermal replacement orebody and vein system. Galena is the primary ore mineral. It is accompanied by lesser amounts of sphalerite, pyrite, and chalcopyrite. The ore is heavily oxidized in the upper parts of the mine—

Right: This 7.3 cm wide wulfenite on calcite—the best of the best—was among the specimens collected in 1939 by Ed Over from the Hilltop Mine, Cochise County. Over's backer Arthur Montogomery elected to keep this specimen in his personal collection, selling the remainder of the lot. Montgomery allowed collector Dick Bideaux to select 50 specimens from the former's collection, and this specimen as well as the Red Cloud wulfenite pictured on page 112 were among Bideaux's choices.

Arizona-Sonora Desert Museum collection
Jeff Scovil photo

Below: Wulfenite on calcite, 6 cm tall Hilltop Mine, Cochise County

Marty Zinn collection
Jeff Scovil photo

Above: A 1993 view of the the Kasper level tunnel opening. The Kasper level is the main level of the Hilltop mine. The Kasper level adit was extended through the mountain to the other side.

Chet Lemanski photo

cerussite and anglesite are common. There are isolated pods of pure sulfide ore at greater depths.

The Hilltop claim was first filed (and named the Dunn Mine) in 1881 by Jack Dunn, the same Jack Dunn who had filed the original discovery claim at Bisbee in 1877. Work began at the Dunn later in the year with capital provided by Chicago interests. E.N. Titcomb was put in charge of the property. He erected a smelter about 1½ miles west of the mine.

In 1897 Englishmen Frank and E. J. (John) Hands purchased the Dunn Mine, and in 1899 it became known as the Hilltop. The Hands brothers shipped hand-sorted ore, which contained 73 percent lead and yielded 36 ounces of silver per ton of ore. Hilltop Metal Mining Company bought the property in 1913, and development of the lead-silver property began in earnest. But when the price of lead dropped in 1926, the Hilltop Mine closed (Brittain 1954). The years 1924-1927 saw good wulfenite specimen production (J.B. King 1940). Miners collected wulfenite with the approval of mine management. The specimens were given in turn to the mine superintendent and were stored underground and then offered for sale to visiting collectors and mineral dealers (Miller and Wilson 1983).

In December 1928 a mineral dealer from New York reportedly opened a cavern that yielded several hundred specimens of wulfenite on white calcite. (King, in Miller and Wilson 1983). Ed Over visited the mine in 1939 and found a pocket that produced about 200 wulfenite specimens, the largest weighing about 70 pounds. Like his Red Cloud find, Over's Hilltop pocket is one of the finest wulfenite pockets ever found in Arizona (Miller and Wilson 1983). Arthur Montgomery handled and kept some of the specimens that Over collected, but Schortmann's also carried some of Over's Hilltop finds.

The mine changed hands several times over the next couple of decades with no real production until 1948 when Piedmont Mines, Inc., picked up the property (Miller and Wilson 1983). American Zinc, Lead, and Smelting Company optioned the property from Piedmont in 1951 and staffed up for further production. After driving additional drifts and raises, they produced galena and silver-bearing cerussite. But production was short lived, and the mine shut down in 1955.

For the next decade the Hilltop appears to have served as a teaching mine, and there was undoubtedly some mineral-collector activity. There was a brief flurry of production from 1968 to 1975 under HRM Investment Company. But after 1975 the mine was only worked for specimens.

The coveted buttery-yellow tabular wulfenite crystals usually occur on soft, white, vuggy calcite. Some of the wulfenite and matrix pieces seem to have been exposed to a fire as they have a "smoked" appearance, which renders them less desirable than the specimens of pure yellow opaque clusters and single crystals. Hilltop wulfenite ranges in color from yellow-orange to amber to brown. Crystals to 12 centimeters (!) have been reported (Miller and Wilson 1983) although the crystals are usually well under 2.5 centimeters. The flattened pinacoid faces are generally satiny to dull but the crystal edges are brilliant.

In the mid-1950s, Richard Brittain, mine superintendent for American Zinc, Lead and Smelting Company, augmented his salary by selling Hilltop

Sunset on the stone pillars in the Chiricahua National Monument, some of the spectacular rock formations visible enroute to the Hilltop Mine

Robert W. Von Mayr, July 3, 2005

wulfenite specimens to Schortmann's (Miller and Wilson 1983). Brittain's specimens reportedly came from football-sized pockets in the Number 1 and Number 3 stopes on the Kasper level of the mine.

Over the last few decades, a number of collectors have spent time attempting to locate the wulfenite occurrence in the Hilltop Mine. Tony Potucek and John Lucking have both made several trips into the mine, and they have concluded that access to the stopes on the wulfenite-producing Kasper level has been blocked by a cave-in.

Right: A geologist (right) samples a fresh face in an underground mine at Bisbee, circa 1927. He is dressed in the professional garb of the time: engineer's high-lace boots, campaign hat, and a field book in the pocket of this jacket. The man on the left is a miner or mucker working as his assistant.

Graeme-Larkin collection

Left: Azurite "roses," to 2 cm wide
54 Stope, 300 level, Holbrook Mine

Colorado School of Mines professor Ben Parker, Jr., had kept these azurite "roses' in a shoebox for nearly 50 years before selling to collectors Jim and Dawn Minette. The label that was kept with them indicated the stope and mine, and based on that information, the specimens were probably collected in the late 1920s, before the Holbrook was sealed.

The "shoebox azurites" are in the collection of Scott Rudolph.

Jeff Scovil photo

Bisbee

The Great Depression necessitated a merger of Bisbee's two copper powerhouses, and the Calumet and Arizona Mining Company was folded into the Copper Queen Branch of the Phelps-Dodge Corporation. Only the richest ores in the Campbell and Junction mines were then profitable enough to mine. The other mines were thus largely leased to independent miners who worked the margins of old stopes in the Czar, Holbrook, and Southwest mines.

These modest efforts brought more of the classic botryoidal malachite and drusy azurite to light. Always daring and very hungry, the independent miners also made several limited attempts to mine the Dividend Fault ores before being beaten back by the incessant movement and unyielding pressure of the clays. Over several years, a few more azurite roses rewarded their efforts, but this was the last series of attempts to win ore from the Dividend Fault as now every new penetration into the zone must be cut through the mass of jumbled timber left by those before. The costs were too high and risks too great to warrant further efforts.

As the Great Depression wound down, Phelps-Dodge redoubled its efforts to mechanize and modernize mining and improve safety and working conditions in the mines. The improved economic conditions of the mid- to late 1930s brought with them years of expansion as Phelps-Dodge opened new areas near the Lowell, Gardner, and Sacramento mines (Mills 1958). With these efforts came large azurite crystals similar to those mined 20 years earlier. Fine specimens of drusy azurite and acicular malachite were mined from oxide stopes between the Sacramento and Lowell mines on the 1200 and 1400 levels.

Work to the east of the Campbell Shaft and on the 1600 level found a series of important oxide orebodies along the Campbell Fault. These specimen-rich stopes stretched from well above the 1600 level to almost the 2000 level. Malachite pseudomorphs after azurite in a wide variety of forms were abundant in the orebodies below the 1600 level. Much of the malachite was of a light hue caused by included zinc. Azurite crystals of good size were routinely found near the limestone walls in the stopes. Here, as elsewhere, the large (up to 15 centimeters long) azurite crystals were in reality second-generation azurite overgrowths on malachite pseudomorphs after azurite. Cerussite as single and as sixling twinned crystals to one

Azurite overgrowth on malachite
pseudomorphs after azurite
7.8 cm wide
1900 level, Campbell Mine, Bisbee
Evan Jones collection
Jeff Scovil photo

Azurite, 7.6 cm tall
Sacramento Mine, Bisbee, Cochise County
Dick Morris collection
Jeff Scovil photo

centimeter were commonly associated with the azurite and, to a lesser degree, the malachite pseudomorphs after azurite.

The Campbell Fault zone on the 1800 level was locally rich with large azurite roses in hematite-rich clay. While often bright, they were lighter in color and more rounded than the classic roses from the Czar and Holbrook mines. And most of the roses had been somewhat etched and then partially overgrown by a later layer of bright crystalline azurite.

Copper prices plunged again as World War II drew to a close and the demand for copper was decreased. The U.S. Government withdrew its subsidies for the production of copper from low-grade ores. At Bisbee, the lessees working the Czar, Holbrook, Uncle Sam, Higgins, and Southwest mines could no longer operate at a reasonable profit; thus, in late 1944 Phelps-Dodge terminated these leases, and the Czar Mine—Bisbee's greatest specimen-producer—closed after 60 years in operation. Off-duty miners did some minor collecting and recovered a few additional specimens over the next several years before the shaft was finally sealed.

Bisbee, Cochise County

Upper right: Malachite after azurite
8.6 cm wide
105-C stope, 1900 level, Campbell Mine
Mark Hay collection, Jeff Scovil photo

Lower right: Calcite, 8.3 cm wide
Freeport McMoRan collection
Jeff Scovil photo

Below: Chalcocite and smithsonite
Crystals to 1 cm, Campbell Mine
Evan Jones collection
Jeff Scovil photo

Facing page, lower left: Main street Bisbee 1940, Graeme-Larkin collection

Facing page, upper left: Calcite
11.1 cm tall, Campbell Mine
Evan Jones collection
Jeff Scovil photo

Facing page, upper right: Anglesite
4.3 cm tall, 1800 level, Campbell Mine
Evan Jones collection
Jeff Scovil photo

Facing page, lower right: Goethite
5.6 cm tall, University of Arizona Mineral Museum collection, Jeff Scovil photo

For a short while (1942–1950), lead and zinc ores were much sought after and great quantities were mined from the Junction, Campbell, and Denn mines (Mills 1958). In the Campbell Mine, some of the lead-zinc orebodies had been partially oxidized to depth along several persistent faults. In the thin oxide zones on the 1800 level, a few outstanding leadhillite specimens were discovered, as were interesting specimens of cerussite and anglesite. Even a few attractive yellow to orange wulfenite specimens, some with copper, were found there.

Exploration for lead and zinc in the Campbell and Denn mines also uncovered several post-ore and hydrothermally developed openings near lead-zinc orebodies. For the most part, these were lined with sharply crystalline calcite. Unless it was brightly colored by inclusions of copper minerals, calcite had been underappreciated at Bisbee and was only rarely preserved. But with calcite being found in forms not previously seen at Bisbee and with specimen-quality finds of other, more desirable minerals decidedly less common, these new finds attracted the attention of the miners, and many good specimens found their way out of the mines. Unfortunately few escaped at least some damage, as collecting was difficult and after all, "it was only calcite."

Clusters of five-centimeter-long, complex white to colorless calcite crystals perched on a dark matrix of goethite or manganese oxides or even barren limestone, were collected from one opening. Another produced aesthetic groups of tabular, centimeter-size, crystals, lightly coated by yellow-brown goethite; the calcite had the added bonus of being highly fluorescent. From yet another void came groups of cream-colored, complex, five-centimeter-long crystals with a faint iridescent sheen imparted by goethite. Many other forms and associations were collected at much the same time, albeit in lesser quantities.

Below: Cerussite "sixlings" on goethite
4.2 cm tall
1800 level, Campbell Mine, Bisbee
Evan Jones collection
Jeff Scovil photo

Right: The massive hoist installed in the Campbell Mine was a state-of-the-art machine in the mid-1930s. The cable on the hoist is almost 5 cm (2 ins.) in diameter. The man pictured in the background provides a sense of scale.
Graeme-Larkin collection

Bisbee miner Angel Salas collected this 16 cm wide pseudomorph of malachite after azurite when he was driving a short raise between the 1300 level of Campbell Mine and the same level of the Briggs Mine in Bisbee. Angel had planned to ask $10 for the specimen, but when he saw the look on collector and barber Esker Mayberry's face, he decided to drive a harder bargain. Esker gave Angel free haircuts for life in exchange for the specimen that is today housed in the Rice Northwest Museum of Rocks and Minerals.

Jeff Scovil photo

In 1947 a short raise was driven from the Campbell 1300 level to connect with the 1300 level of the Briggs Mine to improve ventilation. The very last blast in this working exposed a 6-foot-long pocket lined with incredible malachite pseudomorphs after azurite. The first miners to enter the working after the blast recovered outstanding clusters, exceeding 15 centimeters, of sharp, radiating, bladed crystals on a light brown to tan goethite and associated with the occasional centimeter-sized bleb of malachite. The best of these may well be the best specimens of any type that Bisbee ever produced.

Siderite had long been found in association with the oxidizing orebodies, often in masses of many thousands of tons (Trischka et al. 1929). Occasionally, an attractive specimen of botryoidal material was collected from these occurrences. In the early 1950s, a mass of box-work siderite estimated at more than 30,000 tons was encountered in the Junction Mine. Hundreds, if not thousands of attractive specimens were recovered from this single occurrence, which extended from above the 2200 level to below the 2566 level. Miners continued to collect good specimens long after active mining in the area stopped, as it remained accessible on several levels for many years. These frequently handsome specimens were botryoidal masses on boxwork forms with a drusy, often iridescent surface.

Right: View of Tiger, Arizona, site of the Mammoth-St. Anthony Mine

Charles H. Herbert photo
Arizona Historical Society collection

Below: Gold wires
field of view 3.1 cm wide
Mammoth-St. Anthony Mine, Pinal County
Collected in 1974 by Keith Imoele while on a TGMS-sponsored field trip

Arizona-Sonora Desert Museum collection
Jeff Scovil photo

Left: Wulfenite and cerussite
9.2 cm tall
Mammoth-St. Anthony Mine
Pinal County

Flagg Mineral Foundation collection
Jeff Scovil photo

Mammoth-St. Anthony Mines

These mines in Pinal County were founded in Arizona's territorial days. They saw the territory achieve statehood in 1912 and were consolidated during the Depression. The Mammoth-St. Anthony Mine's final and longest period of continuous production ran from 1938 through 1952. Mammoth-St. Anthony produced significant economic and strategic metal ores as well as a number of mineral species and some fairly unique specimens. The Mammoth-St. Anthony mine is the correct name to use when labeling specimens, especially those collected after the 1938 consolidation.

In 1909 W.P. Blake wrote his classic Minerals of Arizona report to the Territorial governor. In it, he mentioned that the Mammoth Gold Mine at Schultz, Pinal County, yielded probably "the finest crystallized groups of this mineral (wulfenite) and in massive form in the greatest profusion..." Blake went on to say that two or three carloads were washed out of the tailings by sluicing. But very few specimens from the early part of the century were recovered; a devastating mine collapse in 1901 had kept the then-Mammoth-Collins Mine shuttered until 1913, when it was leased and reopened by Great Western Copper Company.

Until 1914 the only metal recovered was gold. But that changed with the increased demand for strategic metals, including the steel hardener molybdenum, during World War I. The veins were rich in wulfenite, a source for molybdenum, and parts of the mines were successfully reworked for the mineral. The tailings were reworked as well—four times between 1914 and

Groups of reticulated cerussite crystals to seven centimeters across are not uncommon at Tiger and can be composed entirely of twinned crystals without matrix. Cerussite pockets were filled with loose, reticulated clusters (Mattison through Thompson, pers. comm.); in fact, most Tiger cerussite has little or no matrix attached. Heart-shape twins to eight centimeters have been recovered.

Above: Cerussite on malachite after azurite, 16.4 cm wide
Mammoth-St. Anthony Mine
Tiger, Pinal County

Les and Paula Presmyk collection
Jeff Scovil photo

1919—before the mines closed again when the price of molybdenum fell after the war. After a few false starts, mining resumed in 1935, at which point more varied milling operations were instituted to accommodate expanded recovery of metals, which by that time included gold, silver, molybdenum, vanadium, zinc, lead, and copper.

Back in the 1880s, the small town that grew up around the mines had been known as Schultz. But Schultz had been abandoned for several decades, so when the United States Postal Service wanted to establish a station to serve the mining area, a name was needed. Two were considered: Saint Anthony and Tiger. The latter, settled on in 1939, is said to have been inspired by a coin purse fashioned from a tiger scrotum that was carried by the mine owner (Panczner 1976). Many of the specimens from these mines are only labeled "Tiger, Arizona."

Peterson (1938) was able to see firsthand what the ore looked like while working on his dissertation in the mines during the early part of 1937. He described the material as being a porous earthy mass in which the constituent minerals were masked by red or brown iron oxides, hydrates, and black powdery manganese oxides. Quartz, occasionally amethystine, occurred in narrow discontinuous stringers, some with comb structures in the center. In the more open parts of the veins, Peterson saw yellow and orange crystals of wulfenite and vanadinite lining cavities or coating breccias fragments. "Scintillating crusts of descloizite and mottramite" occurred in a similar fashion. There were also lenses of black calcite as long as 50 feet in the veins. White calcite occurred as well but, according to Peterson, it was an indicator of lower-grade ore.

Peterson also gave tantalizing descriptions of the minerals. He mentioned "long orthorhombic crystals [of cerussite] projecting into open cavities" and malachite as "emerald green, mosslike masses commonly associated with cerussite." He observed hemimorphite as "slender needles bristling from quartz crystals that formed on the walls of open cavities. One specimen shows tufts of the slender needles growing on wulfenite crystals." Peterson described the vanadinite as stubby hexagonal prisms which were normally orange or coral in color. Where wulfenite and vanadinite occurred together, he noted the vanadinite was deposited on the wulfenite. Wulfenite was crystallized "in thin plates up to 0.8 inch square which bristle from the breccias and line open fractures in a most striking manner. The crystals range from a light yellow through various shades of orange to bright red." Azurite

Malachite most commonly occurs at Mammoth-St. Anthony as pseudomorphs after azurite and is commonly associated with minor cerussite and smaller epitaxial azurite crystals. Although malachite is not abundant from the mines, many suites of Tiger minerals contain good examples.

Below: Malachite after azurite with cerussite, 8.2 cm wide
Mammoth-St. Anthony Mine
Tiger, Pinal County

Scott Rudolph collection
Jeff Scovil photo

and linarite are notably absent from his descriptions of fine crystallized specimens; in fact, linarite is not even mentioned in the descriptive portion of his work.

None of the famous rarities were encountered, or at least recognized, during the 1930s. Certainly wulfenite specimens were produced, given that the mineral was being mined for ore. However, the vast majority of the ore material was lead ore. The veins of wulfenite were tightly crystallized and did not lend themselves to fine specimens. Peterson's thesis (1938) listed a total of 41 species, 3 of which were questionable and one that was later discredited (ecdemite); in fact mimetite was commonly mistaken for ecdemite throughout Mammoth's early history.

In a *Rocks and Minerals* (1949) article entitled "Notes on Mammoth, Arizona," author Francis Wise mentioned some of the rarer species found in the mines. Among them was leadhillite "in pale blue, barrel-shaped crystals up to a centimeter in length."

In 1940 Harvard mineralogist Charles Palache visited the mine and realized the truly unique assemblage of rare species that the deposit represented. But a comprehensive study of the minerals and their paragenesis was not completed until the 1960s. As part of his graduate work at Harvard University, Richard Bideaux made it his mission to examine as many "Tiger"

Mammoth-St. Anthony Mine
Pinal County, Arizona

Right: Miners underground at the Mammoth-St. Anthony Mine

Below left: Leadhillite, 4 cm wide
Evan Jones collection
Jeff Scovil photo

Below right, upper: Dioptase and wulfenite, field of view 3 cm wide
Tony Potucek collection
Jesse La Plante photo

Below right, lower: Smithsonite
4 cm wide
Les and Paula Presmyk collection
Jesse La Plante photo

Facing page, upper: Dioptase
2.2 cm tall
Carolyn Manchester collection
Jeff Scovil photo

Facing page, lower: Wulfenite, dioptase, cerussite, and fornacite (black)
5.8 cm wide
Evan Jones collection
Jeff Scovil photo

Right: Cerussite on dioptase, 3 cm tall
Mammoth-St. Anthony Mine

Les and Paula Presmyk collection
Jeff Scovil photo

Left: Tiger vanadinite occurs in a variety of colors, ranging from tan to root beer-brown to brilliant yellow and orange. Individual crystals can be up to 2.2 centimeters long; specimens do not generally exceed 7.5 centimeters across.

Vanadinite, 5.2 cm tall
Mammoth-St. Anthony Mine

Dick Morris collection
Jeff Scovil photo

specimens as possible, both in private and museum collections. He ultimately viewed more than 2,000 specimens. Bideaux defined the oxidized minerals as having formed as part of two sequences: a "normal sequence" and an "anomalous sequence." The normal sequence created most of the more common minerals: azurite, cerussite, cuprite, descloizite, dioptase, hemimorphite, malachite, mimetite, smithsonite, vanadinite, willemite, and wulfenite as well as rare minerals such as tsumebite, plancheite, creaseyite, and macquartite. All but the four rare minerals result from the "normal" oxidation process seen in many other lead, zinc, and copper deposits.

The remarkable suite of rare species that have contributed to Tiger's fame

Left: Cerussite, 5.7 cm wide

Mark Hay collection
ex Tom Mckee
Jeff Scovil photo

Lower left: Azurite and cerussite, 4 cm tall

Tony Potucek collection
Jesse La Plante photo

Lower right: Smithsonite, 7 cm tall

Smithsonian Institution collection
Jeff Scovil photo

Right: Willemite pseudomorph after cerussite, 10 cm wide

Les and Paula Presmyk collection
Jeff Scovil photo

Lower left: Descloizite on vanadinite 3.5 cm tall

Dave Bunk Minerals specimen
Jesse La Plante photo

Lower right: Wulfenite, 4 cm wide

Tony Potucek collection
Jesse La Plante photo

is the result of a second, "anomalous," oxidation sequence. A number of these species contain two metallic ions. In addition, some of these species are both carbonates and sulfates. Chlorine was vital to the formation of many of the secondary minerals in this sequence. Its source remains in question, having been introduced either by groundwater or hydrothermal fluids. The anomalous sequence of species includes alamosite, bideauxite, boleite, caledonite, diaboleite, leadhillite, linarite, paralaurionite, wherryite, and yedlinite. The solutions that precipitated these one-of-a kind species were trapped in a natural container that had been created by faulting.

Almost 90 mineral species have been described from the Mammoth-St. Anthony Mine. It is the type locality for eight species, four of which are unique to the locality. There are several more unknowns currently under study. Some two dozen species are sought by collectors for their crystallization, rarity, associations, and/or aesthetics. Another ten to twenty species are required to complete a suite of Tiger microminerals.

The mines at Tiger were threatened with closure on several occasions. In 1944 a fire on Labor Day started near the bottom of the Mohawk shaft and burned all the way up the shaft. It spread to the machine shop, coarse-crushing plant, and other structures. Infamous for collapses, fires, and other catastrophes; the mines at Tiger were never a place where non-mining people were able to freely collect. Mineral collectors thus owe a debt of gratitude to the miners who brought specimens home in their lunch pails. The various company officials also took a rather enlightened attitude toward preserving specimens as well, perhaps because they recognized the unique mineralogical character of their mines—or perhaps because they could also generate additional income for the company.

Left: Caledonite, 5.5 cm tall
Mammoth-St. Anthony Mine

Evan Jones collection
Jeff Scovil photo

Right: This specimen (field of view 3 cm wide), consists of bideauxite (yellowish), leadhillite (light blue), and possibly linarite (dark blue), was mislabeled "Sherman Tunnel, Colorado," but the rare species bideauxite is only known from the type locality the Mammoth-St. Anthony Mine, Tiger, Pinal County, Arizona

Terry Wallace collection
Jesse La Plante photo

Known for its rare species, the Mammoth-St. Anthony Mine is the type locality for eight of them:

Bideauxite, $Pb_2AgCl_3F(OH)_2$
(pictured on page 137)
Creaseyite, $Pb_2Cu_2Fe_2Si_5O_{17}\cdot 6H_2O$
Macquartite, $Pb_3Cu(CrO_4)SiO_3(OH)_4\cdot 2H_2O$
Mammothite, $Pb_6Cu_4AlSbO_2(OH)_{16}Cl_4(SO_4)_2$
Murdochite: $PbCu_6O_{6.5}Cl$
Pinalite, $Pb_3WO_5Cl_2$
Wherryite, $Pb_7Cu_2(SO_4)_4(SiO_4)_2(OH)_2$
Yedlinite, $Pb_6CrCl_6(O,OH)_8$
(pictured on page 139)

Upper left: Green tufts of creaseyite commonly occur in association with wulfenite and, in this case, pale yellow cerussite (field of view 4 mm wide). The species was described in 1975 by Sid Williams and Richard Bideaux and was named for USGS geologist Saville C. Creasey.

Robert Meyer specimen and photo

Lower left: Macquartite (crystals here to 1 mm) is known only from Tiger. It is commonly mistaken for mimetite and is found on the same quartz-rich matrix on which dioptase occurs. The species was first described in 1985 by Sid Williams and Marjorie Duggan and was named for French chemist Louis Charles Henri Macquart (1745–1803).

Robert Meyer specimen and photo

Upper middle: Mammothite occurs as bright blue sprays of radiating acicular crystals or as flat plates embedded in anglesite and in association with are caledonite, wulfenite, phosgenite, and leadhillite (Peacor et al. 1985). The type material is part of the National Mineral collection at the Smithsonian Institution. The spray of crystals shown here (field of view 2.4 mm) are on anglesite.

Dick Thomssen collection
Bruce J. Murphy photo

Lower middle: Murdochite was also first described from Tiger. It occurs as black octahedral crystals on fluorite, hemimorphite, willemite and in and on wulfenite. The field of view of the above ex Bideaux specimen is 4 mm wide.

Brent Thorne collection and photo

Upper right: The Mammoth-St. Anthony Mine is the only known locality for pinalite, which occurs both as isolated crystals and divergent sprays in cavities formed from the dissolution of an unknown mineral and are enclosed by chromian leadhillite. The field of view of this example, a group of twinned crystals, is 1 mm wide.

Brent Thorne collection and photo
ex Bideaux

Lower right: Mammoth-St. Anthony is the type locality for wherryite, which was
continued on the facing page

discovered in 1943 by mining engineer E.B. Daggett, who had encountered "a small vug of leadhillite crystals with associated cerussite, anglesite, phosgenite, paralaurionite, hydrocerussite, diaboleite, boleite, matlockite, and quartz" (Bideaux 1980). The light green, fine-grained, dense wherryite surrounding the vug was intimately associated with massive diaboleite. It was described and identified in 1950 by Fahey, Daggett, and Gordon.

In this photo, colorless to light green wherrylite are associated with green caledonite and blue diaboleite. The field of view is 2.5 mm high.

Bruce Murphy collection and photo

Above: Yedlinite was first described from and is only known to occur at Tiger. By far the finest specimen consists of some 200 crystals and is in the Smithsonian Institution. More typically, specimens display just a few 0.5 mm crystals generally in massive diaboleite. The field of view in the specimen pictured here is 1.2 mm wide.

Brent Thorne collection and photo

Upper right: Diaboleite, 3.9 cm wide Mammoth-St. Anthony Mine

Les and Paula Presmyk collection Jeff Scovil photo

Like Phelps-Dodge before it, the Mammoth-St. Anthony Mining Company employed two miners, Archie Griffith and his brother George, to collect specimens and sell them for the company. Archie ultimately retired from Magma Copper Company having concluded his mining career at the San Manuel Mine south of Tiger. George continued on as a mineral dealer after the Mammoth-St. Anthony Mine closed. Even the mine manager, J.J. Strutzel maintained a small collection.

George Bideaux, Richard Bideaux's father, recalled visiting George Griffith in the early 1950s and that the roof of his house was covered with wulfenite specimens. (per. comm.) Dick Jones, a dealer and collector whose day job involved repairing office equipment, had occasion to tour the mine before it closed and was allowed to collect specimens from the conveyor belt bringing ore from the shaft to the mill. He loved to tell of standing at the conveyor at various times selecting dioptase, wulfenite, or whatever the "species of the day" happened to be (pers. comm.).

Finally, in 1952, a young Dick Bideaux sent a clipping to *Rocks & Minerals* (Nov-Dec) advising collectors of the pending closure of the Saint Anthony Mining and Development Company Mine at Tiger. The closure was slated to begin November 1, 1952. Depressed lead and zinc prices were the reasons given for shutting down the world-famous mineral specimen locality. Groundwater overwhelmed the mines, and the last miner left in 1953.

CHAPTER 8

The Post-War Boom

The post World War II boom dramatically changed Arizona's economy. The state had attracted manufacturing investment during the war and the dawning of the 1950s saw explosive growth. Phoenix quickly became a sprawling urban center. By 1975 it had grown from an obscure spot on the nation's map to a metropolis with a population of nearly 1.5 million. In 1949 attracted by cheap land, warm weather, and large military contingent (the summer heat issue was resolved with the invention of air conditioning), Motorola selected the Phoenix area for its Military Electronics division. In 1951 when Howard Hughes built an enormous missile plant in Tucson, the southern part of the state had its manufacturing anchor. In 1950 the number of Arizonians employed in manufacturing was 17,000; by 1960 it was nearly 50,000. Although dwarfed by manufacturing, jobs in mining also increased during this period. The traditional pillars of Arizona's economy—copper and cotton—now accounted for just one third of the total wages in the state (Sheridan 1995).

Above: A circa 1952 YGAR-1 missile, part of the family of Hughes "falcon" air-to-air missiles built in Tucson

United States Air Force photo

Left: Copper, 21.2 cm tall
Pearl Handle pit, Ray Mine, Pinal County

Arizona-Sonora Desert Museum collection
Jeff Scovil photo

Although the relative importance of mining decreased, the copper industry prospered during the boom. The profitability of low-grade copper deposits spurred copper exploration in Arizona in the mid-1950s and led to the discovery and rediscovery of a dozen significant mines including the **Silver Bell**, **Bagdad**, **Sacaton**, and **San Manuel**. One of the most important was a series of deposits 20 miles south of Tucson along the flanks of the Sierrita Mountains east of the Santa Cruz River. The Spanish had prospected for silver in the Sierrita Mountains, and those deposits were rediscovered in the 1880s. The principal mines were **Mineral Hill**, **San Xavier**, **Twin Buttes**, and the **Esperanza** (organized in the Pima Mining District). Most of these mines were dormant after the stock market crashed in 1929. But when the United Geophysical Company, headed by Herbert Hoover, Jr. (son of former U.S. President Herbert Hoover), sought to expand the company's activities from oil to more general mineral exploration in the 1950s, the company targeted the Pima district thinking that the orebodies of all of the mines might be related. A series of surveys found a strong magnetic anomaly, which was interpreted as a buried porphyry copper stock. Drilling revealed an orebody that was larger than 1 million tons that averaged 4.5 percent copper. A lower grade (0.5 to 1 percent), 200 million ton, orebody was also discovered. Three large open-pit mines were eventually established: **Pima**, **Mission**, and **San Xavier**. These mines produced more than three

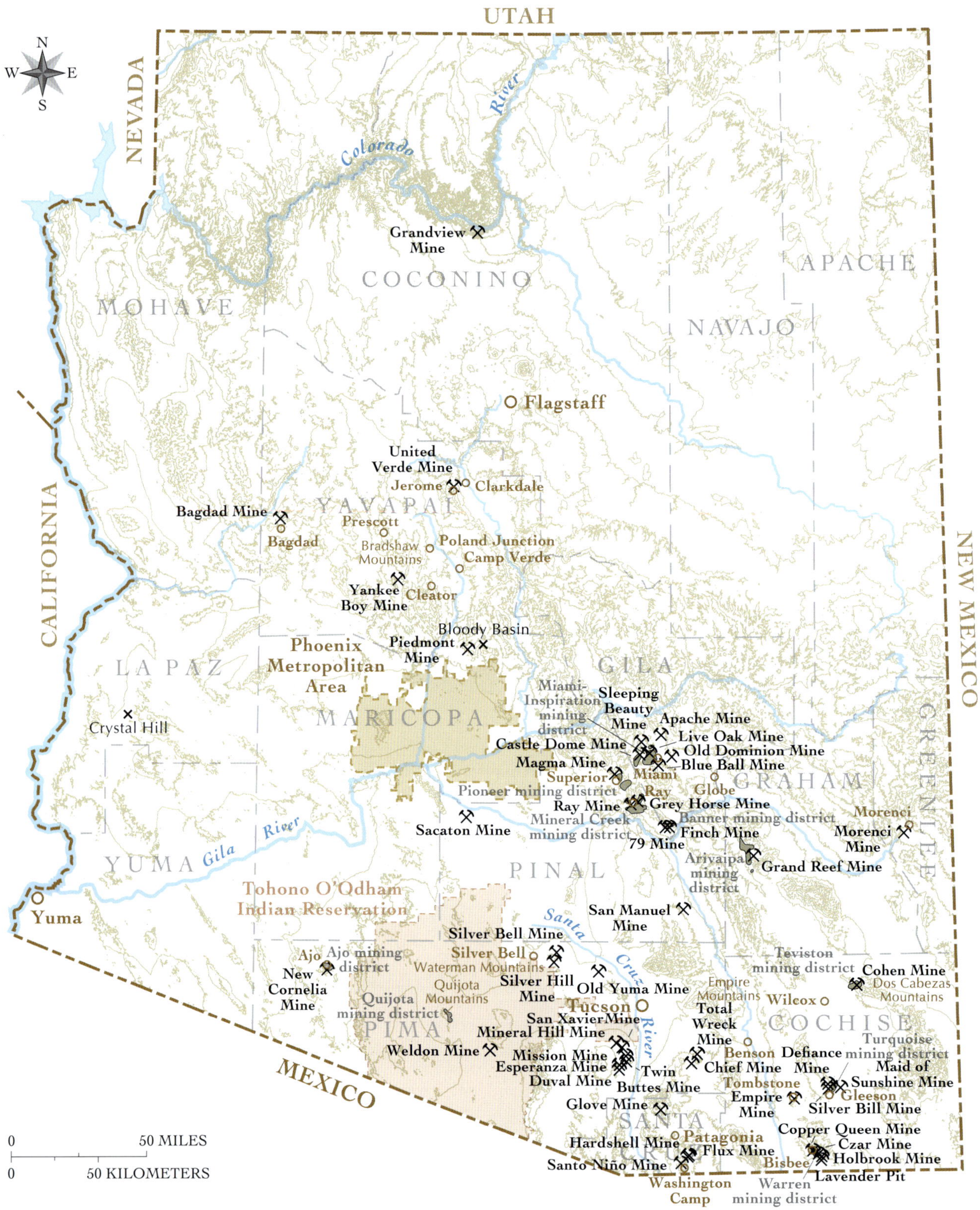
UTAH
NEVADA
CALIFORNIA
NEW MEXICO
MEXICO
N
S
E
W
Colorado River
Gila River
Santa Cruz River
Grandview Mine
COCONINO
MOHAVE
APACHE
NAVAJO
Flagstaff
United Verde Mine
Jerome
Clarkdale
YAVAPAI
Bagdad Mine
Bagdad
Prescott
Bradshaw Mountains
Poland Junction
Camp Verde
Yankee Boy Mine
Cleator
Bloody Basin
Piedmont Mine
Phoenix Metropolitan Area
LA PAZ
GILA
MARICOPA
Crystal Hill
Miami-Inspiration mining district
Sleeping Beauty Mine
Apache Mine
Live Oak Mine
Castle Dome Mine
Old Dominion Mine
Blue Ball Mine
Magma Mine
Superior
Miami
Globe
GRAHAM
GREENLEE
Pioneer mining district
Ray
Ray Mine
Grey Horse Mine
Mineral Creek mining district
Banner mining district
Morenci
Sacaton Mine
Finch Mine
79 Mine
Morenci Mine
YUMA
Yuma
PINAL
Arivaipa mining district
Grand Reef Mine
Tohono O'Odham Indian Reservation
San Manuel Mine
Silver Bell Mine
Silver Bell
Waterman Mountains
Ajo
Ajo mining district
New Cornelia Mine
Quijota Mountains
Quijota mining district
Silver Hill Mine
Old Yuma Mine
Empire Mountains
Teviston mining district
Cohen Mine
Dos Cabezas Mountains
Wilcox
Tucson
COCHISE
PIMA
San Xavier Mine
Mineral Hill Mine
Total Wreck Mine
Turquoise mining district
Weldon Mine
Mission Mine
Esperanza Mine
Duval Mine
Twin Buttes Mine
Benson
Chief Mine
Defiance Mine
Maid of Sunshine Mine
Tombstone
Gleeson
Empire Mine
Silver Bill Mine
Glove Mine
SANTA CRUZ
Copper Queen Mine
Czar Mine
Holbrook Mine
Hardshell Mine
Patagonia
Flux Mine
Santo Niño Mine
Bisbee
Lavender Pit
Washington Camp
Warren mining district
0
50 MILES
0
50 KILOMETERS

Below left: Brothers Wayne and Evertt Thompson were active collectors in the 1950s and 1960s. Based out of Phoenix, their families frequently accompanied them on field trips. The Thompson's nephew Wayne went on to study mineralogy ultimately becoming a well known player on the international mineral market. Wayne's father, the Thompson's brother Carl, was not a collector. Wayne and Evert are pictured here in the field with their wives in 1957.

Below right: Gene Schlepp has been a presence on the Arizona mineral-collecting scene since he was a teenager in the 1950s. He has worked virtually every specimen-producing locality in the state and has been active in the Tucson Gem and Mineral Society for more than 50 years. He and his wife Jackie are highly regarded American mineral dealers. This photo of Gene holding a recently killed rattlesnake at the Flux Mine dates to 1971.

Left: The Principal Landmarks of Arizona Mineral Collecting and Mining During the Post-War Boom (roughly 1950 to 1975)

John Callahan cartography

billion pounds of copper but almost no specimens of collector interest. In the mid-1970s Pima was the largest copper district in the United States, producing 30 percent of the country's copper—nearly 17 percent of the world's production (Bideaux and Wallace 1997).

The period from 1950 to 1975 was a halcyon time for mineral collecting in Arizona. Many of the richest copper mines were moving large amounts of rock and mineral discoveries were frequent. The growth of the mineral collecting hobby created awareness among miners and mine geologists to preserve the minerals finds—recovering specimens meant extra cash! The mines in **Bisbee** (page 162), **Ajo** (page 148), **Jerome** (page 231), **Miami** (page 186), **Magma** (page 201), and **Ray** (page 205) all yielded remarkable specimens during these years.

Collectors visited long-abandoned prospects all across Arizona and discovered collectible mineral specimens, which had been of no interest to previous generations of miners. Passionate and dedicated collectors who today cast a long shadow in the lore of Arizona minerals recovered many of Arizona's best mineral specimens during this time. Collectors such as Gene Schlepp, Dick Jones, Monnie Speck, Wayne Thompson, Sr., and Dick Bideaux pored over old mining records and reports and looked for hints that might lead to mineral treasure. And many more hitched rides and spent weekends moving rock, turning over dumps, and restoring infrastructure in exchange for flats of ore and minerals.

Remote mines that were modest economic producers found fame as classic Arizona specimen localities. The world's best cyanotrichite was found at the **Grandview Mine** (page 240), and linarite came from the **Grand Reef** (page 182). Vanadinite from the **Old Yuma** (page 170), wulfenite from the **Glove** (page 223) and **Defiance** (page 170) mines, and large orange scheelite crystals from the **Cohen Mine** (page 176) were all found and carefully mined by intrepid collectors.

Many localities were worked for specimens for more than three decades! The **79** (page 193) and the **Finch** (page 200) mines trickled out wulfenite; the **Flux Mine** (page 212) produced hundreds of bright white jackstraw

Fifteen-year-old Dick Bideaux's mineral collecting prowess was highlighted in a Stookie Allen drawing that appeared in an April 1952 issue of *The Phoenix Gazette*.

Wayne and Dona Leicht collection

cerussite specimens; and the **Washington Camp** area (page 219) pumped out wonderful Japan-law twin quartz. And a number of oddball localities enjoyed a healthy following of collectors: **Camp Verde** (page 228) turned out literally tons of calcite after glauberite pseudomorphs, the **Blue Ball Mine** (page 192) yielded azurite nodules, and the old gold diggings of **Weldon** (page 161) produced wonderful barite and calcite. Mineral specimens recovered in this era form foundations for many of today's top collections of Arizona minerals. The 1950s, 60s, and 70s were without a doubt boom years both for the state and for Arizona mineral collectors.

With the end of World War II, rockhounding swept the country and hundreds of mineral clubs formed. In December 1946 a group of 28 enthusiasts gathered in Tucson and founded the Tucson Gem and Mineral Society (TGMS). In many ways the TGMS was like clubs elsewhere—they had meetings, lectures and field trips—but the TGMS was also in a unique position. In the 1950s Tucson was an international center for mineral exploration, and the University of Arizona had strong academic programs in economic geology and mining engineering. There were better than one thousand mines and prospects within 100 miles of Tucson, and many of them hosted colorful secondary minerals. The TGMS drew hobbyists and professionals and was a strong club by 1955.

In early 1955 TGMS President Clayton Gibson and Vice-President Dan Caudle visited the Phoenix Gem and Mineral show and floated the idea of a show in Tucson. The society met the idea of a show with criticism and told Gibson and Caudle that no club funds could be used for a show. Gibson and Caudle were undeterred. They recruited a group of society members and held a show on the weekend of March 19-20 in Helen Keeling Elementary School. Nine dealers set up at that first show, which drew more than 1,500 attendees! The University of Arizona Mineral Museum put in an exhibit and launched the concept of bringing museums to shows. In the 56 years since, more than one hundred different museums have displayed at the TGMS show. Many of the participants in the first show have figured prominently on Arizona's mineral-collecting scene. Fifteen-year-old Gene Schlepp (Western Minerals) put in a case of self-collected minerals and won a first place ribbon. Twenty-year-old Richard Bideaux not only put in a display of his minerals but also launched his career in the mineral business by setting up a card table and selling minerals.

The success of the 1955 show sparked interest in and dedication to a TGMS show, which moved the following year to a cavernous Quonset hut on the Pima County Fair and Rodeo grounds. Attendance swelled to 3,000, and the special exhibits included a first for the state—a fluorescent mineral display. Richard Bideaux won Best Exhibit in the show, and Gene Schlepp repeated as the best junior exhibitor.

After a successful 1960 show, the society decided to go national. Show Chairman Clayton Gibson invited a dozen museums to display at the 1961 show; only one curator replied, but that reply changed the TGMS show forever. Curator Paul Desautels committed the Smithsonian Institution. He brought two cases: "Uncommon Gemstones" and "Single Crystals." That same year saw the first satellite show, held at the Holiday Inn South.

Right: Smithsonian curator of minerals Paul Desautels (left), Best-in-Show award winners Bob and Becky Fordham (middle), and Show Chairman Clayton Gibson (right) were pictured in the *Tucson Citizen* during the 1961 TGMS show.

Below: Beginning in the early 1960s, the legendary Desert Inn ("DI") was the center of "off-site" mineral collector activity surrounding the TGMS show.

Bob Jones photo

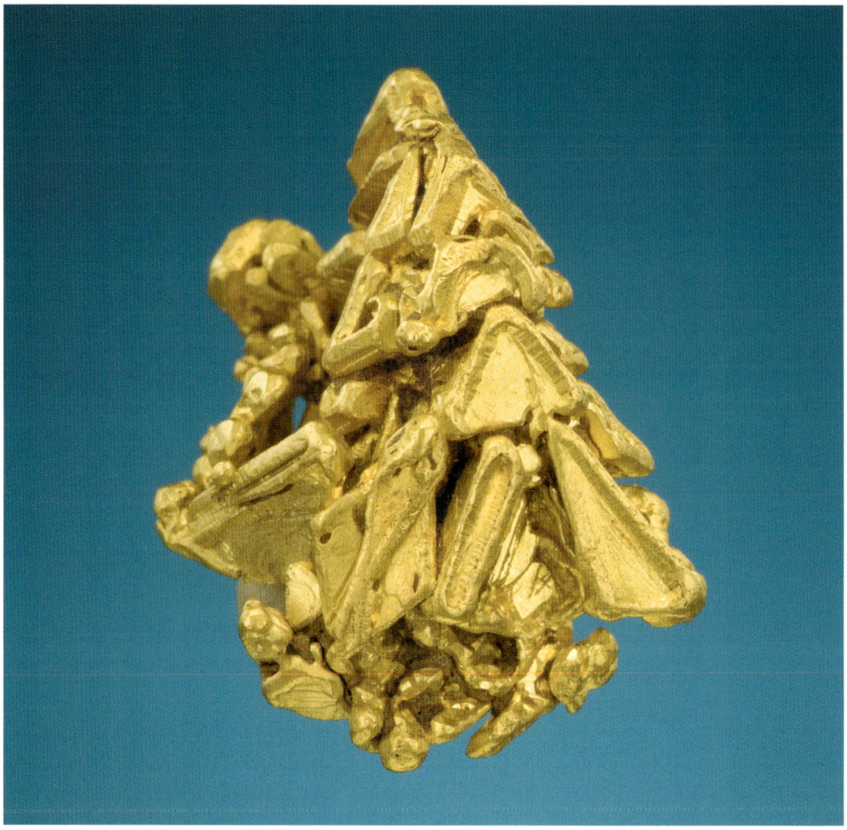

Right: This crystallized gold specimen weighs 15.5 grams and is 2.3 centimeters tall. The specimen was purchased in 1980 from Dick Moore, a geologist from the old Arizona Bureau of Mines. Mr. Moore was the author of *One Hundred Arizona Minerals*. He had obtained this specimen and others from prospectors who were working near Quartzsite, La Paz County.

University of Arizona collection
Cat. No. UAMM 11400
Jeff Scovil photo

Above: The U of A Mineral Museum was housed in the Mines and Engineering Building from 1958 until 1993, when it was moved to the Flandreau Center

University of Arizona photo

University of Arizona Mineral Museum

Arizona's mineral museums were also thriving. At the University of Arizona, Frederic "Fritz" W. Galbraith had become curator of the Mineral Museum in 1941. He had a vision for the museum that included a "mineral hall that the public could enjoy." It took more than 15 years but in 1957 the Mineral Museum opened its doors in a new building. The displays included numerous floor and wall cases with nearly 2,000 specimens on display. Galbraith sought the support of local mining companies and alumni to expand the scope of the collection with an eye for display materials. Several private collections were donated; most notably, the Percy G. Beckett and Ellen A. Brophy collections of Bisbee minerals. Beckett was Vice-President and General Manager for Phelps Dodge's western mines and smelters. He had spent time in the early twentieth century supervising the Copper Queen Mine. In addition to a fine suite of Bisbee minerals, Beckett donated some unique mining art to the museum. In 1923 he had commissioned William D. White to paint miners and mines at Bisbee and Nacozari (Sonora, Mexico); 16 of these paintings adorned the New York offices of Phelps Dodge until 1958 when they were sent to the Mineral Museum.

During the 1950s the Mineral Museum began acquiring specimens through purchase or by trading from its extensive stock of Arizona material. One of the most significant trades was with Boodle Lane, the famous dealer in Tri-state minerals from Galena, Kansas. The museum received a dozen large specimens of galena, sphalerite, calcite, and dolomite that were placed on open pedestals and were extremely popular with children visiting the museum! Susie Davis was another of the mineral museum's favorite mineral dealers. Davis grew up in Tombstone where her father was a miner; she began building a mineral collection as a child. In the early 1950s Davis opened Davis Minerals and served as an outlet for the colorful minerals flowing out of Mexico. She donated and sold hundreds of pieces to the Mineral Museum.

In 1941 Galbraith published *Minerals of Arizona* (bulletin 149 of the Arizona Bureau of Mines). Galbraith used the minerals in the museum collection as the basis for his descriptions. The bulletin was extremely popular with the collecting public and was the reference guide for Arizona rockhounds. It was revised in 1947 and 1959. *Minerals of Arizona* became closely associated with the Mineral Museum, and many collectors came to look at examples described in the bulletin, often leaving donations from their collecting exploits.

In 1964 John Anthony became curator of the Mineral Museum. Anthony loved minerals, and approached his curatorship with a passion for both the scientific and aesthetic values of mineral specimens. As the Tucson Gem and Mineral Show rose in prominence in the 1960s and 1970s, the Mineral Museum was visited every year by thousands of mineral dealers and collectors from around the world. The collection grew to nearly 12,000 catalogued specimens with more than 1,300 species represented. Anthony followed the example of Galbraith and, in 1977, published the penultimate tome on Arizona minerals *Mineralogy of Arizona.*

Above: University of Arizona curator John Anthony

Bob Jones photo

Right: The Mineral Museum purchased this 6.5 cm wide Bisbee copper specimen in 1958 from Morris Jesup Elsing's The Collector's Shop in Tucson for $10.

University of Arizona collection
Cat. No. UAMM 5800
Jeff Scovil photo

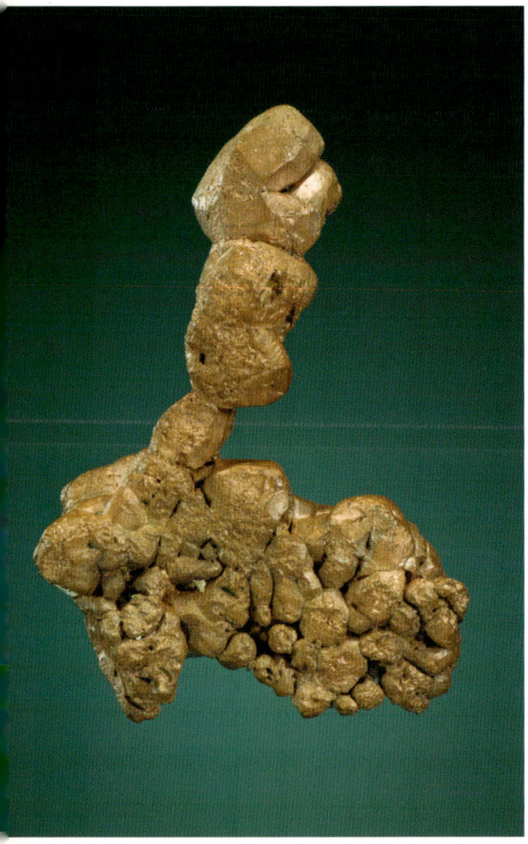

Ajo Mining District

Ajo attracted the interest of some of the earliest prospectors in the state, but the lack of precious metals delayed the development of the massive low-grade copper deposit until 1911 when Calumet and Arizona Mining Company's John C. Greenway secured an option on New Cornelia Copper Company stock, funded a drilling program, and proved that the copper-stained hills and the silica-rich iron rock was the top of a large orebody.

A large-scale leach plant was completed in May 1917. By the middle of June the first copper cathodes were sent east for processing. The leach plant operated until the oxide ores were depleted in 1930. By 1919 the New Cornelia Copper Company had acquired all of the other properties and interests in the area (Rickard 1996). And in 1923, a new concentrating plant began processing 5,000 tons of sulfide ores per day. When the oxide leach plant closed, the capacity of the sulfide mill was expanded to 15,000 tons per day (Thomas and Gibbs 1983). Calumet and Arizona already had a 65 percent stake in New Cornelia when the two companies merged in 1929. When Phelps-Dodge and Calumet and Arizona consolidated in 1931, the **New Cornelia Mine** began part of Phelps-Dodge (Rickard 1996).

While numerous specimens from the rich oxidized zones at Ajo were undoubtedly encountered, none are known to have been preserved. But since 1950, the New Cornelia open pit and surrounding area have produced noteworthy and distinctive specimens. Nearly 100 mineral species have been described from Ajo (Thomas and Gibbs 1983).

The mine is best known for native copper, which generally takes two forms: unique (at least for Arizona) crystals up to 5 centimeters wide; and parallel, feathery groups somewhat reminiscent of herringbone. Of the former, specimens weighing more than 200 pounds and a 6-foot-wide string

Above, left: Copper, 6.8 cm tall
New Cornelia Mine, Ajo, Pima County
John Lucking collection
Jeff Scovil photo

Above, right: Azurite after malachite pseudomorphs after azurite, 4 cm wide
New Cornelia Mine, Ajo, Pima County
Tony Potucek collection
ex Betty Llewellyn
Jesse La Plante photo

Above: "Steam Shovel Mining New Cornelia Copper Co." circa 1916 real photo postcard

John Taylor collection

Left: Azurite, 3 cm tall
New Cornelia Mine, Ajo, Pima County

Dave Bunk Minerals specimen
ex Jim and Dawn Minette
Jesse La Plante photo

Right: Copper, 4.9 cm tall
New Cornelia Mine, Ajo, Pima County
collected in 1968

Grant Gibson collection
ex Terry Wallace
Joe Budd photo

New Cornelia Pit, Ajo, Pima County

Lower left: Ajo is the type locality for ajoite, $(K,Na)_3Cu_{20}Al_3Si_{29}O_{76}(OH)_{16}\cdot{\sim}8H_2O$, and papagoite, $CaCuAlSi_2O_6(OH)_3$. This 5 cm tall vug of ajoite and shattuckite was in the collection of Harold Michel.
Dave Bunk Minerals specimen
Jesse La Plante photo

Lower right: Native copper and copper pseudomorphs after cuprite 6.9 cm tall
Les and Paula Presmyk collection
Jeff Scovil photo

Facing page upper: "Herringbone" copper 14.3 cm wide
Les and Paula Presmyk collection
ex John Sinkankas
Jeff Scovil photo

Facing page, lower left: Azurite after malachite after azurite, 7.6 cm wide
Les and Paula Presmyk collection
ex Tom McKee
Jeff Scovil photo

Facing page, lower right: Copper and cuprite, 4.1 cm tall
Dick Morris collection, ex Monnie Speck
Jeff Scovil photo

of crystals have been recovered. The deposit also yielded pseudomorphs of copper after cuprite.

Calcite is common in the area. The most noteworthy crystals are made bright red by cuprite inclusions. Individual crystals to 1.5 centimeters and clusters up to 8 centimeters have been recovered.

An interesting suite of copper silicates is found at the New Cornelia Mine, which is the type locality for two: ajoite and papagoite. The original samples of ajoite were discovered in 1941, but the description was not published until 1958. The species occurred as fracture-coatings and vein-fillings and as well-formed crystals (elongated, bladed or acicular) in association with shattuckite, quartz, pyrite, conichalcite, kaolinite, barite, olivenite, and pyrolusite. Papagoite occurred in narrow veins and fractures deep in the pit (Thomas and Gibbs 1983). Shattuckite is relatively common at the mine, and the occasional vug of radiating crystals is found, although it most commonly occurs in masses. Shattuckite, copper, and cuprite in chrysocolla are highly sought lapidary materials.

Although uncommon, New Cornelia Mine azurite specimens are significant and noteworthy. From 1958 to 1962 pseudocubic crystals as singles or in clusters were recovered from the mine. In some cases, these specimens evidence several periods of pseudomorphing and numerous overgrowths of azurite. A 1972 a pocket in the upper west side of the pit contained crystals up to 2.5 centimeters on edge in clusters up to 12 centimeters wide (Thomas and Gibbs 1983). The crystals were randomly lustrous or dull. At Ajo malachite is generally associated with the azurite, and some of the 1958 find of azurite crystals are associated with balls of malachite, usually on one side of the specimen.

Right: Vanadinite, 8 cm tall
Old Yuma Mine, Pima County
Les and Paula Presmyk collection
ex Lazard Cahn
Jeff Scovil photo

Above: Vanadinite, 5.9 cm tall
Old Yuma Mine, Pima County
Les and Paula Presmyk collection
ex Lee Price
Jeff Scovil photo

Old Yuma Mine

with contributions by Gene Schlepp

Charles Stevens located the original Old Yuma claim in 1885 and vanadinite and wulfenite pockets were periodically encountered from the 1880s into the early twenty-first century when the mine closed. Ed Over collected a wulfenite pocket at the mine in 1929, and collector Fred W. Schmeltz briefly wrote of his experience at the Old Yuma in 1931. Advertisements in the early 1940s give an indication as to the types of specimens being found at the mine during World War II, when Grady Wilson was leasing the property.

On a 1949 visit to the mine, teenagers Dick Bideaux, Hugh Thompson, and Jonathan Browne visited the mine and exposed an important pocket although they did not realize its extent. Tucson collectors Clayton Gibson and Dan Caudle made a subsequent, independent visit to the mine and began excavating the vanadinite pocket that the boys had opened. The vug

> *... proved to contain a great deal of vanadinite. Loose crystals 6 to 12 millimters were everywhere, and rained down by the dozens from up above every time a hammer blow was struck anywhere near the pocket. Caudle scooped up a bottle full of loose crystals that amountd to half a kilogram or so.* (Wilson and Schlepp 2008)

Other collectors learned about the find, and many collectors recovered its contents before the pocket was finally worked out. Al Haag probably recovered the best specimens, some of which were later purchased by George and Dick Bideaux.

The strike at the Old Yuma Mine stimulated collecting which continued at a sometimes-frenetic pace into the mid-1950s. Al Haag, Dick Jones, and George and Dick Bideaux were among those who frequented the locality in the 1950s. Dan Caudle and Clayton Gibson continued to visit the Old Yuma as well. Dick Jones recalled a "washtub-size" vanadinite pocket that he and Kermit Lee found on a ledge in the glory hole. "The pocket produced about 35 flats of cavernous, skeletal vanadinite crystals on white calcite." One remarkable single crystal exceeds 2.5 centimeters in length (Wilson and Schlepp 2008).

Gene Schlepp remembers one Sunday in 1953 when he and his buddies Burton Miller and Larry Meyer ignored their parents' warnings and rode

Below: This 2.1 cm wide specimen of vanadinite on calcite from the Old Yuma Mine, Pima County, was collected by (and belonged to) long-time Phoenix collector Edna Andregg (1904–1985). A hand-written note in her field catalogue states that it was found in 1946 "near [the] surface with calcite and dendrites."

Les Wagner collection
Jeff Scovil photo

Right: Ed Over recovered this 9 cm wide Old Yuma Mine wulfenite specimen during his one day collecting at the locality in 1929. The specimen was in the collection of Art Montgomery and was later acquired by Dick Bideaux.

Wayne and Dona Leicht collection
Wendell Wilson photo

Right: Vanadinite, 7 cm tall
Old Yuma Mine, Pima County
Collected in 1949 by Al Haag

Mark Hay collection
Jeff Scovil photo

Facing page, upper: Old Yuma Mine, Pima County, circa 1939

Mineralogical Record collection

Facing page, lower: Vanadinite
8.8 cm wide
Old Yuma Mine, Pima County
Collected in 1955 by Don Caudle

Les and Paula Presmyk collection
Jeff Scovil photo

their bikes out to the Old Yuma Mine. They knocked around the glory hole, which at that time had beautiful square-set timbering in it and went down the main decline to the 100-foot level. The boys tied a clothesline onto a piece of rail and dropped on down to the 200-foot level of the mine. The kids found a seam of vanadinite about 5 inches above the floor at the far eastern end of the level. It was getting late, but before heading back home, they took a last crack at the rock and opened the vein at the base of the rib. Gene stuck his head into the hole and peered up into a 12-foot high chimney filled with vanadinite. The red crystals were like candy to the kids, and they collected as many as they could carry. The boys collected about three flats of floater crystals, but they were going to need help to get at the matrix specimens.

They took their finds to the Tucson Gem and Mineral Society meeting where several adults offered to take Schlepp, Miller, and Meyer out to the mine the next weekend to finish the pocket. The adults went, but the boys were left sitting on their doorsteps. As with Bideaux's 1949 pocket, a collecting frenzy ensued as Dan Caudle and Clayton Gibson, Dick Bideaux, Al Haag, Dick Jones, and others collected the pocket over the next several weekends. The vanadinite chute turned out to be about 30 feet long and produced countless vanadinite matrix specimens and loose crystals.

In 1955 Dick Jones struck a pocket of wulfenite near the surface by the east side of the main decline of the mine. He and Al Haag had been collecting vanadinite on the 200 level. Jones went outside to take a break and spotted a yellow-orange area on the west side of the decline. He hit it with a rock, the wall gave way, and wulfenite began falling out of the pocket. Jones retrieved Haag, who hung over the ledge while Jones held onto his feet. Haag collected the equivalent of a 5-gallon bucket of wulfenite specimens to 6.5 cm wide from the pocket.

Gene Schlepp hit his first big wulfenite pocket at the Old Yuma in 1968. He was working then at the top of the glory hole in a crack that was producing big single red vanadinite crystals. The singles just kept coming from the crevice, so Schlepp rigged up some scaffolding and a ladder in the glory hole to gain better access. But when rocks began rolling out of the crack, he started to feel uncertain about the stability of the overhang and decided to approach it from a different angle. Schlepp enlisted the help of John Cesar and his metal roll-up ladder and they began to work the crack from the top of the glory hole. The pair returned a week later with their wives to see that the glory hole had changed dramatically: the huge slab that was the overhang had caved in. Schlepp's tools and scaffolding and Cesar's ladder were buried beneath tons of muck.

Wulfenite, 8.8 cm tall
Old Yuma Mine, Pima County
Collected in 1968 by John and Karen Cesar and Gene and Jackie Schlepp
John and Karen Cesar collection
Jeff Scovil photo

The large broken boulders that had been the overhang were lying in the bottom of the glory hole and were covered with loose red vanadinite crystals. The four began scouring the boulders for crystals. Then Jackie Schlepp spotted a pocket with yellow-orange wulfenite crystals near the top of one of the boulders. Cesar crawled down between two of the boulders and found an exposed wulfenite pocket, which had partly spilled its contents onto the ground between the boulders. Both pockets were large and contained numerous wulfenite crystals. The couples agreed to work the find together, the Schlepps focusing on the upper pocket and the Cesars collecting the lower pocket. It took Schlepp and Cesar three days to collect the whole thing. The largest specimen—a multiple stacked orange-yellow crystal—is 8.8 centimeters tall and is pictured left.

Collectors including Wayne Thompson opened significant wulfenite pockets in the late 60s and early 70s. But if one person had a knack for collecting Old Yuma, it was Gene Schlepp. Around 1970 Schlepp found a 1.5-centimeter-wide, zoned orange and yellow, beveled tabular wulfenite crystal on matrix that is associated with silky white cerussite sixlings (pictured on page 158). He saw a bit of wulfenite in an overhead crevice and, using a long, sharpened screwdriver, "tickled" the base of the crystal. Schlepp caught it as it dropped. The specimen was undamaged.

Gene Schlepp was back at the Old Yuma in 1971, this time with Jerry Meyer. They climbed down the decline and worked their way to the westernmost stope on the 65 level. At the base of the winze, Schlepp caught a glimpse of yellow and dug 8 to 10 inches into the soft ground. A pocket opened, and Schlepp removed 3 sticks of wet dynamite! After recovering his composure, he focused his efforts below the dynamite where he encountered brecciated rubble. Schlepp prepped for a pocket, spread a cloth, and gently moved his fingers through the rubble, which brought out loose breccia fragments coated with wulfenite. All told Schlepp removed around 25 specimens of lustrous, yellow-orange wulfenite, some of which had just fallen out onto the cloth undamaged. Some of the crystals exceeded 4 centimeters on an edge. A few days later, dealer Dave Wilbur called Schlepp to see what was new. Wilbur wanted the whole pocket. They met on the steps of the Ramada Inn in Tulsa, Oklahoma, a couple of weeks later and consummated the deal.

A month or so later Schlepp returned alone to the Old Yuma Mine. He noticed that the vein that had produced the previous pocket continued across the back (ceiling) and down the other side of the stope. Schlepp had located another pocket essentially 180 degrees from the first. This second pocket was about 75 centimeters high by 15 centimeters wide, and went back into the wall about 75 centimeters. It produced a cabinet-size specimen of wulfenite on matrix and a number of smaller pieces. The wulfenite was a rich yellow-orange and was highly lustrous. The cabinet-size specimen is a stout 5 by 2.5 centimeter crystal on an altered volcanic matrix.

In 1972 Chris Panczner, son of former Arizona-Sonora Desert Museum curator Bill Panczner, found a fine pocket of complex red vanadinite crystals to 4 centimeters. The pocket was near the western edge of the mine. Collectors continued to visit the mine throughout the 1970s, and many came away with good specimens of vanadinite or wulfenite.

Upper: Wulfenite, 3.3 cm wide
Old Yuma Mine, Pima County
Collected by Gene Schlepp in 1971
Mark Hay collection, Jeff Scovil photo

Lower: Color-zoned wulfenite, the main crystal is 1.5 cm wide, with cerussite
Old Yuma Mine, Pima County
Collected circa 1970 by Gene Schlepp
Scott Adams collection
ex Arizona-Sonora Desert Museum collection, Wendell Wilson photo

Right: Wulfenite, 4.5 cm wide
Old Yuma Mine, Pima County
Collected in the early 1970s by Gene Schlepp
May Hay collection, Jeff Scovil photo

Above: Vanadinite, 9.8 cm wide
Old Yuma Mine, Pima County
Collected in 1949

Dick Morris collection
ex George Feist
Jeff Scovil photo

Right: Gene Schlepp at the Old Yuma Mine in 1966

Jackie Schlepp photo

Right: The azurite in this 8.9 cm wide specimen on malachite has an etched appearance as is typical for the Silver Hill Mine, Pima County. It was collected by Dick Jones.

Evan Jones collection
Jeff Scovil photo

Above: While unusual for the locality, lustrous azurite crystals, such as this 4.6 cm wide specimen of blocky, lustrous azurite on malachite is known from the Silver Hill Mine in Pima County. Collected by Dick Jones in the early 1960s

Les and Paula Presmyk collection
Jeff Scovil photo

Silver Hill, Empire and Quijotoa Mines

The **Silver Hill Mine** is in the Waterman Mountains, Pima County, northwest of Tucson. Gene Schlepp first became interested in the locality when he visited with his science club in 1955. During that trip, he and his buddies sneaked underground and collected azurite druses.

With fond memories and a matured interest, Schlepp made a solo trip to the mine in 1962. He went down the drift to the first ore chute and up the raise where he entered a small stope. He quickly located a 15-centimeter-wide vein composed of dark blue drusy crystalline azurite with some banded vivid green malachite. The vein had never been worked and was loaded with specimens. Over five days, Schlepp removed 650 pounds of azurite and malachite. He wrapped the specimens, loaded them in sacks, and hauled the sacks, two at a time, out of the mine and a half-mile down the road to his car. He sold the specimens to Tucson dealer Suzie Davis.

A few years later, Dick and Roy Jones along with collecting partner Kermit Lee visited the mine. They decided that not only were there specimens to be collected but there was enough copper oxide material (azurite, brochantite, and malachite) to ship as ore. The three of them proceeded to clean out the adit, reinstall missing rail and ties, and begin mining. They ultimately produced several tens of tons of oxidized copper ore, all the while saving anything that could be sold as specimens.

Azurite and malachite from the Silver Hill Mine are fairly easy to identify. The malachite tends to form light to medium green crystals up to about 3 millimeters long covering flattish plates up to 25 centimeters across. The azurite formed in crystallized patches and bunches scattered on top of the malachite. Much of the azurite has a slightly etched appearance although at least one small pocket, recovered by Dick Jones, contained blocky, lustrous azurite crystals associated with malachite and aurichalcite.

Also in Pima County, the Empire Mountains are home to a number of argentiferous lead and copper deposits in Paleozoic limestone that have produced a number of good specimens. These deposits were first discovered in the 1870s. The principal mining camps were the **Total Wreck Mine**, **Copper Camp** (**Hilton Ranch**), and **California Camp** (**Andrade's Ranch**). None of the mines were particularly large, and all produced oxide ore. Sulfide minerals are rare. The district was in its heyday in the late 1800s. Gene Schlepp recovered many flats of light blue fibrous aurichalcite mats on altered limestone over a number of years from the Empire district's **Lone Mountain** and **Chief** mines.

The Quijotoa Mountains and mining district are on the Tohono O'odham Indian reservation, a vast tract of land between Tucson and Ajo in western Pima County. Spanish miners were working gold placer deposits in the area as early as 1774. F.N. Guild's 1910 mention of altered barite encrusted with drusy crystals of calcite, siderite, or other similar minerals in area is probably what enticed collectors in the 1950s and 1960s to investigate. Bob Jones made three trips to the district's **Weldon Mine** in the late 1960s and early 1970s and recalls,

> *A trail to the adit cuts around and across the considerable dump. Scattered remains of small animals and a large depression in the dirt—a den—made it obvious that a large animal, probably a mountain lion, was residing in the mine. A little way into the adit, a small tunnel accesses an area where bladed white to pink barite clusters and simple galena cubes occur. The barite is typical of the tightly compacted sub-parallel tabular blades from several European localities. The pink color may be due to an iron oxide. The galena is interesting in that the cubes show no modification. Most are a little more than a centimeter on edge. They are covered with a thin sparkly coating of either anglesite or cerussite.*

Calcite from the Weldon Mine is very similar to that from Santa Eulalia, Chihuahua, Mexico, although the Weldon Mine crystals formed on pink barite and are thus easily distinguishable from their Mexican counterparts.

Pale pink barite crystal group with colorless calcite blades at the base of the specimen and drusy quartz crystals on the back. The specimen is 9 cm long and is from the Weldon Mine, Pima County.

Bill and Diana Dameron collection
ex Charles Lawrence
Bill Dameron photo

Right: Motor crew on the 2833 level of the Campbell Mine

Pete Kresan photo, 1973

Left: This and several similar malachite specimens were collected from the 600 level of the Cole Mine in 1957 as a crosscut was being widened. This 32.5 cm wide specimen is composed of a loose mass of tiny crystals and is quite light.

Freeport McMoRan Copper & Gold collection
Jeff Scovil photo

Bisbee

In 1951 Phelps Dodge opened the large **Lavender** open pit on the site of the Sacramento Hill stock. This mine exploited the lower-grade porphyry stock (averaging about 0.7 percent copper). An enormous amount of rock was moved to uncover the ore. With the opening of the Lavender Pit came the discovery of some of the world's best turquoise. For the next 20-plus years, the periodic stripping of the post-ore Glance Conglomerate along the east pit wall brought ever more turquoise to a hungry lapidary market at the peak of a turquoise craze. But in spite of the huge quantity of turquoise produced, high quality matrix specimens are rare as the lapidary value of the material typically exceeded its specimen value.

The **Cole Mine** had been developed by the Calumet & Arizona Mining Company in 1904. The Cole is south of most of the large mines and was well beyond the known limits of economic mineralization. From early on though, it was a good ore producer (Calumet & Arizona 1905). Following the 1932 merger with Phelps Dodge, the Cole Mine was operated off and on until it was fully reopened in 1942 (Mills 1958). This important mine then operated until 1975, when all mining ceased. While both oxide and sulfide ores had been mined over these many years, it only produced notable specimens during its final 25 years in operation.

In early 1952 development work on the 800 level hit clay zones, which when mined, produced many hundreds of interesting azurite roses and attractive spherical masses. Many showed a partial alteration to malachite. Both frequently contained fragments of Devonian era fossils. While these roses are attractive, they are somewhat dull and nowhere near match the quality of those found in the Czar and Holbrook mines.

Some of Bisbee's very best chalcotrichite (a variety of cuprite) was discovered in a stope above the 1100 level of the Cole Mine in 1954. The delicate acicular crystals had formed along relic bedding planes in altered and oxidized clay-rich Abrigo Limestone. Impressive mats of lustrous, hairlike crystals to 3 centimeters in length were recovered in specimens as large 15 centimeters across. To the credit of the miners who found them, a good number of these specimens were recovered with minimal damage.

Impressive quantities of large clusters of radiating connellite crystals in massive cuprite were found on the 1200 level of the Cole in 1958. One fist-size piece is almost pure connellite. Mineralogist Dick Bideaux suggested that that specimen alone may well contain half as much connellite as all of the rest of the connellite specimens in the world combined (Anthony et al. 1995). Numerous smaller but important pieces of connellite in massive cuprite were recovered at about the same time. Also on the 1200 level and in 1959, author Richard Graeme collected a specimen of cuprite and malachite as well as teineite and a mineral that turned out to be new to science and was Bisbee's sixth type species—graemite (Williams and Matter 1975).

"Native" copper had long been mined at Bisbee, but was under

Above: "Native" copper
12.5 cm wide
2200 level, Campbell Mine, Bisbee
Collected in 1955
Les and Paula Presmyk collection
Jeff Scovil photo

Facing page, upper: Malachite pseudomorph after azurite, 3 cm wide
105-C stope, 1900 level
Campbell Mine, Bisbee
Collected in 1970
Graeme collection
Kevin Dixon photo

Facing page lower: Cuprite, variety chalcotrichite, 8.2 cm wide
1100 level, Cole Mine, Bisbee
Collected in 1954
University of Arizona collection
(UAMM Cat. No. 4008)
Jeff Scovil photo

appreciated by the miners, as it paled alongside the spectacular and much coveted copper carbonates. Thus little was saved early on, but as good azurite and malachite specimens became less common, native copper gained a degree of respect. A spectacular orebody consisting almost entirely of copper was encountered on the 2200 level of **Junction Mine** in 1955, just as the mine was being closed. This fine orebody was exploited through the nearby and more efficient **Campbell Mine**. Over a period of five years, more than 14,000 tons of native copper were mined from the several stopes opened on this orebody. Thousands of good specimens were saved by a constant stream of miners, including one of the authors (RG), who frequented these stopes.

The copper frequently occurred as large, crystalline masses in a friable, very yellowish goethite or as spectacular spinel-law twins, clusters, or doubly terminated single crystals in the soft iron-rich clays. Any hard lump felt when running one's hands through the soft mud in the stope was invariably a piece of native copper.

More than a few huge masses, some exceeding several hundreds of kilograms, were left behind as they would have required too much effort to cut into pieces small enough to pass through the ore handling system. Several meter plus boulders of pure copper sat on the 2200 level station ore pocket *grizzlies* (sizing rails to prevent rocks too large for the ore handling system from being put into the chutes) for years in mute testimony of how difficult it was to mine "native."

Above: Hemimorphite, copper and calcite, 9 cm tall
Cole Mine, Bisbee
Freeport McMoRan Copper & Gold collection
Jeff Scovil photo

Facing page, upper: A natural gold film coats this 12.8 cm wide native copper specimen that author Richard Graeme pulled from the 700 level of Bisbee's Cole Mine in the 1960s.
Graeme collection
Jeff Scovil photo

Facing page, lower: Rosasite
13.2 cm wide
26 J stope, 1200 level, Cole Mine, Bisbee, Cochise County
Evan Jones collection, ex Al Viorin
Jeff Scovil photo

At much the same time, a couple of small stopes on the 700 level of the Cole Mine were being mined for ores largely composed of copper, which occurred exclusively as closely spaced individual masses in a gray claylike material, which was streaked with hematite. This occurrence was unusual for the nature of the crystals. While the individual crystals were typically small, they were uncommonly sharp and bright. Many had an imperceptibly thin veneer of elemental gold that made them subtly iridescent and ensured that they would never tarnish.

Malachite replacements of copper were abundant in the 26-J stope on the 1200 level of the Cole Mine in the 1960s. The area also exposed boxwork pockets of goethite with light green, zinc-rich malachite in aesthetic forms, as though it had been poured, molten, into the voids. Rosasite was occasionally found as striking botryoidal linings of hematite vugs containing colorless, gemmy calcite crystals. The rosasite had in fact replaced malachite when zinc sulfate-rich solutions passed through the area after the malachite had been deposited. Hemimorphite crystals to 5 centimeters were also found, although clusters of centimeter-size copper and/or chalcophanite-included hemimorphite on hematite were more common.

The 26-K stope on the 1300 level of the same orebody produced exceptionally long spinal twin copper crystals, some exceeding 20 centimeters in length. A slightly altered cuprite coated all of them and caused them to appear almost black. These copper twins were associated with delicate masses of acicular malachite. The ores continued to below the 1300 level and were mined through 26-L stope, where a few large, superb specimens of spherical clusters of malachite crystals on white calcite were found. These specimens were mistakenly referred to as "primary" malachite.

The late 1960s saw the expansion of the Lavender Pit to the west. It passed through the **Sacramento Pit**, which had been opened a half century before, and into the ground that the specimen-rich **Holbrook Mine** had exploited with great success some 80 years prior. Anticipation among collectors was high as samples recovered from exploration drilling frequently contained cuttings of azurite, malachite, and/or cuprite. An exploration adit into the pit wall intercepted the 400 level of the **Gardner Mine**, and when extended, this adit also hit small patches of rich oxide ore containing fine chatoyant malachite specimens.

Phelps Dodge anticipated encountering mineral specimens and made author Richard Graeme responsible for mineral recovery. The first major discovery came late on Thanksgiving Day afternoon in 1969 when a large electric shovel in a dropcut hit a 40-centimeter thick malachite mass. Active mining in the dropcut was stopped and the extent of the malachite determined by selectively removing the soft oxide surrounding the material. The malachite was nearly horizontal and measured a bit over 36 feet long and more than 3 feet wide. It was wedge-shaped and ranged from 60 to 20 centimeters thick. An estimated 5.5 tons of malachite were recovered from this find. The mass was between two stopes mined in the late 1890s, both of which had caved in and rendered the stopes too dangerous to mine further. Post-mining ground movement had conveniently broken the material into

This 8.5 cm wide malachite on calcite is the largest of the five specimens collected by Melvin Elkins from the 26-L stope, 1300 level, Cole Mine, Bisbee in the 1960s.

Les and Paula Presmyk collection
Jeff Scovil photo

pieces generally weighing no more than 200 pounds.

Afterward large pieces of boxwork malachite with small spheres of azurite were recovered in large quantity. One of these pieces weighed nearly 1,500 pounds, but only small sections were of specimen quality. Over the next eight months or so, several thousand specimens of azurite and malachite, reminiscent of those mined 80 years earlier, were recovered and sold to the general public thus affording local collectors an opportunity to add the classic material to their collections.

The area being mined by this extension of the Lavender Pit had been the source of the type material for mineral chalcoalumite, and sure enough many hundreds of fine specimens of this rare mineral were uncovered. Some of the more interesting chalcoalumite specimens exhibited tiny clusters of sharp, bright crystals, which had not been seen before. Williams and Khin (1971) used this new material to accurately determine the crystallography of this species, thereby completing Larsen and Vassar's 1925 description of the type material.

The huge, modern machinery easily sliced through the high-grade plastic clays along the **Dividend Fault** that had thwarted the efforts of their underground predecessors. When the first old workings of the 200 level Holbrook were exposed in the pit wall, Richard Graeme and another company geologist crawled in to explore. Within ten minutes Graeme had a classic, albeit clay-covered, 5-centimeter-wide azurite rose in his pocket. While nirvana was certainly at hand, he never saw another in either the limited Holbrook workings or in the pit.

As gold and silver prices rose and copper prices fell in the 1970s, Phelps Dodge prospected areas known to be rich in precious metals. The **Shattuck Mine** had long been recognized as an important source of gold, and it was for sale. Phelps Dodge purchased the small mine and began the labor-intensive process of reopening it, as the shaft timbers had burned out in the

A miner on the 2966 level of the Campbell interior shaft in a cage, 1962

This shaft connected the 2966 level of Campbell Mine to the 3100 level of the Denn Mine and allowed for the movement of men and materials during the unsuccessful exploration phase of the lower Denn area

Richard Graeme photo

early 1950s. As work progressed down the shaft, areas that had long been inaccessible were now open. While a few fine malachite specimens were found, the effort was notable for a short while for an unusual find of shattuckite as pseudomorphs after malachite and as small spheres of acicular crystals that were recovered from the type locality on the 200 level. A few specimens of the questionable species bisbeeite (today bisbeeite is thought to be a mixture of other minerals) were also recovered.

The noted "Lead Cave" on the 300 level produced silica breccia fragments that were richly adorned with yellow, waxy crystals of bromargyrite. A small oxidation cave on the 500 level yielded numerous irregularly shaped masses of cuprite covered with bright acicular malachite crystals. A body of hematite on the 600 level contained crystals of that mineral to 3 centimeters. These crystals were so sharp and distinct that they were initially thought to have been pseudomorphs after something else, perhaps calcite. The 600 level also yielded surprisingly good specimens of botryoidal hematite.

The end to mining at Bisbee came in June 1975; nearly a century after it began, but the story of the mineralogy of this fabulous deposit is still being written as new species for the district are discovered in the study of the many specimens saved from this, the best of Arizona's mineral occurrences.

Wulfenite, 20.8 cm wide
Defiance Mine, Cochise County
John Lucking collection
ex Smithsonian Institution collection
Jeff Scovil photo

Defiance Mine

with contributions by Scott Lewis and Roy Jones

Certainly no Arizona mineral collection would be complete without a Defiance Mine wulfenite. Located in the Turquoise Mining District, Cochise County, the Defiance Mine was a late bloomer. Wulfenite was not discovered there, or was at least not well known, until the mid-1950s. The Defiance Mine exploits a series of irregular base and precious metal replacement deposits that are controlled by a fault that cuts through the district.

Rich silver-lead ore was taken from the oxidized portion of the vein near the surface from 1923 to 1924 (Wilson 1927). The workings were enlarged considerably by other companies that later mined the orebody. In 1956 one of those mining companies took a hard look at the mine and discovered the wulfenite-lined watercourse in limestone that produced the now classic specimens. The watercourse was only worked for a short time before good specimen finds became increasingly scarcer.

Above: Advertisements for Martin Costello's St. Louis Beer Hall appeared regularly in the *Tombstone Daily Prospector*, this one on February 12, 1889 (vol. III, no 36)

Library of Congress Chronicling America collection

Below: Gleeson, Arizona, circa 1908
John Taylor collection

The Defiance Mine along with the nearby Silver Bill Mine has been a part of Scott Lewis's family for over 100 years. The claims were bought and patented by his great-grandfather Martin C. Costello in the late 1800s and were mined by his grandfather Jim Giacoma, who discovered the famous 1957 wulfenite watercourse while mining lead, silver, and zinc.

Costello arrived in Tombstone in 1879. An Irish immigrant, he learned carpentry in Philadelphia and applied his craft in the Arizona Territory by building the St. Louis Beer Hall on Tombstone's Fremont Street in 1880. Costello owned the saloon, which featured Anheuser-Busch beer. It was the only place between Denver and San Francisco where one could enjoy a cold "Bud." A natural entrepreneur, Costello invested his earnings in real estate and mining claims.

He bought several claims, including the **Irish Mag**, from Bisbee prospector Jim Daley's common-law wife, who had purportedly grubstaked Daley. But title to the claims was disputed after Daley disappeared from Bisbee, having shot and killed a deputy sheriff (Douglas 1909). It took 10 years and the United States Supreme Court to finally award Costello clear title. He sold the Irish Mag and four other claims to the Lake Superior and Western Development Company in 1898. The Bisbee property was ultimately mined by the Calumet and Arizona Mining Company (see page 91).

With his Bisbee holdings tied up in court, Costello shifted his focus to the Gleeson-Courtland district, 12 miles to the north. He staked the Defiance and surrounding claims in the early 1880s and patented them before the turn of the century. He ultimately bought and patented 50 claims in the district. Costello mined high-grade silver, lead, and copper from the near surface oxide deposits from 1883 to 1885. Production history is sketchy from 1890 to 1920, but in the mid-1920s, reworking of the dumps at both the Silver Bill and Defiance produced an average of 150 to 900 tons of ore per month. The 1930s saw only intermittent production, but in the early 1940s, brothers Tony and Jim Giacoma (Martin Costello's son-in-law and Scott Lewis's grandfather) mined and shipped over 2,000 tons of high-grade lead ore to aid in the war effort.

Mining resumed in 1952 and 1953 as metal prices were on the rise. The Giacomas and two other miners were producing 75 to 100 tons of ore per month. During this time, 80 feet of incline and 425 feet of drifting were mined on and below the main adit haulage level. As the mine deepened and tramming became more difficult and costly, a secondary haulage point was chosen. In 1956 a surface shaft at a 60-degree incline was sunk to the depth of 90 feet. More than 450 feet of new drifting attempted to connect with the lower workings of the main adit haulage level. Heading directions were determined each day after listening to blasts from the respective drifts. This "Italian Sonar" method proved remarkably accurate. The two workings were within 50 feet on one another when the brothers hired Wally Platt, a University of Arizona student working summers at the nearby **Shannon Mine**. Wally surveyed both headings and suggested a slight gradual westerly turn in the drift heading north from the new incline. When the two headings merged, both the Giacomas each saw light at the end of the tunnel—the other's carbide lamp.

During the survey, Platt collected several specimens of wulfenite and calcite and showed them to his friend Richard A. Bideaux. Defiance wulfenite had been discovered! Jim and Tony Giacoma did not spend much time marveling at the slabs of honey-colored wulfenite crystals lining the 45-foot watercourse. The wulfenite was 14 percent lead and would result in larger settlement checks from the ASARCO smelter in El Paso. It was not until later that they realized that wulfenite had a specimen value far greater than that of its lead content. Scott Lewis recalls,

> *The Giacoma brothers began accumulating specimens, and even as a young boy I realized how prized and valuable the wulfenites were. I spent most every Christmas, Easter, and summer vacation with Grandpa Jim in Tombstone. Numerous collectors from all over the country, including miners from Bisbee and Nacozari, stopped by to purchase wulfenite specimens, which were stored in a shed containing several hundred wooden produce crates full of wulfenite and calcite specimens.*
>
> *Pricing was simple: size of his fist, $10; bigger than his fist, $20. Larger pieces ranged from $20 to $100. Miniatures, thumbnails, and shards were free for the asking. He took cash or specimens in trade and would place anything acquired in trade on an upright piano in his living room. I could not wait to visit him and see what new rocks had been added to his collection; many of these specimens are in my collection today.*

Wulfenite on calcite
Field of view 2 cm tall
Defiance Mine, Cochise County
Tony Peterson collection and photo

Probably the best description of the wulfenite zone was written by Richard Bideaux, and published in March-April 1957 *The Mineral Explorer*:

> *... On January 29, 1957, Wally, Richard Jones, and I visited the mine. We were very surprised when we found a large solution cavity exposed in the working face. This cavity was in the bedded limestone, and had a flattened, horizontal aspect. Some of it had been removed during mining, but the portion left was about 35 feet long, 12 feet wide, and from 2 to 4 feet high. The surprising thing was that the walls, floor and ceiling were encrusted with wulfenite crystals in addition to calcite crystals. These wulfenite crystals were similar to those that Wally had collected previously, but some were almost one inch along an edge.*
>
> *We were unprepared for the quantity of material exposed, so we confined our collecting this day to a smaller pocket beneath the large one. This smaller pocket produced a number of transparent thin tabular wulfenite crystals of a beautiful yellow color. Many of these crystals were an inch wide. Although the specimens are fragile, they are superb examples of wulfenite. The matrix is limonite. Plumbojarosite and plattnerite have been identified on some specimens.*
>
> *We returned the following weekend, in the company of Ben Drew and Tom Tucker, to find that about three-fourths of the pocket had been destroyed by mining operations. However, another remarkable cavity had been penetrated adjacent to the first one, and this new pocket was entirely unblemished. Like the first, it was a solution cavity in the limestone even*

Defiance Mine wulfenite crystals frequently show a twisting morphology not unlike saddle-shape dolomite crystals. The wulfenite blades are composed of subcrystals which are slightly overlapped, rotated, and tilted with respect to each other. The subcrystal arrangement resembles a screw-like pattern. The vertical axes of subcrystals are skew lines. This 10 cm wide specimen is in the collection of Tony Potucek.

Jesse La Plante photo

more heavily encrusted with wulfenite and calcite crystals. This one was roughly circular in cross section, and about 45 feet of its length was accessible.

It was an unforgettable experience to crawl into this pocket and be completely surrounded by millions of glittering crystals. Many hundreds of pounds of wulfenite were in sight, so easy to collect that specimens could be pried from the walls without even using tools in many cases. These specimens were so easily collected that I doubt if one percent of all the crystals we handled were broken on our return to Tucson.

The best wulfenite in this pocket occurred as thick crusts either on the limestone walls or on a layer of mixed iron and manganese oxides. Both calcite on wulfenite, and wulfenite on calcite were obtainable. All wulfenite crystals in the pocket were the same brown tabular (type) as the original crystals. A large proportion of the specimens we collected showed a curious crested development in that the large individual crystals would develop into numerous smaller plates on that part of the crystal farthest from the matrix [an is example shown above]...

... Some plattnerite was noted on these specimens as very tiny crystals near the bottoms of the wulfenite crystals, which were on the iron-manganese oxides. Part of the characteristic brown color of these crusts is due to a thin film of iron oxide, which in some cases is lightly overgrown

Wulfenite, 12 cm wide
Defiance Mine, Cochise County
Collected by Clayton Gibson in the 1960s
Joe and Susan Kielbaso collection
Jeff Scovil photo

with manganese oxide in a distinctly dendritic pattern.

Drifting continued along this series of fissures for 130 feet before the last traces of wulfenite were passed. Several other good-sized solution cavities were encountered during this time, also lined with wulfenite, but the contents of these were shipped with the ore. Very likely exploratory mining in this area of the mine will discover other cavities. The largest crystal recovered was 70 mm on an edge.

Roy Jones related some of the back-story to Bideaux's Defiance Mine exploits, recalling that Dick Bideaux got a call encouraging him to return to the mine as a new wulfenite-lined watercourse had been discovered. Dick's father George had just purchased a brand new Buick. He was out of town for the week and had warned Dick not to drive his new car. But the lure of collecting overruled George, and in Dick's mind there would be no harm done as his collecting partners were taking a trailer to haul the specimens home and there would be no telltale red stain in the car. Dick felt safe.

They spent the weekend collecting and had many flats of wulfenite to show for their efforts. They loaded everything onto the trailer and began driving back to Tombstone.

The road was dusty, and Dick did not realize that his partner had stopped the trailer to check the load—Dick rear-ended the trailer. After the dust (literally) settled the boys had two concerns: was the wulfenite okay (not a single specimen was broken or damaged); and how were they going to

Above: Blocky wulfenite crystals on matrix, field of view 2 cm tall
Defiance Mine, Cochise County
Tony Peterson collection and photo

Below: Dipyramidal wulfenite Crystals to 3 mm
Defiance Mine, Cochise County
Peter Cristofono collection and photo ex Neil Wintringham (No. 82-148)

get the caved-in grill of George's new car repaired?

They headed back to Tucson to raise some quick cash and get the car fixed. They figured Tucson dealer Claude Motel might be interested in buying some of the wulfenite. Although Claude was not inclined to buy the fragile crystals, he loaned them the money to get the car fixed. If George suspected that Dick had wrecked his car, he never let on.

Bernie Haag, widow of famed mineral collector Albert Haag, recalled the friendly but intense rivalry between dealers Dick Bideaux, Dick Jones, and Al Haag. And when Al heard that Bideaux and Jones had found a wulfenite locality somewhere in the southern part of the state, Haag struck up a conversation with Jones to try to dislodge some information. Dick told Al that the mine was somewhere north of Tucson. Al immediately said, "I happen to know the mine is south of Tucson." Al of course had no idea where the guys were working, but the briefly startled look on Jones's face clued him in.

Al and Bernie found a secluded spot on the highway east of Tucson and waited just a little while before Bideaux and Jones went zipping by heading east. Tailing them on Benson highway was not a problem, but when they turned south at Benson it became more difficult for Al and Bernie to follow them undetected. Shortly before Dick and Dick reached Tombstone, they sped up and Al lost them. But by that point the Haags knew that Bideaux and Jones were heading for the Defiance Mine in Gleeson. Al and a friend then visited the Defiance Mine and also got many specimens before the vug was depleted.

Gene Schlepp visited the mine in the early 1960s. He explored the first level cutting west at the split and hit a stope at the end of the drift where he collected 8 to 10 beer flats of dipyramidal wulfenite up to 1.2 centimeters wide on oxidized limestone. The wulfenite is butterscotch in color and has a silvery sheen. Gene brought the material out of the mine thinking that they were unusual but disappointed that he didn't locate a good pocket of the classic thicker wulfenite. But Dick Bideaux had a more favorable view of the specimens. He noted that they were the first dipyramidal wulfenite crystals found in Arizona.

Mineral collectors recovered a significant amount of wulfenite during the rest of the 1960s. Thompson (1980) reported that most of the wulfenite was the more fragile, thin-bladed variety. In an effort to stem collecting, the mine owners erected a large concrete cap over the shaft with a steel door held closed with a large steel chain. The chain and door were compromised within a few weeks and the mine was again open, *sans* chain.

According to mine owner Scott Lewis, the Defiance was open to collectors during the 1970s when another oxide zone extending off the bottom of the 1950s watercourse produced several hundreds of flats of commercial material. The property was permanently closed in the 1980s after a teenager's attempt to broad jump a nearby shaft resulted in his death. After liability lawsuits and increased family exposure, the Defiance, Silver Bill, and all of the Costello mines were closed to collecting until almost 20 years later when Scott Lewis organized a mineral collecting operation in the Defiance Mine.

Cohen Mine, Dos Cabezas Mountains

The Cohen Mine in Cochise County produces Arizona's best scheelite crystals and has even yielded a few that are world class. The mine has been known by several names: Comstock, Adams, Cowboy, Rogers, and one that seems to have stuck—the Cohen Mine. The deposit is located about 25 miles east and south of Wilcox in the Teviston Mining District.

The property was just a few gold prospect pits until about 1944 when F.W. Clark and his wife were prospecting and found tungsten. They located 13 claims, which they sold to Aichis Arthur Cohen in 1945. Arizona Department of Mining and Mineral Resources records indicate that the mine operated during World War II as the "Coin" Mine, likely a misinterpretation of "Cohen."

Cohen and his wife reportedly drove a 100-foot long adit into the vein beginning at the east end of the deposit where the vein outcropped. Along the way, they encountered a series of open pockets that produced deep orange scheelite up to 7 centimeters long. While the crystals were obviously more valuable as specimens than as tungsten ore, the pockets were not plentiful enough to make the operation profitable, and the Cohens let the claims lapse about 1950. R.B. Adams relocated the property a few years later.

Adams leased the claims to Tyone Mining Company in 1954, and they constructed a mill for concentrating the scheelite. But expenses exceeded revenue and the operation closed. The property, at this point called the Comstock Mine, seems to have changed hands at least twice in 1955 until it was acquired by C.C. Rogers of Safford, Arizona. Rogers found scheelite crystals, which just like the Cohens, he deemed were more valuable as specimens than as ore.

Scheelite, 11.1 cm tall
Cohen (Rogers) Mine
Dos Cabezas Mountains, Cochise County
Flagg Mineral Foundation
ex Loris P. and Colleen Woolery collection
Jeff Scovil photo

A June 6, 1957 Arizona Department of Mining and Mineral Resources report filed by engineer Axel Johnson stated that owners Donald C. and Pearl C. Rogers were working an 18-inch-wide quartz vein. The Rogers had struck scheelite 10 to 15 feet below the surface, and while they could provide no estimate of the average value of the ore, they showed Mr. Johnson some "beautiful scheelite crystals, which they found in the vein."

In the July-August 1957 *The Mineral Explorer*, Dick Bideaux reported on the Roger's scheelite finds:

> *Occasional pockets of quartz crystals with associated scheelite crystals were encountered in mining operations, but the amount of specimen material reaching collectors was very limited, most of it being shipped as ore. Scheelite crystals donated to the University of Arizona geological museum by A.A. Cohen reach a size of two inches, and are with peculiar bluish-gray translucent quartz crystals. An inclined shaft was recently sunk on a new area of the property, and a particularly fine pocket of crystals was found only 12 feet beneath the surface. The scheelite crystals from this pocket are a deep amber color, with a resinous luster. They are commonly composite in character. Various combinations of first, second, and third order bipyramids have been noted. The majority of the crystals recovered were of medium size, and about 2 to 3 pounds in weight. Several much larger crystals were found, however. A crystal in the writer's*

Right: Scheelite and quartz, 7.6 cm tall
Cohen (Rogers) Mine
Dos Cabezas Mountains, Cochise County
Arizona-Sonora Desert Museum
Jeff Scovil photo

> *collection at the time* [RAB] *measured 4½ x 4½ x 6½ inches, and weighed 12 pounds, 15½ ounces. It is only one half of the original crystal; if it ever existed, would have weighed an estimated 22-23 pounds. It is about 60% anhedral.*

The largest known crystal from this occurrence is in the Arizona-Sonora Desert Museum. It is 25 centimeters wide and weighs about 22 pounds! While the color of the specimens is consistently orange, the larger crystals tend to have just a few good faces and broken or contacted sides where they grew into the quartz vein.

The Rogers family learned too late that they should have kept the scheelite crystals on the quartz crystals. When Les Presmyk visited Mrs. Rogers in 1973, all she had left was one 4-centimeter scheelite crystal and a 25-pound bag of crushed scheelite. She also had a bookcase full of clear, lustrous quartz crystal clusters. Some of the specimens were double sided, with quartz crystals coating both sides. The crystals were up 7.5 cm long.

Mrs. Rogers also talked about the pockets they dug and the crystals they sold. She recalled taking some of their best crystals to the Tucson Show. They met Paul Desautels of the Smithsonian, and he got first pick. Senior Folch

Upper and lower left: Scheelite
10.3 cm wide
Cohen (Rogers) Mine
Dos Cabezas Mountains, Cochise County
The upper photo was taken under normal lighting conditions, the lower photo under shortwave ultraviolet light

Les and Paula Presmyk collection
Jeff Scovil photos

Upper right: Scheelite, 6.1 cm tall
Cohen (Rogers) Mine
Dos Cabezas Mountains, Cochise County

Les and Paula Presmyk collection
ex Flagg Mineral Foundation
Jeff Scovil photo

Girona was very interested in two or three of the crystals until he found out that Desautels had already selected one. By his own admission, the remaining crystals were far superior to anything he had in his collection; still he refrained from buying one because he did not get first choice.

Mrs. Rogers also mentioned one pocket of scheelite that had been stolen from them. Mr. and Mrs. Rogers held weekday jobs in Safford and did their mining on the weekends. Late one Sunday, they hit into a good pocket of quartz crystals and scheelite that they had planned to collect the following weekend. A man stayed at the property during the week to act as a watchman. When the Rogers returned, the scheelite was gone. The watchman explained that he had left one day during the week to get supplies and returned to find that someone had removed the crystals.

Not long after and unbeknownst to the Rogers, a prospector approached Loris P. Woolery (see page 180), who was well known in southern Arizona as a collector who would pay good money for crystals. The prospector offered Woolery a group of scheelite crystals of various sizes and several bags of scheelite ore. "He had to take it all or nothing, with the ore at one price per pound and the crystals at a considerably higher price" (Getsinger 1965). Woolery bought the lot, and the eight specimens that he kept were probably the best group of Arizona scheelite crystals in existence. His largest specimen stands 11.1 centimeters and is the best example known. These specimens all came out of one pocket.

No significant pockets or specimens have been found at the property since the Rogers family stopped their operation. Tucson Show attendees got to see the Woolery crystals on display in 1962. Loris Woolery exhibited the best of his collection, including the top two or three Cohen Mine scheelite crystals. The Flagg Mineral Foundation later acquired Woolery's collection, which included these fine crystals.

The Flagg Mineral Foundation

The Flagg Mineral Foundation, an all-volunteer, non-profit corporation, was formed in 1962 by members of the Mineralogical Society of Arizona as a memorial to honor Arthur Leonard Flagg. The original vision of the foundation was to establish an earth science museum in the Phoenix area. Early on the foundation fortuitously acquired the large and well known Loris P. and Colleen Woolery (Bisbee) mineral collection. The foundation's goal was expanded to become the curator and guardian of this important collection. Although the Woolery collection was worldwide in scope, more than 70 percent of the original 1,600 specimens were from Arizona, the majority from Bisbee. Shortly after the Woolery acquisition, members of the Mineralogical Society of Southern California donated a fine collection of Tiger minerals in honor of Arthur Flagg. The foundation's finest leadhillite (pictured on page 358) was part of that donation.

A few years later, the foundation purchased Ray and Hazel Wright's collection, which included an extensive array of Bisbee calcite specimens. It also contained worldwide specimens, Bisbee rhodochrosite, and the best malachite after azurite known from the Maid of Sunshine Mine at Courtland. Other donations followed including the Harry and Irene Hill's Tiger collection, the Neavitt Memorial Collection of Gem Carvings and Bowls, and the Cherry Memorial Sphere Collection. Every major Phoenix-area collector, as well as a number from the larger community, have supported the foundation's efforts.

From the Collection of the Flagg Mineral Foundation (est. 1962)

Above: Calcite "Bird's Nest"
11.6 cm wide
Bisbee, Cochise County

ex Colleen and Loris P. Woolery collection

Right: Malachite pseudomorphs after azurite on cerussite
9.7 cm wide
Mammoth-St. Anthony Mine, Tiger
Pinal County

Facing page, upper: Vanadinite
14 cm wide
Grey Horse Mine, Pinal County

Facing page, lower: Copper with calcite, 8.8 cm tall
Campbell Shaft, Bisbee, Cochise County

The photos in this spread are by Jeff Scovil.

Grand Reef Mine

The Santa Teresa Mountains are remote, wild lands about 60 northeast of Tucson. The Aravaipa Mining District is in the southeastern Santa Teresas and is home to numerous small, lead-zinc mines, which also recovered modest amounts of gold and silver in the first half of the twentieth century. One of those mines, the Grand Reef, has produced a suite of wonderful secondary lead minerals including seven type species: aravaipaite, artroeite (named for the famous micromounter), calcioaravaipaite, grandreefite, laurelite, pseudograndreefite, and shannonite (named for the Arizona collector Dave Shannon). The Grand Reef Mine has also produced Arizona's best linarite and some very fine leadhillite.

The mine is named for the massive silicified breccia and rhyolite pipe that it exploits. The dike cuts across the landscape and almost forms a dam across Laurel Canyon. It looks like a giant fin rising above the surrounding terrain. Adjacent to and within the dike, a series of nearly vertical faults localized the lead ores.

The Grand Reef was first worked in the 1890s by J.W. Mackay. Eventually a small mill was built to concentrate the ores. In the 1920s the concentrates were hauled a few miles away to the town of Aravaipa, located on the only flowing stream in the area, to be processed in a floatation mill. In 1929 and 1930 the Grand Reef was the second largest producer of lead in

Linarite on quartz, 6.7 cm tall
Grand Reef Mine, Graham County

Arizona-Sonora Desert Museum
Jeff Scovil photo

The vertical sheet of silicified breccia from which the Grand Reef Mine takes its name
Jolyon Ralph photo, Feb 2009

the state. The mine has been idled by the early 1940s.

Early references to the localities exotic lead minerals appear to have prompted Ed McDole to visit, but the Grand Reef withheld its collector treasures until the late 1960s. In 1968 Dick Jones, his son Roy, and Roy Lee did some research and determined there was a linarite locality near the ghost town of Klondyke in the Santa Teresa Mountains. The trio made a trip to the mine and was surprised to discover rocks in the creek bed stained blue by linarite. Apparently someone had blasted out the drill pads and in the process had scattered a number of blue-stained rocks.

Cuprite variety chalcotrichite, 5.4 cm tall
Grand Reef Mine, Graham County
Dick Morris collection
ex Tom McKee
Jeff Scovil photo

Although the mine most certainly encountered numerous pockets of cerussite and linarite in the early days, all of the known specimens from the mine were recovered by collectors . Dick, Roy, and Roy spent the next several months exploring and collecting a number of fine linarite specimens. And they found other oddball minerals including a pocket of cuprite crystals, a creedite specimen, as well as leadhillite and olivenite crystals.

Over the next three decades, the Grand Reef Mine became a destination for collectors. While appearing to be discontinuous and perhaps friable, the breccia is very difficult to break. Collectors Wayne Thompson and Dick Miller spent hours in the spring of 1971 trying to dislodge a 30-centimeter wide rock that seemed to have roots to the center of the earth. They could move it but it was wedged into the rock face. As they tried to work it out, the occasional specks of blue came out of the bottom of the rock leading them to believe there were linarite crystals in the vicinity. Linarite is fragile, so they did not want to start pounding away indiscriminately. They finally left the mine with plans to return.

Thompson mentioned his frustration to collector Lee Bridges that week and without Thompson's knowledge, Bridges headed up to the mine with collector Bob Dryer. Within 30 minutes, they had dislodged the rock and were amazed to see a vug filled with 2.5-centimeter (1 inch) crystals (pictured below). Joy cannot adequately express their feelings. They brought the specimen back, showed it to Thompson, and ultimately traded it to him. It was the best Grand Reef linarite that either Bridges or Thompson have ever seen.

Thompson and Les Presmyk discussed trimming the specimen for more than a year. Today it would be a relatively easy to do, but none of the sophisticated trimmers and saws were available to them in the early 1970s. All they knew was that the matrix was some of the hardest they had ever encountered and linarite has perfect cleavage. Thompson finally decided that rather than risk trimming the piece, he would offer it for trade.

Thompson took the specimen to the Tucson show and showed it to Australian mineral collector Albert Chapman who told Thompson that he would bring an undamaged, crocoite specimen with terminated crystals to Tucson the following year in exchange. They made the trade and the linarite was taken to Australia. Chapman's collection was ultimately acquired by the Australian Museum in Sydney, where it is currently on display.

Below: The Albert Chapman Mineral Collection exhibit at the Australian Museum, the Grand Reef linarite recovered by Lee Bridges in 1971 is pictured in the bottom center of this James King/Australian Museum photo

Right: Linarite in quartz breccia matrix
12 cm tall
Grand Reef Mine, Graham County
Collected in 1971 by Lee Bridges
Australian Museum collection

Miami-Inspiration District

Butte copper magnate William Boyce Thompson founded Inspiration Copper Company in 1908. The first shafts—the **Joe Bush**, **Scorpion**, and **Bull Dog**—were sunk near Miami, a small silver town just west of Globe and its **Old Dominion Mine** (page 51). The **Colorado** shaft came next.

A year earlier, the Miami Copper Company had formed to mine ore reserves including the Oaks-Newman group of claims and the Keystone property, both of which were accessing the same orebody as the Inspiration mines. In 1912 Inspiration Copper Company merged with the Live Oak Development Company to become the Inspiration Consolidated Copper Company. The properties were later purchased by Cyprus and then Phelps-Dodge, now Freeport McMoRan Copper & Gold, Inc.

The **Live Oak Mine**, the district's most important specimen producer, had been a source for drusy quartz on chrysocolla and gem silica since underground production began in the early twentieth century. The few miners who enjoyed the beauty of the pieces (or knew their value) saved them. But when the open-pit operation began in 1947, shovel operators, truck drivers, drill operators, pit bosses, and management all had a better opportunity to spot and save specimens. It was not necessarily a flood of material that started hitting the collector market in 1947, but almost.

Collector and writer Bob Jones (pers. comm.) recalls being told of miner/collector Harold "Curly" Talbott's experience working on the 550 level of that mine on the very last shift of the underground operation. Talbott and his crew were having trouble with a plugged draw point. When they shined their lights up the chute to see what was blocking it, they were astonished to see a large boulder of pure gem chrysocolla with drusy quartz. Curly's nephew Layton recalled his uncle describing the boulder as being as big as a Volkswagen. When Curly shined his light up the chute, the whole boulder seemed to glow. With the underground operation shutting down and the open pit starting, management told the crew, "Let 'er lay."

Years later contract collector-dealer John Mediz recovered a 150-pound

Above: Globe-based collector Fred Jones assembled a strong collection of specimens from the Globe-Miami mines

Les Presmyk photo, December 1983

Right: A circa 1930s real photo postcard of the Live Oak Mine, part of the Inspiration Consolidated Copper Company mine complex

John Taylor collection

Facing page: Chrysocolla after azurite
7.1 cm tall
Live Oak Mine, Miami, Gila County

Dick Morris collection
Jeff Scovil photo

Right: Malachite after azurite, 5 cm tall
Live Oak Mine, Gila County

Tony Potucek collection
Jesse La Plante photo

Left: An icon of both the Live Oak Mine and the Les and Paula Presmyk collection, this 6 cm wide quartz encrusted malachite pseudomorph after azurite was recovered from the Live Oak Mine near Miami in the Miami-Inspiration district, Gila County, just before 1950.

Jeff Scovil photo

boulder of gem chrysocolla. When Curly heard about the find, he went to see if it might be the same boulder that had been lodged in the draw point 40 years prior. Although it would have been a neat ending to the story, Curly said that it was not the same piece.

A number of locals began assembling mineral collections after the open-pit operation was underway. Layton Talbott recalled that by around 1955 Curly was paying $19 a pound for mine-run gem silica. Curly was one of the few guys back then that the miners knew they would pay money for the material—he was just crazy for it. They would sometimes show up with their pickups loaded when they got off shift. Curly would invite people into his little trailer to look at his showcase, and if he liked the people, he would take them out to the "rock shed" and show them the tonnage. By buying great quantities of material, Curly was able to assemble one of the best local collections of Live Oak material.

Curly's brother-in-law Bob Watts was the mine's general manager and a collector. Watts built a large collection with help in no small part from Curly. A portion of the collection was displayed in the Globe office of the First National Bank and was later donated to the Arizona Mining and Mineral

Upper left: This interesting, 8.1 cm tall chrysocolla pseudomorph after azurite was from a unique pocket recovered in the early 1960s from the Schulze Claim, about 3 miles west of Miami in Gila County. The Schulze became part of the Oxhide Mine, which was operated by Inspiration Consolidated Copper Company

Wayne and Dona Leicht collection
Jeff Scovil photo

Upper right: Gypsum and azurite
6.3 cm tall
Castle Dome Mine, Gila County

Les and Paula Presmyk collection
Jeff Scovil photo

Museum in Phoenix. But the bulk of the collection—the part that was not on display—was stolen from Watts's home in the 1970s.

Fred Jones of Globe built a collection boasting at least 50 Live Oak Mine specimens ranging in size from 5 to 50 centimeters. The specimens that came out of the vugs and veins in the open pit tended to be larger than those that could have been reasonably extracted through the underground operation. Chunks of gem silica weighing up to 18 pounds were found in the pit. Crystallized specimens of azurite and malachite after azurite with and without drusy quartz were also recovered. Average individual crystals were smaller than 2.5 centimeters (1 inch), but some of the pseudomorph specimens consisted of crystals as large as 6 centimeters wide. The finest known chrysocolla pseudomorph specimen is 25-by-12 centimeters and is in the Arizona-Sonora Desert Museum; it is composed of 4-centimeter wide individual pseudomorphs.

Exploration of the Castle Dome copper porphyry deposit started in 1941, at a time when the U.S. Government was subsidizing raw material production as part of the war effort. Located 6 miles west of Miami in Gila County, the **Castle Dome Mine**, an important turquoise producer, came on line in 1943, and mining continued until 1953. The most significant specimens the Castle Dome copper porphyry produced were azurite that came out of an open fissure in the smaller **Red Hill** pit at the southern side of the deposit. The lustrous blue-black crystals are small, up to 3 millimeters, and occur on a distinctive iron-stained quartz porphyry matrix with interspersed blebs of pyrite and chalcopyrite. Plates and ribs up to 25 centimeters across were recovered, but most of the specimens are in the 5- to 8-centimeter range.

Large quantities of cutting-grade turquoise were recovered at the **Castle Dome Mine**; in fact, shovel operator L.W. Hardy had amassed enough material while on the job to go into the business of treating and cutting turquoise full time after the mine closed. When the mine reopened as the **Pinto Valley Mine** in 1974, Cities Service Company took bids on a contract to recover turquoise, and Hardy was the successful bidder.

The **Copper Cities** or **Sleeping Beauty Mine** has been famous for light to medium sky-blue turquoise since the mid-1950s. Nicknamed for Sleeping Beauty Mountain north of the mine, the operation closed in the late 1970s. Like the Pinto Valley Mine, Copper Cities was under the auspices of Cities Service Company and Leonard Hardy was awarded the contract to mine "Sleeping Beauty" turquoise. Hardy sold the contract to Monty Nichols in 1988. Since that time, the operation has mined both turquoise and "chalk," which needs to be treated before it can be cut.

Right: Turquoise, 5.4 cm tall
Copper Cities (Sleeping Beauty) Mine
Globe, Gila County

Houston Museum of Natural Science
Jeff Scovil photo

Below: Turquoise cabochon, 3.9 cm long
Copper Cities (Sleeping Beauty) Mine,
Globe, Gila County

49'ers Minerals specimen
Jeff Scovil photo

Bottom: Pinto Valley (Castle Dome) Mine
as seen in August 1974

Les Presmyk photo

Blue Ball Mine

About 8½ miles southwest of Miami in the Summit Mining District, the Blue Ball Mine exploited a copper deposit in a clay gouge along a split of the Williamson Fault, which encloses a wedge of sheared and mineralized Precambrian-age Pinal Schist (Grant 1989).

Azurite nodules are collected from the oxidized zone of the deposit and are used as display specimens and in jewelry. When broken open, the nodules display drusy azurite and the occasional feathery spray of lustrous bright green, or even solid crystalline malachite. The most prized of these geodes are those which form a five-arm star, the star being an opening lined with sparkly bright blue azurite crystals. Most of the nodules are 1 to 3 centimeters in diameter but the occasional, larger specimen has been found.

The area was reportedly known in the 1930s, and in 1951 David P. Record reported finding 3 azurite geodes. When split open, "one was lined with lustrous azure-blue drusy azurite; another lined with lustrous drusy azurite and a mesh of minutely thin needles of green malachite; the third was lined with greenish xline brochantite and tiny whitish calcite crystals."

Local collectors and Boy Scouts collected nodules, the latter to raise money for equipment and trips. While working in the Globe-Miami area in early 1970s, author Les Presmyk tracked down the Blue Ball Mine, which had been made obvious by some earlier bulldozer work.

Presmyk collected a few small nodules that day. Later in the 1970s, collectors Bob Johnson and Jimmy Vacek drilled a four-foot hole into the conglomerate and montmorillonite clay at the Blue Ball Mine. They set off a charge and Jimmy said nothing happened. But in 1977 a chance meeting between Vacek and a couple of local diggers initiated a spate of commercial production that spread the little blue nodules all over the world.

Top: Les Presmyk collecting at the Blue Ball Mine near Miami

Hugh Bynum photo, September 1971

Above: Azurite geode half
3.7 cm in diameter
Blue Ball Mine, Gila County

Les and Paula Presmyk collection
Jeff Scovil photo

Right: A pair of azurite and malachite geode halves to 3.7 cm in diameter
Blue Ball Mine in Gila County

Les and Paula Presmyk collection
ex Jimmy Vacek collection
Jeff Scovil photo

79 Mine

with contributions by Stan Keith and Bob Lane

The 79 Mine is on the border of Gila and Pinal counties in the Banner Mining District, several miles northwest of the smelter town of Hayden. The mine sits on the southwest side of the Dripping Springs Mountains in the shadow of Tam o'Shanter Peak, and has produced world-class specimens of several species including, wulfenite, aurichalcite, and smithsonite.

Mike and Pat O'Brien located the claim in 1879. The 79 began as a base metal producer, changing hands a number of times until it was purchased by the Seventy-Nine Lead-Copper Company in 1928. That year the company shipped more than 2,300 tons of lead ore to Douglas, and the 79 Mine ranked fifth in the state for lead production. The following year it was in first place, having produced nearly 15,000 tons of oxidized lead ore. It was thirteenth in silver production. The mine continued production and was developed until January 1938, when the price of lead dropped, and the 79 Mine was idled. A series of companies and lessees operated the mine over the next three decades until Inspiration Consolidated Copper acquired it in the fall of 1969.

The 1971 death of the property watchman ostensibly opened the 79 Mine to collectors, who endured some interference on the part of the Pinal County Sheriff Department. In August 1971 Inspiration tried to seal the mine off by setting a thick concrete plug in the inclined shaft and collapsing

Upper: Aurichalcite
field of view 6.5 cm tall
79 Mine, Gila County

Canadian Museum of Nature collection
Jeff Scovil photo

Lower: The 79 Mine site in 1972

Casey Jones photo

Right: Wulfenite 4.7 cm wide
79 Mine, Gila County

Dick Morris collection
Jeff Scovil photo

a dump over the access through the second level adit. But collectors found other ways to access the mine and collecting continued.

No specimens are known from the 79 Mine's active production periods, as in general, the orebodies did not yield much collectible material. The 79's specimen sweet spots are the low-grade oxidized mineral occurrences which are associated with post-sulfide fractures. Once collectors gained access to the mine they found an abundance of hemimorphite, smithsonite, aurichalcite, and wulfenite (with minor amounts of azurite), cerussite, chrysocolla, mimetite, and pyromorphite. In addition the lower levels—the orebodies—produced crystallized pyrite, quartz, hematite, chalcanthite, halotrichite, and a suite of rare minerals including clinoclase and olivenite.

Stan Keith underground in the 79 Mine
Thomas Trebisky photo

Dick and Roy Jones from Casa Grande and Cy Mack, a Southern Pacific railroad employee stationed in Hayden, were the first modern collectors known to have had gotten into the mine to explore the workings. The trio gained access after 1967 when the original wooden telephones were still in place on the shaft stations.

They agreed to focus on minerals rather than antiques, thus appearing harmless to local law enforcement should they be caught in the mine. But the lure of the antiques was too much for Cy, who ventured into the mine alone and was met by deputies when he emerged with a wooden phone. The team of Jones, Jones, and Mack ultimately produced hundreds of flats of specimens, mostly hemimorphite, aurichalcite, and smithsonite, with some wulfenite. But their focus was admittedly on quantity rather than quality.

Stan Keith and his collecting partner also had an encounter with law enforcement in the spring of 1970. As Stan recalls,

> *We were still in the process of exploring the mine and had that day enjoyed good success collecting several flats of sulfates oxidizing from sulfides on the fifth and sixth levels. We had been underground for hours and decided to come up for air with several flats of chalcanthite goodies. On the 470 level we encountered a gentleman who we assumed was a fellow collector. But instead of showing us minerals, he showed us a loaded revolver and introduced himself as Norm Grisson, the mine watchman. We were unceremoniously ushered up the decline at the point of a gun and introduced to the Pinal County Sheriff who had been waiting on the surface for over six hours while Mr. Grissom painstakingly searched the collecting levels. We were cuffed and transported to the Sheriff's office in Winkleman where we were locked in a holding room.*
>
> *They held that we were guilty of mineral trespass. We threatened, if they fined us, to sue Grisson for the value of the equipment that we had been forced to leave in the mine. We were not fined and were released with a warning not to return. We were back in the mine three weeks later with new equipment, conducting night raids until Grisson's death about one year later. We never saw Grisson or the law again.*

While hemimorphite, smithsonite, and aurichalcite were abundant at the 79, the Holy Grail of the 79 are its rare, trademark "red-dot" wulfenite crystals (example pictured facing page). The red-dot occurrence is difficult

Right: This superb, 4.9 cm tall, "red-dot" wulfenite on mottramite was collected from the wulfenite room on the 4th level of the 79 Mine in Gila County by Reg Barnes and Rick Jackson in 1971 and was the first pick from the pocket.

Mark Hay collection, ex Rick Jackson
Jeff Scovil photo

Below: View toward the southeast of the main fault in the wulfenite area on the 4th level of the mine as it was exposed in 1972

Stan Keith photo

to access, and the specimens are difficult to collect. The wulfenite occurrence is on the fourth level of the mine. It is accessed through a short timber-supported area. White mold is growing on the timbers. The room is on a fault: anyone wearing glasses has to wipe them off because the humidity in the area is so high. During wet seasons, the fault that cuts the room can move several inches. The red-dot crystals originate in a rarely frequented, much smaller room on the other side of the fault.

The best single mimetite and wulfenite pocket was about the size of a football. Stan Keith recovered the contents: a few specimens of botryoidal orange-red mimetite with yellow wulfenite crystals. In 1969 or 1970 Sam Rua of Superior collected a fine pocket of thin, clear yellow wulfenite crystals on black mottramite matrix. The best cabinet specimen is 15 centimeters wide. The specimen, which went to Tom McKee, consists of three clusters of 2-centimeter-wide crystals. It is now in the Arizona-Sonora

Top: Chrysocolla pseudomorph after wulfenite, the crystal is 1.2 cm wide
79 Mine, Gila County

Joyce and Jimmy Vacek collection
Jeff Scovil photo

Middle: Susan Keith in an emptied 79 Mine mimetite-wulfenite pocket

Stan Keith photo, April 1972

Lower: Hemimorphite, 2.6 cm wide
79 Mine, Gila County
Collected by George Stevens

Shirley Fiske collection, Jeff Scovil photo

Desert Museum collection.

Many young collectors including Andy Clark, Gary Fleck, Casey Jones and Tony Potucek cut teeth at the 79 Mine in the early 1970s, coming away with good specimens and great memories. Bob Lane had met Wayne Thompson at Glendale Community College, where they were both taking Bob Thompson's mineralogy course. Lane and (Wayne) Thompson made dozens of trips to the 79 Mine. And according to Lane, they collected

> *hundreds of flats of various specimens, including wulfenite, hemimorphite, aurichalcite, smithsonite, cerussite, mimetite, and chrysocolla.* [Lane's] *collection of 79 minerals was rapidly growing and eventually became the best array of self-collected specimens from the locality.*

College students Les Presmyk, Wayne Thompson, Reg Barnes, and Rick Jackson made a series of trips to dig the 79 Mine in the spring of 1971. A rich wulfenite pocket kept the interest of several configurations of the group for two consecutive long weekends until all four convened at the site and tensions arose as to ownership. Barnes and Jackson left Presmyk and Thompson in the wulfenite room and went off to check out the aurichalcite room where they immediately broke into a pocket. Not realizing the extent of this discovery and because of the tension, everyone left the mine.

A few days later and unaware of the finds of the prior weekend, Wendell Wilson directed new Arizona State University graduate student Kent Knock and a friend to the 79 Mine to dig. The students came back the next day with dusty beer flats filled with astonishing specimens of turquoise-blue aurichalcite specimens. The two had opened a pocket in the back (ceiling) of the "aurichalcite room" on the fourth level, but they had only collected half of the pocket before they had to leave. They suggested that Wilson go back and finish it.

Wilson located the seam in the back of the aurichalcite room, removed his hard hat, and stuck his head into what he later described (1987) as, "a dark, horizontal opening perhaps 5 or 6 cm high and 20 or 25 cm wide." He wasn't able to see them, but Wilson felt loose pieces and pulled one out. This first piece was 12 centimeters long and was later sold Tom McKee for $125. It is certainly one of the finest aurichalcite specimen ever collected from the 79 Mine, or anywhere. The specimen has no matrix but is a hardened crust of aurichalcite, with individual crystals so large that the terminated laths of aurichalcite are visible. Wilson removed three flats of specimens only a few of which exhibit the coveted dark turquoise-blue.

John Mediz hit another aurichalcite pocket in the 79 Mine in 1979. European mineral dealer and collector Jean Pierre Cand was at the mine and purchased the entire pocket on the spot. Cand presumably sold the specimens to a private collector, and they have yet to surface, but Mediz later described some of the pieces as black and brown stalactites scattered with balls of deep blue aurichalcite.

At one point, Wayne Thompson and Lee Bridges collected a pocket of green smithsonite with radiating sprays of aurichalcite up to 1.5 centimeters long. Unfortunately, a deputy sheriff greeted them as they left the mine, and they wound up spending the next several hours waiting for a judge to set

79 Mine, Gila County

Right: Wulfenite on mottramite
5.9 cm wide

Les and Paula Presmyk collection
Jeff Scovil photo

Lower right: Aurichalcite, 9.3 cm tall

Bruce Carter collection
ex Sam Elbin
Jeff Scovil photo

Lower left: Rosacite and aurichalcite
7.2 cm tall

Arizona-Sonora Desert Museum
Jeff Scovil photo

Facing page, upper left: Casey Jones underground in the 79 Mine in 1972

Bob Lane photo

Facing page, lower left: Smithsonite with calcite and aurichalcite
Field of view 5 cm wide
79 Mine, Gila County
Collected in 1971

Les and Paula Presmyk collection
Jeff Scovil photo

Facing page, right: Aurichalcite
7.6 cm wide
79 Mine, Gila County
Collected in 1971 by Wendell Wilson

Les and Paula Presmyk collection
Jeff Scovil photo

Below: Aurichalcite
Field of view 3.6 cm long
79 Mine, Gila County

Arizona-Sonora Desert Museum collection
Jeff Scovil photo

bail. While they waited, they heard the deputies and other employees exclaiming how beautiful the specimens were every time someone opened one of the flats.

Finally in 1975 Pacific Specimen Recovery, a group of collectors, leased the mine from Inspiration. Pacific Specimen Recovery included Tom McLean, Jim and Joyce Vacek, Bobby Lane, Mark Sullivan, Andrew Rogers, Jim Herr, Mike New, and Grant Richards, the group's organizer. Former State Mine Inspector Verne McCutchen brokered the deal. After the Pacific group was granted the lease, they reopened the blocked entrance, cleaned and refurbished the mine, and collected aurichalcite and green smithsonite from the fourth level. Pacific hit a wulfenite pocket in the wulfenite room. The top specimen was a 4-centimeter wide, transparent, orange-yellow crystal. The group also collected red dot wulfenite on mottramite from an opening near the wulfenite stope on the fourth level.

Bob Lane remembers his brief stint contract collecting with Pacific Recovery,

> *I got a call from Grant Richards, who wanted to know if I was interested in working the 79 and Christmas Mine open pit for specimens. Was I interested? Heck, yes! I recruited Andy Rogers and his friend Mark, friends from Northern Arizona University to help.*
>
> *Grant had done most of the prep work, but before we could haul the drilling equipment into the mine, we had to muck out the aurichalcite room, which had an unbelievable amount of backfill in it. The backbreaking job took weeks. When we are almost done, Grant and I went to Phoenix for some rest and supplies. We left Andy and Mark to watch the camp.*
>
> *We had been gone a few days when Verne McCutchen called to tell us that Andy had had an accident and was in the hospital. Evidently he and Mark had been joyriding in Verne's Jeep when they rolled it off the edge of the mine road and into a ravine. The Jeep had rolled over Andy, the knobless stick shift had pierced his stomach, and his intestines were protruding from his belly. Somehow Mark got Andy back to the camp trailer, covered the wound with a wet towel, and ran all the way to the highway near Hayden to get help. A rescue team took Andy to the hospital. He ultimately recovered.*
>
> *Grant and I came back to the mine and finished the mucking out of the aurichalcite room. Before we started drilling, we poked around the newly exposed areas where Kent Knock had found his famous aurichalcite specimens. We uncovered several pockets of beautiful green smithsonite, and on that first night, we hauled out several flats of very nice specimens. As we continued exploring, I noticed a 5-centimeter wide light blue seam of pure aurichalcite tufts in the floor. I dug at it, and it started to open up. Within a few hours, we had two flats of velvety blue-green aurichalcite similar to what Kent found but a little more blue. We could see back in the pocket for at least another 30 centimeters, and both sides were coated*

with a thick velvety fur. We collected at least four more flats of primo stuff but the zone went down into the floor of the stope. Grant and I decided to drill and blast some rock away to further expose the pocket. We put in a hole and set off the charge, planning to come back in the morning to see what we had.

We were excited to get back into the mine the next morning. We cleared away some of the muck and started pulling out nice chunks of aurichalcite. A few minutes later, we noticed a light—Inspiration Copper Company officials were walking down the drift toward us. They asked us to come to the surface with them. When we got there, they informed Grant and me that the project had been turned over to mineral dealer John Mediz from Globe. We needed to pack up and leave.

It was such a rude and disappointing experience that I went back to Flagstaff and gave up collecting. I had a fantastic collection of specimens many of which I had collected from the 79 Mine, United Verde pit, and the Red Cloud Mine, but I had lost interest. I sold the whole collection to Andy Rogers for $200 and have regretted it for the rest of my life!

The 79 Mine was under Mediz's watchful eye for the next 15 years.

Finch aka Barking Spider Mine

with contributions by Paul Harter

The Finch Mine is a favorite Arizona collecting locality. In the Banner Mining District in the Dripping Springs Mountains, Gila County, the Finch Mine is just east of the road into the 79 Mine. It is comprised of the workings on the Brick #3 claim, which was first located around 1880. After which and until about 1920, various owners and lessees periodically mined lead ore and hauled it to the smelter in Hayden.

In the 1950s Dick Jones, Roy Jones, and Roy Lee met with moderate success digging for wulfenite at the Finch. But the mine had always given up its treasures with reluctance, and the team was not equipped to do much drilling and blasting at that point. They did, however, find a few nice pockets, with wulfenite crystals up to 1.5 centimeters across in clusters up to 6.5 centimeters long. They also found yellow-green vanadinite crystals with multiple faces along their *c*-axes. Vanadinite clusters as large as 6 centimeters wide were recovered as were mottramite-coated vanadinite crystals.

In 1973 when Pacific Specimen Recovery was working the 79 Mine, Joyce and Jimmy Vacek decided to take a leave of absence from the project and, along with Bob Johnson, began to methodically collect the nearby Finch Mine. Bob drilled four, 4-foot holes, used one stick of dynamite in each hole, and shot a crosscut between the east and west adits. Fine pockets were found in the rib as well as the floor. Wulfenite from the rib was generally orange to apricot in color. Wulfenite from the floor was firmly attached to a limestone matrix and was coated with a blue-gray hemimorphite, a combination that is common at the Finch Mine. Virtually all of the wulfenite, regardless of color, is coated with drusy quartz crystals. Jimmy recalls that, except for a coating of blast dust, the wulfenite was in every respect pristine.

Above: The "Magma Mine, Superior, Ariz." circa 1910 postcard

John Taylor collection

Facing page, upper left: Wulfenite encrusted with quartz
The central crystal group is 1 cm wide.
Collected in 1973
Finch Mine, Gila County

Joyce and Jimmy Vacek collection
Jeff Scovil photo

Facing page, lower left: Wulfenite on hemimorphite and encrusted with quartz
The central group is 1 cm wide.
Finch Mine, Gila County

Jeff Scovil collection and photo

Facing page, right: Wulfenite encrusted with quartz, 8 cm wide
Finch Mine, Gila County
Collected in the 1950s by Roy Lee

Les and Paula Presmyk collection
Jeff Scovil photo

The locality gained its moniker over the course of the Vaceks' 1970s wulfenite operation when Jimmy had been drinking volumes of milk. Flatulence was the music of the day and with every "tune," John Garsow, who was working with the Vaceks, announced yet another "Barking Spider."

Magma Mine

The Magma Mine, known best for fine barite and pyrite specimens, was an underground metals mine in the Pioneer Mining District in Pinal County. Magma began the Irene and Hub claims, which worked veins in the Silver Queen area (see page 33). Those operations closed in 1893 when the price of silver collapsed, but as the infrastructure in the area developed, new technologies were introduced, and the demand for the various base metals increased, the prospects of the Silver Queen area regained viability.

George Gunn and Butte copper magnate William Boyce Thompson purchased the property in 1911. They deepened the mine and brought it online as Magma Copper Company, a copper and silver producer. By 1913 ore production was continuous enough to warrant constructing a concentrator. A 200-ton per day mill was built in 1914. And in the same year, Magma began construction on a narrow-gauge railroad to the railhead at Florence, 31 miles away.

The mine was eventually deepened to the 4000 level on the Magma as well as on the Koerner vein 1,200 feet to the south. Sphalerite was

Right: Gemmy barite put the Magma Mine on the mineral specimen map. This pretty 9.1 cm tall cluster was collected from the "A-bed" replacement orebody on the 3000 level in the 1950s.

Les and Paula Presmyk collection
Jeff Scovil photo

Facing page, upper: The Boyce Thompson Arboretum in Superior was founded by mining magnate William Boyce Thompson, owner of the Magma Mine. The arboretum houses a collection of desert plants from around the world, but the saguaro (*Carnegiea gigantea*), pictured here in glorious bloom, is native to the desert around the Magma Mine. Saguaro is Arizona's state plant; the saguaro's blossom is the state flower.

Dean Hueber photo; April 2007

Facing page, lower left: As specimens from the ore veins at Magma were infrequently encountered and even less frequently attractive, few were preserved; thus this 7.9 cm tall tetrahedrite with quartz and pyrite collected from the Magma Mine before 1920 is a true rarity.

Les and Paula Presmyk collection
Jeff Scovil photo

Facing page, lower right: In the early to mid-1970s, a miner working on the 3300 level of the Magma Mine encountered a clay-filled zone that produced thousands of euhedral, floater pyrite pyritohedrons, some as large as 3.5 centimeters. The crystals varied in luster from satiny to very fine. The miner, Anderson, carried the clay-covered crystals out by the bagful and cleaned them at home. He made the local rock show circuit also selling them to collectors who visited him at his home in Kearney. This 6.6 cm tall cluster from the find is in the collection of Evan Jones.

Jeff Scovil photo

significant enough in some of the stopes to justify a zinc circuit, which continued throughout World War II.

The company worked the Magma and Koener veins through the mid-1950s when replacement beds in the limestone above the quartzite boundary were discovered. The mineralization along the contact was rich in copper, which became the economic driver for the operation.

One stope in the Koerner vein was very rich in silver. With one of the first blasts, miners reported it looked like sheet metal hanging down from the back of the stope. Those sheets turned out to be native silver. When management discovered the mineralization, they had the ore bagged and hauled directly to the smelter. These unusual measures may have been due in part to the richness of the ore and the apparent need for security, but the reality is if they had put the material into the ore chute and sent it through the mill, virtually none of the silver would have been recovered. By bagging the ore, 90 percent of the silver was recovered (Tom Kerich pers. comm.).

Voids were scarce in the Magma and Koener vein mineralization, and there was thus no space in which crystals could form. As a result there was no culture at the Magma Mine for preserving whatever crystals the miners might have occasionally encountered; thus, mineral specimens were almost unknown from the Magma Mine until the late 1950s. But when an incline was driven to mine the new replacement bed horizons, numerous pockets of lustrous, golden, green, reddish, or black barite crystals and pyrite crystals up to 5 centimeters across were exposed. Magma miners were suddenly enticed to collect mineral specimens (Ricardo Campos pers. comm.).

The area known as the "A-bed" (the lowest of these replacement bed orebodies) produced Magma's best barite and pyrite. Unfortunately this part of the mine spontaneously caught fire in 1962. The fire spread to the timbered areas and that section of the mine was sealed (Tony Johnson pers. comm.). Barite pockets were occasionally encountered while mining A-bed pods down to the 3600 level, but these were nothing like what was found in the stopes above the 3000 level. The barite crystals reached 5 centimeters across and clusters up to 20 centimeters were recovered. Pyrite specimens were of similar size. At the time, they were some of the best being produced anywhere in the world (Dick Jones pers. comm.).

Numerous calcite crystals were recovered from watercourses in the limestone as the most easterly replacement beds were being mined. But because the miners did now know how collect and wrap the crystals, collector-quality calcite specimens from Magma before 1975 are rare.

Magma miner Richard Campos became enamored with crystals when he began hitting the barite pockets, so much so that he eventually left the mine and became a full time mineral dealer. For 20 years, his rock shop was a required stop for any mineral collector who passed through Superior.

Above: Calcite, 6.6 cm tall
Magma Mine, Pinal County

Les and Paula Presmyk collection
Jeff Scovil photo

Right: Barite, 5 cm wide
A-bed, 3000 level
Magma Mine, Pinal County

Tony Potucek collection
Jesse La Plante photo

A 1917 certificate for 1,000 shares of stock in the Arizona Ray Copper Company

Mark Hay collection
Jeff Scovil photo

Ray Mine

with contributions by Bob Jones

The Ray Mine is located in Pinal County, about 65 miles north of Tucson near the Gila River in the Mineral Creek Mining District. This important mine enjoys a long production history, but to collectors it is known for fine twinned copper crystals, cuprite, and chrysocolla.

Native Americans probably mined deposits in the Mineral Creek area but members of the Army of the West were the first to encounter them in modern times. In 1846 soldiers found gold and copper ore and subsequently named the stream Mineral Creek. Mining activities occurred around 1870 when a local prospector established the first mining claim. Mineral Creek Mining Company built a five-stamp mill on Mineral Creek in 1880 and numerous other claims were subsequently filed.

By 1883 the Ray Copper Company was organized in New York and operated a 30-ton copper smelter to handle mine production (Ransome 1919). The primary ore from the underground mines was native copper. In 1898 the Globe Mines Exploration Company, a London group, bought the claims and reported reserves of 190,000 tons of sulfide ore containing 4 to 5 percent copper. A 75 ton-per-day concentrator was built along the creek.

A rail line was established between Phoenix and the mining town of

A circa 1908 real photo postcard of the "No. 1 Copper Mine, Ray, Ariz." The mines in Ray were consolidated under the Ray Consolidated Copper Company in the early twentieth century. The towns of Ray and Kelvin were both moved in the 1950s and the ground under the former towns mined via the Pearl Handle open pit.

John Taylor collection

Kelvin in 1905, and shortly thereafter Ray Consolidated Copper Company purchased the mine and appointed Daniel C. Jackling supervisor. While at Utah Copper Company, Jackling had been instrumental in developing methods for mining low-grade sulfide ores through open pits. He also instituted the first low-grade ore flotation process, at Bingham Canyon, Utah. Jackling brought that process to Ray, thereby turning the mineralized rock into ore.

A number of mines in the Mineral Creek Mining District were all producing ore from the porphyry copper system, and by 1927 all had became part of the Nevada Consolidated Copper Company (Metz and Rose 1966), which in 1933 merged into Kennecott Copper Company.

Mining was accomplished by a combination of shrinkage stoping and block caving, which was producing 50,000 tons of copper annually by 1942. The mine began to transition to an open pit operation in 1948, a move that was completed in 1955. The towns of Kelvin and Ray were moved, and the former town sites were mined. By 1966 production was more than 23,000 tons of ore per day.

Open pit mining continued sporadically at Ray until 1982, as production was dependent on the price of copper. The operation shut down temporarily between 1982 and 1986, when ASARCO purchased the property and restarted the mine. From the start of underground mining in

1880 until 1979, the Ray deposit produced over 2.25 million tons of copper, 1,500 kilograms of gold, 225,000 kilograms of silver, 5,400 tons of lead, 3,000 tons of molybdenum, and 100 tons of zinc (Department of Mines 1981). Mining continues under the auspices of ASARCO Grupo Mexico.

The Ray porphyry deposit is fairly unique, and the primary ore is chalcopyrite. Other primary minerals include pyrite and molybdenite. When Ransome (1919) investigated Ray, he noted native silver in some of the "ordinary ores," in particular in a small vein with cuprite and chrysocolla in the No. 3 (Ray Central) Mine. He also made note of small particles of native copper and a cuprite occurrence in the same mine. Ransome's early works on the geology and mineralogy of ore deposits evidence his appreciation for crystallized minerals; thus the paucity of such descriptions in his work on Ray is telling.

While stripping overburden from above the Pearl Handle open pit around 1950, miners uncovered a 3-foot-thick caliche zone (Jones 2008). A common product of Arizona's arid environment, caliche is a result of continuous groundwater evaporation and concentration of mineral compounds (mostly calcium carbonate). The caliche formed at or near the surface and created a hard, relatively impermeable horizon. The zone held rich deposits of crystallized native copper in long wires and herringbone clusters (Jones 2008), many of which are coated with bright red to dark maroon cuprite. Individual twinned copper crystals faithfully coated by cuprite reached 16 centimeters.

Above: Copper encrusted with cuprite
11.5 cm tall
Pearl Handle pit, Ray Mine, Pinal County
Tony Potucek collection
ex Harold Michel
Jesse La Plante photo

Right: Copper encrusted with cuprite
5.1 cm wide
Pearl Handle pit, Ray Mine, Pinal County
Dick Morris collection
Jeff Scovil photo

Miners collected many of the crystallized copper specimens, selling and trading them for extra income. Phoenix-based collector Fred Burr began travelling to Ray in the early 1950s to trade citrus grown on his property for the cuprite-coated copper wire clusters.

Although less common, acicular cuprite (chalcotrichite) was also present in this zone. It was typically found as heavily felted masses of bright red to orange-red needles up to 6 centimeters in length. Specimens commonly show needles twinned at 90 degrees to one another to form a delicate angular lattice. They occurred on limonitic matrix and in association with cubic cuprite (Jones 2008).

Cubic cuprite crystals seldom reach 1.5 centimeters at Ray although some superb, unblemished 2.5-centimeter cubes have been recovered. Bob Jones recalls one Ray Mine specimen composed of several perfect cuprite cubes, each exceeding 2.5 centimeters.

Ray is best known for superb, spear-shaped twins of native copper. Undoubtedly the most important specimens from this deposit, these twins are also Ray's rarest and most desirable copper crystals. Native copper specimens also occur as two-dimensional arborescent groups, which range in size from 2.5 centimeters to more than 50 centimeters across. Many of the specimens have pieces of matrix material adhering to the copper. Depending on the degree of oxidation, the crystals range from a dull brownish red color (usually caused by oxidized coatings of cuprite) to fairly bright, uncommonly coated with malachite. Individual crystals usually take the form of small dodecahedra.

Above: Bob Jones collecting sheet copper in the Pearl Handle pit of the Ray Mine

John Mediz photo

Facing page: Spinel-law twinned copper 13.1 cm wide
Pearl Handle pit, Ray Mine, Pinal County

Daniel Trinchillo collection
James Elliott photo

Most of the native copper from Ray occurs as compact, arborescent sheets that formed in thin cracks in the host rock. Collecting is easy: smash the rock with a hammer since the malleable copper is not negatively affected. Sheet copper can be feet long but usually lacks visible crystals. The material was a nuisance at the mine as it jams the crushers.

Beginning in late 1972 and early 1973, long-bladed, spinel-law twins of native copper coated by malachite were found in an area between the 1860 and 1780 levels of the pit. Miners collected these specimens, which found their way to the Tucson 1973 show. The periodic loose, twinned crystal was still being found late into 1973.

As exciting as those copper twins were, the find was eclipsed in March 1974. There were no indications of copper in the immediate vicinity of this remarkable occurrence. Specimens were encountered suspended in clay in a pocket about the size of a refrigerator. The best crystals were 12 centimeters long by up to 2.5 centimeters wide with the longest twins reaching 18 centimeters! Unlike the previous find, this pocket contained large, spectacular groups, with an occasional specimen on fragmented breccia. Numerous single floater crystals were also recovered. Many of these exceptional specimens were lightly coated with bright green malachite. Others are a dark reddish color—coated by cuprite—and were also fairly bright. The pocket produced several thousand wonderful clusters and singles. In some cases, small sharp cubes of later-formed copper were epitaxially aligned along the twinning contact (Jones 2008).

Susie Davis bought most of the specimens from this pocket. Bob Jones

From the Pearl Handle Pit

Upper left: Copper encrusted with the chalcotrichite variety of cuprite
5.4 cm wide
2020 Bench, Pearl Handle pit, Ray Mine
Pinal County

Dick Morris collection
Jeff Scovil photo

Lower left: Twinned copper, 6.2 cm wide
Ray Mine, Pinal County

Don and Gloria Olson collection
Jeff Scovil photo

Right: Twinned copper, 6.2 cm tall
1740 bench, Emperor Fault
Pearl Handle pit, Ray Mine, Pinal County
Collected in March 1974 by Ray miner
Fred Liebert

Mark Hay collection
Jeff Scovil photo

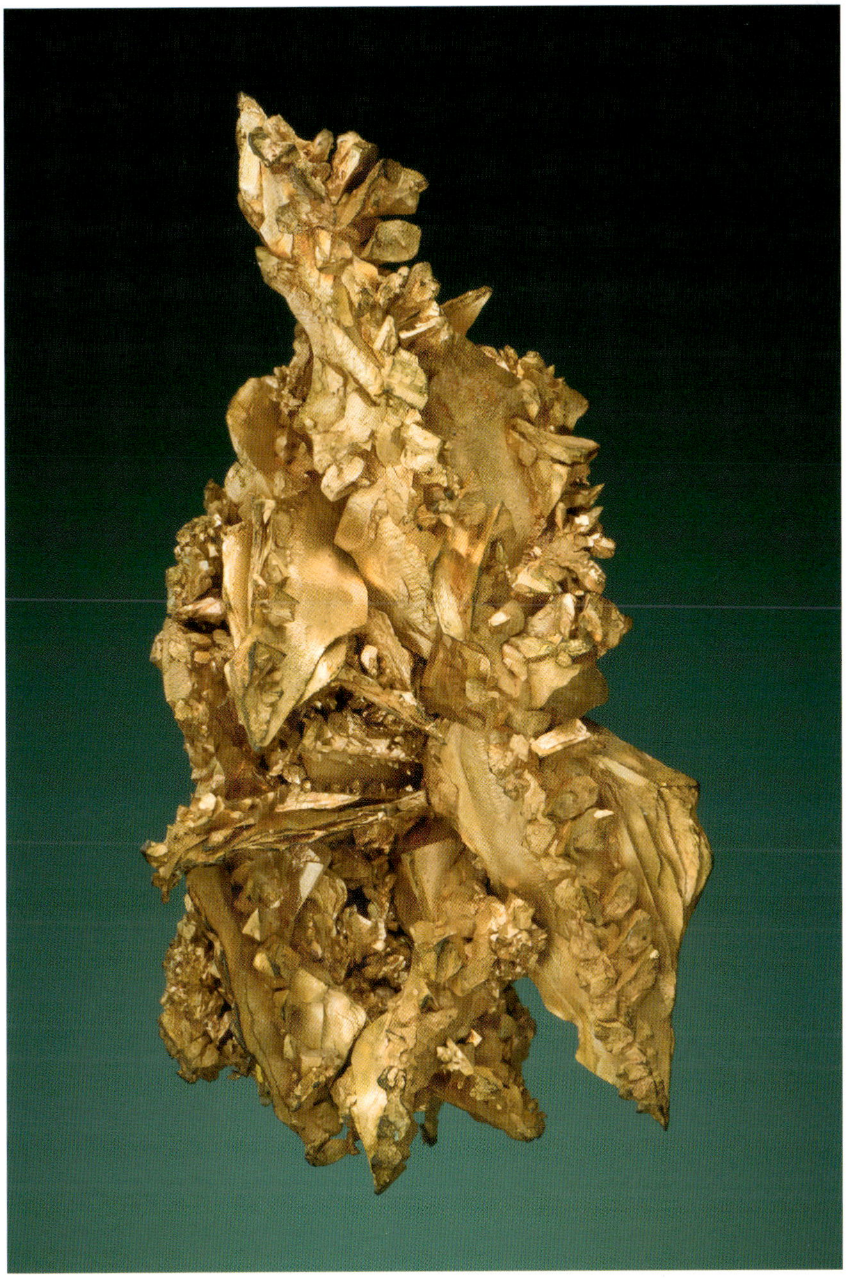

happened to visit Mrs. Davis while she was processing the copper. Susie had a number of small plastic children's wading pools in her back yard. Each was full of twinned coppers in groups and singles that were being acid soaked. Lying in the sun were an equal number of bright specimens that had been soaked and washed and were drying. More pieces had been washed and were awaiting their acid baths. Bob impressed upon Susie the importance of preserving many of the malachite-coated coppers. She was agreeable, and the Arizona-Sonora Desert Museum specimen pictured on page 140 is a great example of one of the untreated specimens.

INCORPORATED UNDER THE LAWS OF ARIZONA

No. 3 Shares 12,200

The Flux Mining Company

Capitalized at $1,500,000.00

1,500,000 SHARES PAR VALUE $1.00

This Certifies that Roger T. Felton is the owner of Twelve Thousand Two Hundred Shares of One Dollar each of the Capital Stock of The Flux Mining Company, fully paid and non-assessable, transferable only on the books of the Corporation by the holder hereof in person or by Attorney upon surrender of this Certificate properly endorsed.

In Witness Whereof, the said Corporation has caused this Certificate to be signed by its duly authorized officers and to be sealed with the Seal of the Corporation this 9th day of July A.D. 1919

SECRETARY PRESIDENT

SHARES $1.00 EACH

The Flux Mine has produced outstanding specimens of jackstraw cerussite. On a visit to the mine in 1909, U.S. Geological Survey geologists F.C. Schrader and J.M. Hill (1917) witnessed a spectacular sight:

> *From the roof in the inner part of the south or 260-foot tunnel and adjacent parts of the crosscuts hang great masses of closely spaced acicular or filiform silky white cerusite* [sic] *about a foot in length.*

This Flux Mining Company stock certificate, dated 1919, is in the collection of Dick Morris.

Flux Mine

Collectors discovered spectacular cerussite in the Flux Mine in the 1950s. The upper workings of the mine consist of a small open-pit "glory hole" and numerous adits up Flux Canyon in the Harshaw mining district near Patagonia, Santa Cruz County. Mineralization in the entire district is peripheral to the Red Mountain copper porphyry deposit hydrothermal system. The Flux Mine is situated in an intrusive rhyolite with silicified and brecciated veins/dikes. It exploits a silver-bearing lead deposit, the upper reaches of which are heavily oxidized, and sulfides such as galena have been completely altered to cerussite. Sulfides become more abundant with depth and consist of galena, pyrite, sphalerite, and lesser amounts of chalcopyrite.

Ore from the Flux was smelted in adobe furnaces along Alum Gulch near Sonoita Creek in 1858. Schrader (1915) reported that the mine furnished lead for ammunition used during the Civil War. The mine apparently languished until 1878, when it was relocated. It and the Hardshell Mine were shipping ore to a smelter at Benson, and the mines were referred to collectively as the Hardshell mines.

After 1896 the mines were bonded to the Arizona Gold and Copper Company, which built a smelter at Patagonia. The company mined about 2,000 tons of ore from the Flux, smelted it, and then relinquished the mine. In 1904 the mine shipped 10 carloads of high-grade galena ore. Benjamin Heney took over the Flux Mine in 1905. He sank the mine workings to 260 feet and expanded them. He passed through the oxides (which he was apparently mining) at a depth of 250 feet and stopped when he encountered the sulfides. Heney suspended operations in 1909. In 1914 a consortium of Californians bonded the property. By 1915 the Flux Mine had produced 50,000 tons of ore and had more than 5,000 feet of workings on the surface and on the 70, 100, 125, and 260 levels. Lessees worked the mine throughout the 1920s and 30s.

ASARCO purchased the Flux Mine about 1939 and worked the lower levels for lead and zinc sulfides containing silver and copper. By 1944 the workings extended to 430 feet. The Flux workings reached a depth of 590 feet in the 1950s. Ore was processed upstream at the Trench Mill in Alum Gulch. The mine ceased operations in 1963 under a lessee. An estimated 850,000 tons of ore were produced (Keith 1985).

When quizzed about the upper workings of the mine where the cerussite was found, one of the mining engineers recalled that they were encouraged to stay out of them. He did, however, recall one night when the headlights of the car that he was in hit the apricot tree growing in the mine

Galena with siderite and pyrite
9 cm wide
Flux Mine, Santa Cruz County
Collected by Dick Jones in 1955
Mark Hay collection
ex Tom McKee
Jeff Scovil photo

superintendent's yard, and the tree "rained" coatimundis, which had been up in the tree feasting on apricots.

In 1956 Tucson collectors Clayton Gibson and Dan Caudle discovered white and rust-colored "jackstraw" cerussite crystals up to 12 centimeters long in one of the adits of the upper workings. They recovered a number of fine specimens. One 30-centimeter-long piece is made up almost entirely of 10- to 12-centimeter-long crystals.

Dick Jones found two specimens of a strong yellow-green chromium-rich cerussite along with other minerals in the mine. Striated cubic pyrite crystals were commonly collected there in the 1950s.

In the 1960s Gene Schlepp and his wife Jackie went to the Flux Mine to collect sulfides. They entered the concrete portal at the base of the hill and went down the winze on the west side of the drift. Reaching the first level below the haulage level, the Schlepps noticed standing water on the level just below them. They walked north and encountered a large stope in the center of which stood a 25-foot tall pile of muck. The Schlepps found striated pyrite cubes up to 5 centimeters on an edge embedded in the wet clay. After collecting a number of them, they shined their car headlights at the ceiling above the pile. Schlepp remembers it was like looking at a summer night sky with light reflecting off of the pyrite crystal faces exposed in a pocket in the ceiling.

The coatimundi or coati (*Nasua narica*) is a relative newcomer to Arizona, probably migrating into the state from Mexico. The first documented sightings were in the late 1800s. Coatis along with raccoons and ringtails are in the Procyonidae family of mammals. They are a social species and frequently live in bands of 30 or 40 individuals. Coatis are omnivorous and inhabit mountain canyons (Flux Canyon, for example) in the summer. Much of the habitat in the vicinity of the Flux Mine is preserved as past of the Coronado National Forest, Patagonia Lake State Park (est. 1975), and the Patagonia-Sonoita Creek Preserve, the latter owned and managed by the Nature Conservancy. The Sonoita Creek Preserve is one of the southwest's birding hotspots. The Elegant Trogon (*Trogon elegans*) pictured **above** was sighted and photographed on the preserve in February 2006 by George Vlahakis.

The white-nosed coati pictured **right** is in the Sonora-Desert Museum live collection.

Photo © 2007 Paul and Joyce Berquist/Arizona-Sonora Desert Museum

Right: Cerussite, 9.4 cm tall
Flux Mine, Santa Cruz County
Collected in 1956 by Clayton Gibson and Dan Caudle

Paul S. Harter collection
ex Clayton Gibson collection
Jeff Scovil photo

Gene walked to the other side of the muck pile where he spotted the wooden planking on which the muck pile was resting. Beyond the boards, Schlepp saw a black hole. He dropped a rock off the edge, and it was many seconds before he heard the rock hit the bottom. Gene then realized that the planks and muck were all that was between them and a deep pit. Schlepp did not waste any time getting back over to Jackie, and the two of them beat a hasty retreat out of the stope.

Bob Jones and Bill Panczner spent some time in the deeper sulfide zone of the Flux. Bob remembers pockets containing up to 10-centimeter-wide pyrite crystals. A 30-centimeter wide crystal in the University of Arizona collection is the largest crystal from the mine. While that specimen shows significant damage, unbroken crystals to 15 centimeters are preserved elsewhere. Flux galena crystals can reach 1 centimeter across and exhibit good luster. The somewhat distinctive sphalerite usually forms in knobs of seemingly black, lustrous octahedra.

Gene Schlepp gave up on the sulfides after investigating that precariously perched pile muck and instead focused on the cerussite. The material had been mentioned in Frank C. Schrader's 1917 USGS Bulletin on the Mineral Deposits of the Santa Rita and Patagonia Mountains and was being sold by several dealers. Schlepp enlisted Mexican miner Mateo to accompany him to the Flux one day. They found a likely area and began mining for the brilliant white cerussite on wad. Schlepp spotted some wood and lifted up on the board in the loose muck at their feet.

Déjà vu! The floor began flowing slowly into a void beneath the board. Mateo and Gene hugged the wall as the floor disappeared. But the flow stopped after lowering the stope several feet, and they found some virgin ground that produced about 20 flats of excellent white jackstraw cerussite crystals to 15 centimeters long scattered on and off the wad matrix. And the virgin area produced something else—a half dozen small white porcelain bottles with Chinese writing on them. While Schlepp thought they were interesting, he left all of those relics of very early mining at the Flux Mine.

Gene and Jackie Schlepp were camping with another couple at the Flux Mine on New Year's Eve 1967. The foursome had been digging cerussite before coming out to make dinner. After the meal, the Schlepps went back into the underground workings of the Flux and continued to work. The ground they were digging became richer in cerussite. Schlepp checked his watch just after opening a unique pocket of light lime-green twinned cerussite sixlings; it was just about midnight.

Right: Anglesite, 8.7 cm tall
Flux Mine, Santa Cruz County
Collected in the 1970s by Ted Granillo
Les and Paula Presmyk collection
ex Arizona-Sonora Desert Museum
Jeff Scovil photo

Facing page, upper: The small Flux Mine open pit (now filled in) reveals some of the earliest workings of the mine.
Tony Potucek photo, 1974

Facing page, lower: Striated pyrite
6.8 cm wide
Collected in the late 1950s by Patagonia resident Richard J. Rowe
Flux Mine, Santa Cruz County
Mark Hay collection
Jeff Scovil photo

The twins are elongated like most of the jackstraw cerussite at the Flux, but when viewed down their *c*-axes, these crystals exhibit a six-ray star pattern—they are twinned from top to bottom. They found crystals as large as to 7.5 centimeters growing in bundles on and off the wad matrix. Gene collected the pocket while Jackie wrapped the specimens. The pocket produced three flats of the fine, twinned cerussite.

In the early 1970s Ted "Eddie" Granillo and his partners made a small find of what may well be Arizona's best anglesite specimens. The well-defined, tabular crystals are up to 1.3 centimeters, are milky white with just a hint of yellow, and show silky luster.

Not every trip to the Flux was productive, but some were certainly memorable. Tony Potucek remembers a trip in 1974 with fellow University of Arizona mining and geology students Pete Knudsen and C. George Lynn. Lynn was the fastest collector in the world to jump out of a vehicle, don his gear, and head into the mine. On this occasion, Lynn climbed the hill to the mine opening where the trio had wrested a rattlesnake out of the entrance to the mine just one week before. Into the mine Lynn went. And as Knudsen and Potucek came up on the dump, they heard a roar from the mine. Lynn came running out and right on his heels was a frightened javelina that had apparently gone into the mine to escape the heat.

Washington Camp-Duquesne

The Patagonia Mountains are home to several specimen-producing localities in addition to the Flux Mine. The mining districts—Harshaw, Palmetto, Patagonia, Washington Camp, and Querces—in this historic region of the state were all organized by the 1870s, but they did not become known as specimen producers until the 1950s.

Sylvester Mowry's pre-Territorial **Patagonia** (**Mowry**) Mine (see page 26) is in the Patagonias, but only one specimen is known to have been preserved from the Mowry: a humble lump of cerussite that turned up in a Tucson collection in 1986. Other mines in the area include the **Pride of the West** (formerly the **Washington Mine**), **Bonanza**, and **Holland**.

In the late 1950s Richard Bideaux and others started The Mineral Exploration Society. Its members researched the mining districts in the Patagonias. The following was published in their newsletter *The Mineral Explorer* (1957),

> *...Magnificent sharp highly modified marmatite (Fe-rich sphalerite) crystals up to 1½ inches in diameter were later* [after 1900] *found in the George Westinghouse mine; also, fine botryoidal blue smithsonite with crystallized aurichalcite and malachite, and three-inch pyrite crystals as isolate cubes in the crystalline limestone. At the Santa Niño property, 2½ miles southwest of Duquesne, curious deeply striated and distorted pyrite crystals up to six by six by twelve inches, and weighing upwards of 100 pounds apiece, were found bordering a molybdenite orebody. The molybdenite was also well crystallized as distinctly hexagonal tablets one-half inch in diameter, occurring on quartz matrix...*

Bideaux also noted a more contemporaneous find:

> *Japanese law quartz twins was* [found] *in an open cut near the Pride of the West mine, above Duquesne. On the face of this cut, thin veins and bands of quartz swirl through massive andradite. A dirt and debris-filled pocket in the face was excavated which must have contained many excellent specimens when first broken into. The quartz in this pocket occurred as very much flattened crystals, partly transparent, and usually incrusted with secondary silica. About one crystal in ten must have been a twin, as many loose individuals were found which had broken along the twinning plane, which is an apparent plane of weakness in these crystals. One distorted loose crystal consisted of one individual three by one inches and the other one by one-half inches. Several twins on matrix were recovered with individuals two by three-quarter inches. One specimen about two by two inches included 14 small limonite-incrusted twins. These crystals were all one-eighth inch in thickness or less. Only one crystal was found which had the heart-shaped outline characteristic of the twins from the type locality; all the rest have much more elongated individuals.*

Armed with this information, Tucson collectors began prospecting old diggings in the Patagonia Mountains, and their research and legwork paid

Upper: Molybdenite and chalcopyrite on quartz, crystals to 1.3 cm long
Santo Niño Mine, Santa Cruz County

University of Arizona collection
Jeff Scovil photo

Lower: The ruins of the Bullion House at the Mowry Mine site, May 1973

Les Presmyk photo

Left: Japan-law twin quartz, 31 cm wide
Holland Mine, Washington Camp, Santa Cruz County

Flagg Mineral Foundation
Jeff Scovil photo

Right: Rutile and muscovite, 7.2 cm wide
Santo Niño Mine, Santa Cruz County

Evan Jones collection

Facing page, upper left: Quartz
6.2 cm tall
Duquesne, Santa Cruz County

University of Arizona collection

Facing page, upper right: Calcite and quartz, 8.3 cm tall
Holland Mine, Santa Cruz County

Evan Jones collection

Facing page, lower right: Quartz (variety amethyst), 8.3 cm tall
Santo Niño Mine, Santa Cruz County

University of Arizona collection
ex Richard Bideaux

Facing page, lower left: Calcite and quartz (twinned), 13.2 cm tall
2nd level, Holland Mine
Collected in 1952

Mark Hay collection
ex Claude Motel

All photos in this spread are by Jeff Scovil.

off. By the 1960s this area was a favorite collecting destination for various clubs in Arizona such as the Mineralogical Society of Arizona. Quartz was abundant in a number of surface deposits as were garnet and epidote crystals. And collectors could always turn up various sulfide minerals by turning over any of the various old mine dumps. No one went home empty handed.

One mysterious pocket was reported from the era. A hiker traversing Mt. Washington claimed to have picked up a transparent, lustrous Japan-law twin quartz crystal about 5 centimeters on a side. He brought it home to Casa Grande and showed it to Dick Jones, who bought it. He described the place to Dick as having "30 or 40 of these crystals just lying on the ground" (Dick Jones, pers. comm.) and provided detailed directions to the spot, but Dick never did find the pocket.

The Holland Mine was also producing in the 50s and 60s. Quartz and calcite were encountered there as well. Some of the quartz took shape as Japan-law twins, and the pink-peach calcite rhombohedra could reach up to 6 centimeters. All of these were coated with a secondary tan calcite or siderite that had to be etched away.

By 1970 several mining companies including Rosario-Simplot were active in the area. They tied up claims and initiated sampling and exploration programs, and ultimately limited access to most of the underground workings. And in the late 1980s, the patented mining claims began being bought by private entities, which closed many of the specimen-producing localities. During a 1994 U.S. Department of the Interior assessment (1994), landowners refused to allow Bureau of Mines personnel onto more than half of the acreage in the Washington Camp-Duquesne area to sample and inspect their property.

Glove Mine

with contributions by Gene Schlepp

One of Arizona's most prolific wulfenite localities is located in the western foothills of the Santa Rita Mountains about 40 miles south of Tucson. Over a 20-year period beginning in the 1950s, the Glove Mine yielded thousands of fine wulfenite specimens, most with a very distinctive habit. Many of these specimens came from remarkable pockets—walk-in grottos of dense, lustrous, butterscotch-color wulfenite crystals.

The Glove began as a relatively undistinguished lead claim first staked in 1907 by Edward T. Sheehy and associates in the Tyndall Mining District, just north of Cottonwood Canyon. In the early years a total of 16 claims were filed although only two proved economic. The sites were prospected and developed from 1907 to 1911, and for the next six years they enjoyed almost continuous production.

The ore, a mixture of argentiferous galena, smithsonite, and sandy-textured cerussite, was hand-sorted and then hauled by mule teams to the railroad station at Chavez, Arizona. Only about $15,000 in ore was shipped during this period as the main orebody was yet to be discovered. The Glove deposit consisted of a series of irregular, replacement pipes penetrating limestone and plunging eastward at variable angles. These pipes were emplaced along permeable zones in the limestone that had been created by faulting and brecciation. Massive lead-zinc sulfide ore was deposited first. When these sulfides weathered they released sulfuric acid, which etched out cavities in the soluble limestone, forming a complex system of caverns. Most of these caverns originally hosted sulfide deposits, but in some cases the acid

Left: Wulfenite, 7.6 cm tall
Glove Mine, Santa Cruz County

Mark Hay collection
ex Monnie Speck
Jeff Scovil photo

Right: The Glove Mine shaft provided collectors access to the mine until it was burned in early 2011. A U.S. Forest Service welder who was installing a bat-friendly gate over the mine opening sparked the fire.

Wendell Wilson photo, circa 1985

Wulfenite, 28.3 cm wide
Glove Mine, Santa Cruz County

Lyda Hill collection
James Elliott photo

resulting from weathering ate into the surrounding limestone to form extended caverns, many of which subsequently became lined with secondary minerals such as wulfenite.

Having shipped just two cars of ore to the smelter between 1918 and 1950, Sheehy and his partners sold the operation to the Sunrise Mining Company in 1951. Sunshine restarted operations a year later and chanced into the main orebody while extending the adit. More than one million dollars in ore was shipped to the smelter in the eight years that followed. Much of the fine wulfenite was recovered as Sunshine worked down through the heavily oxidized ore shoots and into the sulfide zone, which was below the 300-foot level.

As a high school student in 1954, Tucson collector Dick Apfel and his father were among the earliest collectors to discover the Glove Mine. Gene Schlepp recalls that he and Dick attended junior and senior high school together and that they, along with Larry Meyer and Burton Miller, were members of the Science Club. One Monday, Gene and his friends noticed that Dick's hands were stained brown—evidence of having been digging over the weekend. Apfel confessed that he and his father had found wulfenite in various shades of yellow and brown all over the Glove Mine dumps. The mine was operating, so the Apfels struck a deal with mine manager Eddie Mack. The agreement allowed for the Apfels to collect and sell wulfenite and for Mack to receive a percentage of their sales.

Dick and his dad spent 16 successive weekends at the mine, taking out matrix pieces of large wulfenite crystals. Some of the specimens weighed as much as 50 pounds. Mr. Apfel built a 10-by-12-foot storage shed to house the wulfenite. And Gene remembers visiting the Apfel house and seeing every one of the shelves in the shed loaded with wulfenite specimens. Mr. Apfel had arranged to sell the specimens to New York dealer and mineralogist Hugh Ford. Apfel charged $50 to $75 for a powderbox full of specimens. As Gene recalls, the elder Apfel was amused about a letter Ford sent after receiving the first shipment. He wondered whether Apfel would consider reducing the price!

On the Sunday of their sixteenth weekend of collecting, Apfel and his son breached a large cavern containing superb caramel and butterscotch-colored wulfenite crystals. As it was late in the day, Dick and his dad left the mine, and when they returned the following weekend, the mine manager told them that he had reconsidered their deal and they were no longer allowed to collect in the Glove Mine. Shortly thereafter, professional collectors Dick Jones and Al Haag struck a deal with mine management, and they went on to take out many fine specimens.

Dick Bideaux had the good fortune to have collected at the Glove Mine between 1955 and 1959. He reported on his activities in the local newsletter *The Mineral Explorer* (Bideaux 1959):

> *My first visit to the mine in the company of Richard Jones was in January, 1955, when the new development work was just beginning in earnest. A few fragile yellow crystal groups of wulfenite on limonite were found in the older workings at this time. As the new shaft, now a transfer raise and airway inside the mine, was encountering only sparse cerussite, we assumed that we had heard the last of the Glove mine as a wulfenite locality. Thus, I was quite surprised in October of the same year to receive an invitation to see a large pocket of crystals which had been found on deepening the shaft. This invitation was offered through Carl Richardson of Tucson by Jack Splane, the then chief engineer. Mr. Splane stated that the pocket had been found about 100 feet below the adit level and that the crystals were quite large but so fragile that it might prove impossible to collect any. As Mr. Splane had also been an engineer at Tiger, I realized that something rather unusual must have been uncovered, as he would not be easily impressed by wulfenite.*
>
> *This first pocket was typical of many found during the life of the mine. The inclined shaft had narrowly missed it; a miner had noticed a rubble-choked watercourse and had cleared it out, thus making the pocket accessible. Two other channelways were seen which led from the pocket*

Upper left: Vanadinite
field of view 4.2 cm tall
Glove Mine, Santa Cruz County

Les Presmyk collection
Jeff Scovil photo

Lower left: The orange-brown mineral overgrowing the central zone of these Glove Mine wulfenite crystals has been confirmed by SEM/EDS analysis as descloizite, field of view 1.7 cm wide

Tony Peterson collection and photo

Right: Wulfenite
9 cm tall
Glove Mine, Santa Cruz County

Dick Morris collection
Jeff Scovil photo

into the surrounding rock and necked impassably after a few feet. Other pockets were later uncovered by following these. The central cavity was about six by ten feet wide and four feet high. The deposition of wulfenite crystals had reached it is maximum thickness in the main pocket. The walls were covered with a confused aggregate of wulfenite plates up to 14 inches thick in many places. Near and in the channelways out of the pocket were large isolated single crystals and groups attached to the walls. The color of the crystals on the walls and top of the pocket ranged from brown to lemon yellow in spots, while the isolated crystals were a light brown. On the floor of the pocket, a thin layer of mud-covered crystals which were nearly black in color; on the mud layer had been deposited a thick druse of tiny brilliant black needles of plattnerite. Over the entire pocket the crystals were separated from the walls by a thin layer of hard manganese oxide.

Enough collecting had been done in the pocket prior to my arrival that the floor was thickly littered with broken wulfenite plates, a foot thick in places, which proved somewhat precarious and heart-rending to crunch about on. The management was convinced that there was ore around the pocket (there wasn't) and had miners drilling about the mouth of the pocket preparatory to blasting it at the end of the shift during my latter hours of collecting. Due to this deadline, collecting had to be rather hurried; there was not time to get back into the limestone matrix and remove large groups on matrix as was done in other pockets after the intrinsic value of the wulfenite crystals was realized.

Where the wulfenite crystals were isolated, singles and groups over two inches on an edge could be snapped off the walls, often with a little manganese oxide attached to act as a base. In the main pocket, a good specimen-sized area would be selected, and the crystals about it stripped away down to the wall rock. The specimen could then be removed by chipping away at its base until it fell away from the rock. Most of the groups taken in this way were about six inches thick, although they could have been easily much thicker. One hazard of such collecting comes from the wulfenite crystals themselves; as they are thin tabular, with sharp corners and beveled edges, severe cuts could result unless much care was taken. Also, there was poor ventilation in the pocket, so that my glasses, which I kept on to protect my eyes while collecting from the roof where the best specimens were, were continually fogging over from my breath, requiring attention.

Wulfenite, 12.1 cm wide
Glove Mine, Santa Cruz County
John Lucking collection
Jeff Scovil photo

Camp Verde

with contributions by Tony Jones

Jerome is perched high up on the east face of Mingus Mountain, which is part of the northerly trending Black Mountains; the mining camp over looks the Verde Valley, today home to the towns of Cottonwood, Cornville, and Camp Verde. Between 10 and 2.5 million years ago this valley was Lake Verde. This lake filled in and dried out a number of times before it finally disappeared leaving behind a thick sequence of lacustrian sediments composed of limestone, mudstone, and evaporates. The deepest portion of the lake was in the south, and it is there that a unique salt deposit formed. Millions of years later, that deposit was highlighted by *Gem Trails* collecting guides as the place in Arizona to go to collect glauberite and calcite pseudomorphs after glauberite.

About a mile due west of Camp Verde, a small mesa juts out from the Black Mountains along Copper Canyon. The face of this mesa is a layer cake of Glauber's salt (hydrated thenardite), gypsum, and halite along with limestone and a fine-grained clay. The salts were first reported in the fort records for Camp Verde when it was founded in the early 1860s. It seems likely that the fort was intentionally located near the salt deposit as salt is essential for livestock.

Starting in the 1920s several companies mined the deposit for "salt cake" used in paper pulp processing. In 1926 Western Chemical was mining the deposit for thenardite when they uncovered stone tools, torches, and sandals along with collapsed tunnels that indicated that the deposit was mined by Arizona inhabitants long before the sixteenth-century arrival of Spaniards

Right: Calcite after glauberite, 8.6 cm tall
Camp Verde, Yavapai County
Collected in 1970

Les and Paula Presmyk collection
ex Marty Zinn
Jeff Scovil photo

Facing page, upper: Circa 1930s view of the sodium plant at the Camp Verde salt deposit

The plant has since been removed.

Photo courtesy of Sharlot Hall Museum Library and Archives

Facing page, lower: In the early 1980s a large number of green stained calcite pseudomorphs after glauberite crystals such as this 9.1 cm tall example appeared in mineral dealers' stocks. They were often marketed as naturally occurring and a rationale was supplied that stated that the Verde salt deposit was at the mouth of Copper Canyon. However, several dealers later admitted to dying the crystals in a secret solution, probably copper sulfate. These colorful pseudomorphs commonly sold for up to 3 times the cost of natural white crystal group.

Keith Williams Minerals specimen
Jeff Scovil photo

(Morris 1928). Further excavation uncovered a mummy preserved by the salt—a lasting record of an ancient miner's death in a mine collapse.

The locality is easy to find: take Salt Mine Road in Camp Verde and drive a mile or two toward Copper Canyon, park your car, and start collecting. In the 1950s calcite crystal clusters replacing glauberite littered the ground. Virtually all of the collecting is on the surface. And even though layers containing various types of crystals extend over several square miles, most of the collecting is concentrated around the old salt mine. High-quality specimens were abundant, but it did take some effort to find (and later clean) the best specimens, which might have included thenardite, the various pseudomorphs, or halite. With a little luck, a purple or blue example of the latter might be found.

Collector and dealer Tony Jones visited Camp Verde in 1971, but unlike most collectors, Tony figured the underground was fair game. He and a few

friends met Camp Verde veteran Ken Dunham at the mine in March. Ken had described the underground experience as "hairy" but agreed to take the collectors into the mine. As Jones tells it,

> *Getting into the mine required them to dig out the buried entrance and slither in through a very small hole. Once underground, we fired up carbide lamps and saw what Ken meant by hairy. Most of the roof was composed of large heavy layers of halite and sedimentary strata. The timbers had mostly snapped under the weight of the roof and sat in folded masses with halite crystals growing from them. The ceiling was covered with snow-white sparkling stalactites of thenardite. It was not a large underground working, and it was certainly difficult to stand up in there, so most of our collecting was done from a hunched position. But the thenardite was really abundant. We used clear food wrap to protect the stalactites from getting dirty and placed them into beer flats after a final paper wrapping. My own yield was about 25 or so flats for the day.*

In the early 1970s Mike New, Curt Van Scriver, and Tony Jones spent some time exploring the area. They had a standing order from Filer's Minerals for the pseudomorphs after glauberite for their "Mineral of the Month" offering. The trio discovered that while the specimens near the salt mine tended to be weathered, those under the surface in other areas were more pristine. New, Van Scriver, and Jones collected hundreds, maybe thousands, of pounds of very sharp clusters of tan-colored calcite after glauberite. They recovered single crystals up to 3 centimeters and "pineapples" (clusters) up to 13 centimeters across.

Just across the ravine in lower Copper Canyon the team found an occurrence of aragonite after glauberite. And New discovered a site a few miles from the salt mine that produced pure white gypsum after glauberite pseudomorphs. New's locality yielded individual crystals up to 5 centimeters long and clusters up to 15 centimeters wide. They also worked an exposure along one of the ravines that to this day yields large, gray glauberite crystals up to 10 centimeters long.

Right: Circa 1970 postcard of the "Ghost City" of Jerome, Arizona, home of the United Verde and United Verde Extension (Little Daisy) Mines

Lithographie collection

Below: Azurite, 6 cm tall
United Verde Mine, Jerome
Yavapai County

John Lucking collection
acquired from George Bideaux
Jeff Scovil photo

Facing page, left: Halite, 19.8 cm tall
Camp Verde, Yavapai County
Collected in 1964

Les and Paula Presmyk collection
Jeff Scovil photo

Facing page, right: A view of a section of the surface salt deposit at Camp Verde

Peter Cristofono photo, January 2006

Jerome

with contributions by Mike New

The United Verde Mine (see page 65 and 101) in Yavapai County stood open and unwatched (or at least it seemed so) since it shut down in the mid-1950s, and mineral collectors predictably found their way into the mining areas and found azurite, malachite, native copper, cuprite, and a lot of chalcopyrite. Mike New was an early visitor to the "Big Hole."

For ten years beginning in the early 1950s, New's family lived in Phoenix. His dad had hunted for uranium during the rush in the mid-1950s and enjoyed prospecting and visiting old mines. The News had relatives in Utah and travelled on the main highway north from Phoenix through Sedona to visit them. After noticing a sign for Jerome on one trip, Mike began pestering his dad to take him to the "old ghost town."

They left Phoenix early one Saturday morning and arrived in Jerome about 2 o'clock in the afternoon. They drove right into the floor of the open pit of the United Verde Mine, and Mike started collecting. He got some chalcopyrite from an ore pile on the pit floor and started checking the sides of the pit. New's prize of the day was a piece of schist with native copper in the center of a vein of crystallized cuprite, azurite, and malachite.

New started field collecting again in the late 60s after his tour in Vietnam. He spent a lot of time doing shows and prepping rocks to help Jean Ryan and her husband Riley, proprietors of a rock shop near where he and his wife lived in Yorba Linda, California.

> *Jean asked me one day if I knew where to collect native copper. I remembered Jerome from my high school days and promised her that I would revisit the locality. I talked my friend Keith Vanderbosch into going with me. Jerome was a ghost town, with a population of no more than 40. Mingus Mountain High School had closed several years before. The new freeway from Phoenix to Flagstaff had bypassed the town, and it was dying.*

Copper on calcite, 6.8 cm wide
United Verde Mine
Jerome, Yavapai County

Evan Jones collection
Jeff Scovil photo

Keith and I drove right into the open pit just as my dad and I had 12 years before. There was no one there—Big Hole Mining Company had shut down its sulfide mining operation. Keith and I started by walking along the base of the walls in the area where I had found the native copper in the past. We didn't find one piece. We did, however, see pieces of azurite and a few complete roses on one stretch of the haul road in the pit. We started climbing up the vertical schist walls trying to locate the source of the azurite. In hindsight that was a really stupid thing to do as the schist walls often gave way in a spectacular manner with hundreds of tons of rock hurtling to the bottom of the pit.

We got near the top of the east wall and found a white clay seam studded with decomposing azurite rosettes. The ones near the surface were frost damaged but 5 or 10 centimeters into the clay, we were getting nice rosettes to about 4 centimeters in diameter. We collected for about three hours, and when we quit for the day, we had eighty or ninety decent pieces and two or three really nice ones.

On a subsequent trip, while I was walking up the road from the pit to the collecting spot, two men arrived: Phelps-Dodge's Jerome representative (Who knew they had a presence there?) and the town marshal. They asked me to join them in the Phelps-Dodge office in town where I was lectured about private property and high wall safety. I was told to cease and desist collecting. So thereafter, I entered the property at the end of the alley by the Catholic Church. There is a nice gate in the fence and a nice storm drain to access the open pit. I was invisible once I arrived at the collecting spot and could come and go at will.

Over his many trips, New discovered several areas in the mine that produced azurite. Each area yielded a slightly different type: round balls with a sort of twist (pictured page 231); bundles of wheat sheaf-shape sprays; and roses with larger blades (pictured below). The best ball New found went to Jean Ryan. He sold and traded hundreds of azurite specimens; many went to Ricardo Campos, who had a rock shop in Superior, and to Mrs. Lemon of Mother Nature's Art Shop in Apache Junction.

New invited Robert Johansing, a young collector from Pasadena (who was also a fellow member of the Mineralogical Society of Southern California) along on one trip. They collected in the main area and did quite well. New extracted a promise from Robert to not show the locality to others, so imagine his surprise when he returned to California after having lived several years in Hawaii and learned that Robert had disclosed the locality to Bob Thompson, a teacher at Glendale Community College. Thompson and his friends had collected heavily in the main area. Fortunately New never revealed the six or seven other places in the mine that produce azurite roses.

Below upper: The slag pile at Clarksdale, a remnant of decades of mining at Jerome, can be see from aboard the Verde Canyon Railroad Wilderness Route. The slag is currently being reprocessed

Candace Porth photo, Nov. 2011

Below, lower: Malachite (polished)
Field of view about 4 cm tall
Jerome, Yavapai County

Arizona-Sonora Desert Museum collection
Jeff Scovil photo

Right: Azurite, 4.2 cm tall
United Verde Mine
Jerome, Yavapai County
Collected by Mike New

Les and Paula Presmyk collection
Jeff Scovil photo

Right: Molybdenite on quartz
6.5 cm wide
Quartz outcrop near Cleator
Yavapai County
Collected by David Shannon
Les and Paula Presmyk collection
Jeff Scovil photo

Above: The first post office at Cleator was established in 1869 and was then called Turkey Creek. The post office served the placer gold mining community in the area for 3 months. In 1925, James P. Cleator acquired the post office, general store, and many of the buildings in Turkey Creek from his partner Levrett P. Nellis. Cleator renamed the town for himself.

The Cleator post office closed in 1954, but Cleator (pictured far left) had by then sold the town to his son Thomas (far right). After Thomas's death in 1996, his sister Eleanor and her son inherited the ghost town. Several of the town's original buildings, including the old James P. Cleator Post Office (now a general store) are still in use.

United States Postal Service photo

Cleator, Bloody Basin, and the Yankee Boy Mine

with contributions by Bob Lane and Ray Lasmanis

There are a number of interesting and little known collecting localities near the ghost town of **Cleator** in Yavapai County. In addition to the Yankee Boy and Piedmont mines, a schist not more than a few miles from "downtown" produces staurolite crystals up to 2.5 centimeters long, and one stingy outcrop produces Arizona's best molybdenite crystals.

Collector Bob Lane was one of the lucky few who have found success at the molybdenite outcrop. Lane was taking Professor Bob Thompson's *Introduction to Mineralogy* at Glendale Community College. He remembers Thompson as an avid collector who was, "able to teach the subject in a way that would also turn students into mineral collectors." One evening as they were studying specimens, collector and later mineral dealer Wayne Thompson walked into the classroom. His long hair was dirty from a trip to the 79 Mine where he had collected a nice pocket of aurichalcite. Wayne was carrying a flat of specimens, which he wanted to show our instructor. One thing led to another and Professor Thompson scheduled a field trip to Cleator so the class could get a taste of field collecting.

Molybdenite occurs in the quartz core of a small pegmatite. After securing permission to collect, the group climbed to the quartz outcrop and began pounding on the quartz for specimens. They found a few small molybdenite crystals and some ferrimolybdite. It was fun but not what they had been hoping to collect. Then as Lane was climbing back down the hill to leave, he kicked over a boulder and noticed something silver-colored sticking out of the quartz. It was the edge of a molybdenite crystal that was nearly 5 centimeters wide! Bob Thompson freaked out when he saw the specimen. He cautioned Lane not to attempt trimming it there, suggesting instead that they use the rock trimmer at the lab. Lane remembers,

> *When we got back it was late. We left the specimen at school and returned the next day. Professor Thompson had already trimmed the specimen and it turned out fantastic! The specimen was now a 5-centimeter-wide molybdenite crystal with another 2-centimeter-wide crystal sitting at its base and perched on a pure white matrix of quartz. To this day, it probably is the best single specimen to ever be collected from Cleator. I was a mineral collector for life after that trip.*

A few miles south of and 10 miles east of Cordes Junction in Yavapai County, **Bloody Basin** takes its name from an 1873 U.S. Army surprise attack on an Apache encampment on Turret Peak. The raid resulted in the death or capture of most of the Indians involved and was a turning point in the Army's campaign to subdue the Apaches in the Bradshaw Mountains. Bloody Basin falls under the Copper Creek Mining District. Its few small mines and exploration pits have historically seen only limited production, but a find in the mid-1950s made one of those old prospects—the **Piedmont Mine**—one of Arizona's Holy Grail-specimen producers.

In 1955 Phoenix area collectors Bonnie and Evert Thompson, Wayne Thompson, Sr. (uncles of mineral dealer Wayne Thompson), and Monnie and Ed Speck discovered beautiful pseudomorphs of malachite after azurite

Quartz coating malachite after azurite
6.5 cm tall
Piedmont Mine, Bloody Basin
Yavapai County
Collected in 1955 by Bonnie and Evert Thompson, Wayne Thompson Sr., and Monnie and Ed Speck

Mark Hay collection
ex Monnie Speck
Jeff Scovil photo

at a small copper prospect in Bloody Basin. The small mine consisted of little more than a 50-foot deep shaft, but there were several large quartz boulders on the dump. Some of boulders contained pockets of drusy quartz on malachite after azurite. At the time, these specimens were unique in the state and quite possibly the world; in fact, only a handful of worldwide localities have produced similar quartz-coated malachite after azurite. Interestingly the Live Oak Mine in Gila County, Arizona (see page 186), is one of them.

They were obviously not going to get the specimens out of the boulders with hand tools or even sledgehammers, but fortunately Evert Thompson knew how to use dynamite. Over several weekends, the group collected everything the dump had to offer including one 15-by-38-centimeter specimen. As no one in the group was willing to concede that largest and best specimen, however, they ended up breaking it into three pieces.

No one remembers exactly how many pieces were collected, but a casual accounting of the specimens in the three collections puts the figure around 50. For some time Sandi Ashton at Carl's Rock Shop had the best collection of Piedmont pseudomorphs after she purchased Monnie Speck's collection. Mark Hay, Dick Morris, and Keith Williams bought Speck's collection in 1997. Les Presmyk purchased a portion of Bonnie and Evert Thompson's specimens about 1985 (Grant 1986). A year later Dave Shannon acquired a few Piedmont specimens from Wayne Thompson Sr., but the Arizona-Sonora Desert Museum had already purchased the bulk of that Thompson brother's specimens, including his two best.

Facing page: Quartz coating malachite after azurite, 4.5 cm tall
Piedmont Mine, Bloody Basin
Yavapai County
Collected in 1955 by Bonnie and Evert Thompson, Wayne Thompson Sr., and Monnie and Ed Speck

Evan Jones collection
Jeff Scovil photo

It took a second purchase for Presmyk to acquire the balance of Bonnie and Evert Thompson's share of the find. The Thompsons' daughter (and Wayne Thompson's cousin) Margaret was ready to sell her parents' collection in the mid-1990s. Presmyk and (Wayne) Thompson went to look at what was left. Although Presmyk was confident that he had already purchased the best of elder Thompsons' specimens, some good pieces remained.

Presmyk walked into the house and saw no sign of the collection.

> *There had been a ten-foot long display case filled with minerals ten years before, but the case was not there. Then Margaret started to unscrew some panels from the wall, and in just a few minutes we were looking at the collection. Had anything ever happened to Wayne's cousin, the house might have been sold with the collection hidden behind the paneling and no one realizing that it was there. The Piedmont specimens that the Thompsons had held back were there in the case behind the paneling.*

Considering that the Piedmont's only collectable mineral is malachite (the pseudomorphing is complete), there is a fair amount of variety in the specimens. Some of the pseudomorphs are light green and others are a very dark green, and in some instances the single crystals appear mottled, showing various green tones. The matrix ranges from deep chocolate brown to almost white. Some of the quartz coatings and crystals are thin and opaque white. Most are clear and transmit the color of the malachite. Individual crystals can exceed 3 centimeters.

Over the years, a number of collectors have visited this well known locality, but none had much success. Interestingly Utah-based collector Bob

Lane's father co-owned the Piedmont Mine in the 1950s.

> *In the 1950s when I was just five years old, my dad was part owner of this mine. He worked with the Wright brothers of Bloody Basin who were mountain lion hunters. My dad and the Wrights had a number of gold and copper claims in the area, and the Piedmont was one of them. I only visited once, for about a week in 1955. Dad had a mining cabin there and a shed where they kept all their supplies and blasting material. There was a tunnel entrance near the cabin at that time, and a couple of shafts. My Dad and some other guys came out of the mine with specimens, but they weren't chrysocolla or malachite; they were dark blue azurite! I remember how they shined in the sun. It wasn't a druse but individual crystals. I don't know why, but at the end of our trip, the tunnel was blasted shut and the cabin and shed were burned down.*

Graham Sutton and Bob Lane leased the claims in the early 1990s. They did some exploration work and found some chrysocolla and malachite pseudomorphs, but nothing great. The state of Arizona made mining and moving dirt difficult, but Lane believes if they had been allowed to seriously mine the locality, they would have found some incredible specimens.

The area is presently protected as part of Pine Mountain Wildness Area and collecting and mining there is no longer an option; regardless, the shaft from which the boulders came is now filled with water (Grant 1986).

Right: Pyrite, 12.7 cm wide
Yankee Boy Mine, near Mayer
Yavapai County
Ray Lasmanis collection
Bill Dameron photo

Facing page, upper: The range of the cougar (*Puma concolor*) extends from the Canadian Yukon to the southern Andes in Chile. Commonly referred to as mountain lions or pumas, the big cats were classified as "predatory animals" by the Arizona legislature in 1919. A bounty of $50 was offered for them until 1970, when they were reclassified as big game animals, and a tag became required to hunt them. In the 1950s the Wright brothers, partners in the Piedmont Mine, hunted mountain lions for that bounty. According to the Arizona Game and Fish Department an average of 300 cougars are bagged annually in Arizona, animal control officials are responsible for 15 percent of the kills.

Arizona-Sonora Desert Museum live collection, Dwight Hanson photo

Facing page, lower: Quartz on malachite after azurite (repaired), 11 cm wide
Piedmont Mine, Bloody Basin
Yavapai County
Collected in 1955 by Bonnie and Evert Thompson, Wayne Thompson Sr., and Monnie and Ed Speck
Les and Paula Presmyk collection
ex Jim and Dawn Minette
Jeff Scovil photo

The **Yankee Boy Mine**, located near Mayer in Yavapai County, consists of a 50-foot deep shaft with 70 feet of drifting at the bottom. The mine was originally developed for gold and silver, but it produced pyrite and chalcopyrite instead and never turned a profit.

In the 1950s and 60s, collectors found lustrous pyrite octahedra up to 5 centimeters on an edge in the dump near the shaft. The form distinguishes these crystals, a number of which have a lustrous brown coating. Some collectors have attempted to remove the brown coating and found the pyrite lackluster. In 1964 the Mineralogical Society of Arizona hosted a field trip to the Yankee Boy and Marcella (Marc) Watson encountered a dozen or so crystals up to 7 centimeters on edge in the dump near the mineshaft.

Geologist and collector Ray Lasmanis visited the Yankee Boy and found octahedral pyrite and chalcopyrite crystals in calcite on the dump. The largest pyrite crystal he collected measures 3.6 centimeters on a side. He also found the specimen of bright, 1.8-centimeter, pyrite crystals pictured above.

Lasmanis's interest in Yavapai County pyrite was first aroused through a trade he made in 1957 in which he obtained a 2.8-centimeter-wide, bright pyrite crystal, which had been part of a 1930s shipment to Tiffany & Co. The crystal showed a conchoidal fracture plane and had unique very step-growth striations on the cubic faces. The location given was "west of Poland Junction." So when Lasmanis's job placed him in Prescott in the mid-1960s, he began to execute a careful search for the pyrite's source. His search eventually led him to a mine dump 2 miles west of **Poland Junction** on the south side of Poland Junction Road where he found morphologically identical pyrite crystals up to 1.8 centimeters wide with cubic, thin step-growth striations and associated 1.8-centimeter-long clear quartz crystals. This discovery eventually led to his locating the Yankee Boy Mine

Above: "Descending the Grand View Copper Mine Trail," a circa 1906 photograph of a tour along the Berry Trail in the Grand Canyon, which at that time was being privately developed. The canyon was protected as a national forest and then a national monument, but it did not achieve national park status until 1919, three years after the National Park Service was established.

Library of Congress Prints and Photographs collection

Inset: The U.S.G.S. Survey placed this marker along the Berry Trail in 1904

U.S. National Park Service photo, 2006

Grandview Mine

with contributions by Bob Bartsch

The Grandview (or Last Chance) Mine in the Grandview Mining District of Coconino County is probably the most spectacular place in Arizona to collect minerals. The mine exploits a mineralized solution-collapse breccia pipe located on Horseshoe Mesa on the south side of the Grand Canyon; hence the "grand view." The deposit consists of uranium and copper mineralization in a chemically altered breccia. The mine is best known among collectors for producing pleasing velvety blue mats and vugs of cyanotrichite and dark green brochantite, but the mine has also produced some 40 different minerals.

In 1890 Pete Berry and brothers Niles and Ralph Cameron located claims at the top of Horseshoe Mesa and two years later discovered rich copper deposits at the mesa's base (Leicht 1971). Berry began constructing a trail to the claims in July 1892. The hand-hewn Berry Trail took a full year to construct. The modern Grandview Trail roughly follows Berry's original path, parts of which are still visible. The trail drops more than 2,800 feet to Horseshoe Mesa. Berry built the Grandview Hotel at the viewpoint and opened it for business in 1892. In spite of the expense of transporting the ore, they worked their Last Chance Mine at a profit until 1902.

Berry and a partner formed the Grand Canyon Copper Company in 1895. The company's holdings included both the mine and the hotel (Leicht 1971). In 1913 William Randolph Hearst purchased the property, but he was interested not in the mining portion of the business but in the hotel and tourist trade. Before he could capitalize on his investment, however, the National Park Service took over and Hearst was forced to sell his interest. Sanctioned tourist accommodations sprouted up on the south rim of the canyon, and the Grandview Hotel and others were abandoned.

Pete Berry continued guiding tourists into the canyon until 1916, when he left Grand Canyon country. The Grandview Mine stopped operating around 1916. Over its relatively short life, the Grandview is credited with 2,200 tons of base and precious metal ore production, primarily copper (936,000 pounds) and silver (14,000 Troy ounces) (Keith et al. 1983).

In the summer of 1964 Bob Bartsch ventured into the Grandview Mine, having been preceded by no more than a handful of others (well known thumbnail and smithsonite collector Jim Minette was among those few). Dick Bideaux suggested that Bartsch check out the mine. Bartsch was 24 when he made his first trip in the summer of 1964. It was hot at the Grandview, so he and his companion slept between rails on a bed of ore sacks in the mine. He remembers the bats skimming over his face while he was trying to sleep.

Cyanotrichite on brochantite
Cyanotrichite groups to 5 mm in diameter
Grandview Mine, Horseshoe Mesa
Grand Canyon, Coconino County
University of Arizona collection
Jeff Scovil photo

The walls in the upper level of the mine were an amazing sight, with generous amounts of crystallized brochantite everywhere, along with azurite, cyanotrichite, smithsonite, and a number of minerals that I could not identify. Most of the best deep colored cyanotrichite was associated with the brochantite we found in solution cavities and in watercourses within what appeared to be a siltstone cemented with a carbonate. We collected massive amounts of exposed brochantite, sorted through it, and picked out only the most attractive specimens associated with cyanotrichite to pack up the trail. After a while, we had brochantite specimens lined up along both sides of the adit for about 40 feet.

Rock Currier went along with Bartsch on one trip and found what is probably North America's finest zeunerite specimen. From the late 1960s to 1971, Wayne Leicht (1971) made half dozen trips down into the Canyon to the Grandview Mine to collect minerals on a research grant from the U.S. National Park Service. This initial study of the mineralogy of the Grandview breccia pipe helped the Park Service identify the minerals occurring in the deposit. Over the four decades that followed, collecting in the mine has been strictly limited to a very few authorized groups.

Above: Zeunerite, 2.5 cm wide
Grandview Mine, Horseshoe Mesa
Grand Canyon, Coconino County
Collected in 1972 by Rock Currier

Rock Currier collection and photo

Right: Smithsonite with malachite (spheres to 0.5 mm), and azurite, 6.7 cm tall
Grandview Mine, Horseshoe Mesa
Grand Canyon, Coconino County

Bob Griffis Minerals specimen and photo

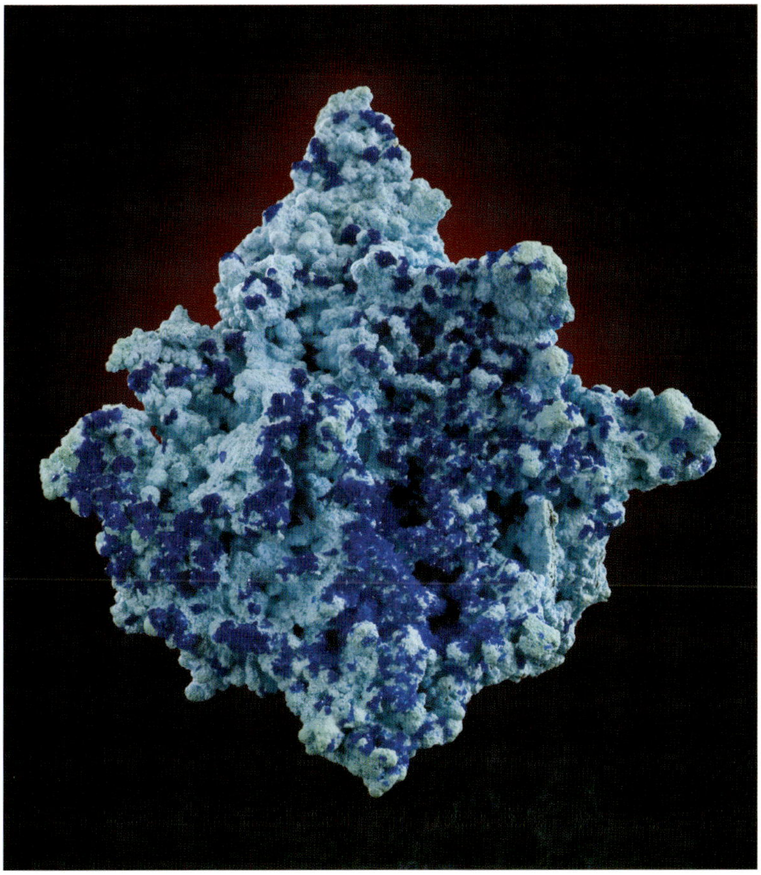

Upper left: Azurite-Chalcoalumite
6 cm tall
Grandview Mine, Horseshoe Mesa
Grand Canyon, Coconino County

Kevin Ward specimen
Joe Budd photo

Upper right: Brochantite, 4.9 cm wide
Grandview Mine, Horseshoe Mesa
Grand Canyon, Coconino County

Kevin Ward specimen
Joe Budd photo

Right: A number of mineral deposits in and around the Grand Canyon were worked before the area fell under the protection of the U.S. National Park Service. This placer mining camp along the Colorado River near Lee's Ferry, in what is today Grand Canyon National Park was photographed in 1913.

Library of Congress Prints and Photographs collection

CHAPTER 9

The Postmodern Era

The final 25 years of the twentieth century saw significant changes in the character of the state of Arizona. The population had continued to explode, growing from 2.2 million in 1975 to over 5.1 million in 2000. Remarkably, this growth was almost exclusively around Phoenix and Tucson. The Phoenix metropolis, which promoted itself as a desert paradise, became known as the "Valley of the Sun," and by the turn of the century was home to over 3.2 million people. In 2006 the population of the Tucson metropolitan area exceeded 1,000,000.

With more than 80 percent of the population living on less than 10 percent of the surface area, a number of environmental issues including air pollution and the depletion of natural resources, particularly water, confronted the state. The very things that attracted people to the Valley of the Sun—the dry desert and a valley surrounded by beautiful mountains—also trapped air pollutants in the city. The visible affects of the rapidly expanding population heightened awareness of environmental issues, spurred conservation programs, and impacted the development of new mining and manufacturing projects throughout the state.

Above: A Central Arizona Project aqueduct cuts across the desert. The project diverts more than one million acre-feet of water from the Colorado River to supply farms and communities across southern Arizona with fresh water.

Central Arizona Project photo

Left: Cerussite, 7.1 cm tall
Flux Mine, Santa Cruz County
Collected in 1982 by Stan Esbenshade

Stan Esbenshade collection
Jeff Scovil photo

In Tucson, where there are no lakes or rivers to supply fresh water, the burgeoning population severely taxed the groundwater. An engineering marvel, the 336-miles of canals and pipelines of the Central Arizona Project provided a solution. The project delivers water from the Colorado River to Phoenix, Tucson, and a number of other communities along the way. However, diverting 1.5 million acre-feet of water from the Colorado River every year has had an impact that was unforeseen by project developers in the late 1960s and early 1970s. By the new millennium, society had begun to rethink the wisdom and sustainability of "conquering" nature.

As for mining, the last two decades of the twentieth century were a time of tremendous copper production in Arizona. Copper prices were solidly above a dollar a pound for much of that period, and output steadily increased, peaking in 1998 with an annual production of more than 2.7 billion pounds. Were it all to have been stacked on a football field, the annual copper tower would have been more than 100 feet high! Morenci was the state's most productive mine. It accounted for about one third of Arizona's copper output during this interval; however, the metal was extracted by moving huge amounts of low-grade porphyry. In addition, by 2000, heap leaching—extracting minerals from the dispersed sulfides with a

A 2009 view of the heap leach at Freeport-McMoRan's Morenci operation

Photo courtesy of Freeport McMoRan Copper and Gold, Inc.

solvent—had become the dominant processing method, so while copper mining boomed in Arizona, there were few mineral specimen discoveries.

Changing attitudes impacted mining (and therefore mineral collecting) as well. Mines, factories, and plants of all kinds became pariahs to many in America. They were seen as scarring the landscape, drawing precious resources, and polluting the environment. The many benefits that mines and manufacturing bring to communities are commonly not recognized until the operations are moved to more politically and economically hospitable foreign climes. This struggle for balance continues today.

As the pendulum swung, the **Tucson Gem and Mineral Society** (TGMS) show continued to grow. It became an anchor for the international mineral collecting world as well as the broader lapidary, fossil, and gemstone communities. In 1981 the American Gem Trade Association (AGTA) formed and quickly became the world's most prestigious organization of colored-stone professionals. The AGTA tied its largest show to the TGMS show, a move that added to the gravity of the Tucson "experience." By 1990 Tucson was hosting more than a dozen large, satellite shows. Tucson's role in the gem, mineral, and fossil communities was difficult to overestimate.

But change was on the horizon as the internet was firmly entrenched in our lives by the turn of the century. Collectors had gained the ability to socialize, and share and trade their finds with the global community of collectors—to get their mineral "fix"—anytime and from anywhere. As was the case with so many institutions, the value of the Tucson experience was being questioned.

In 1970 ***The Mineralogical Record*** (MR) was founded by John White, curator in the Mineral Sciences Department of the Smithsonian Institution, with the help of a newly formed group: Friends of Mineralogy. The MR was aimed at serious mineral collectors and established a high standard for quality in mineral journals that remains today. When Wendell Wilson took over as editor and publisher of MR in 1977, he moved the journal's headquarters to Tucson. Wilson's interest in fine art influenced the magazine's design and photography standards. That ethos in turn inspired a community interest in mineral specimen photography.

One of those inspired was **Jeffrey A. Scovil**. A lifelong mineral collector, archeologist by training, and woodworker by trade, he began dabbling in mineral specimen photography in the mid-1970s. By the mid-1980s, he was a rising star in this specialized field, and had begun bringing his studio from his home in Phoenix to the Tucson Show to document new finds for the various mineral-collecting journals and allow collectors the opportunity to have their treasures professionally photographed. Scovil became a photographer full time in 1990. In 2007 he received the prestigious Carnegie Mineralogical Award for his "outstanding contributions in mineralogical preservation, conservation, and education."

Another stream in the Tucson confluence, the **Arizona-Sonora Desert Museum** (ASDM) opened as a unique facility focusing on the wonders of the Sonoran Desert in 1952 and began planning for an earth sciences center in 1972. The museum hired Bill Panczner as the first mineral-collection curator in 1974. Its Earth Science Center opened in December 1981 and included exhibits on the geology of the region with a hall dedicated to the mineralogy of Arizona and the Mexican state of Sonora.

An Arizona institution, collector, and photographer Jeff Scovil (**above** in a photo by Furenza) collected, cleaned, prepped, and photographed the 6.2 cm wide metatorbernite specimen pictured **right** from near Crown King in the Bradshaw Mountains, Yavapai County.

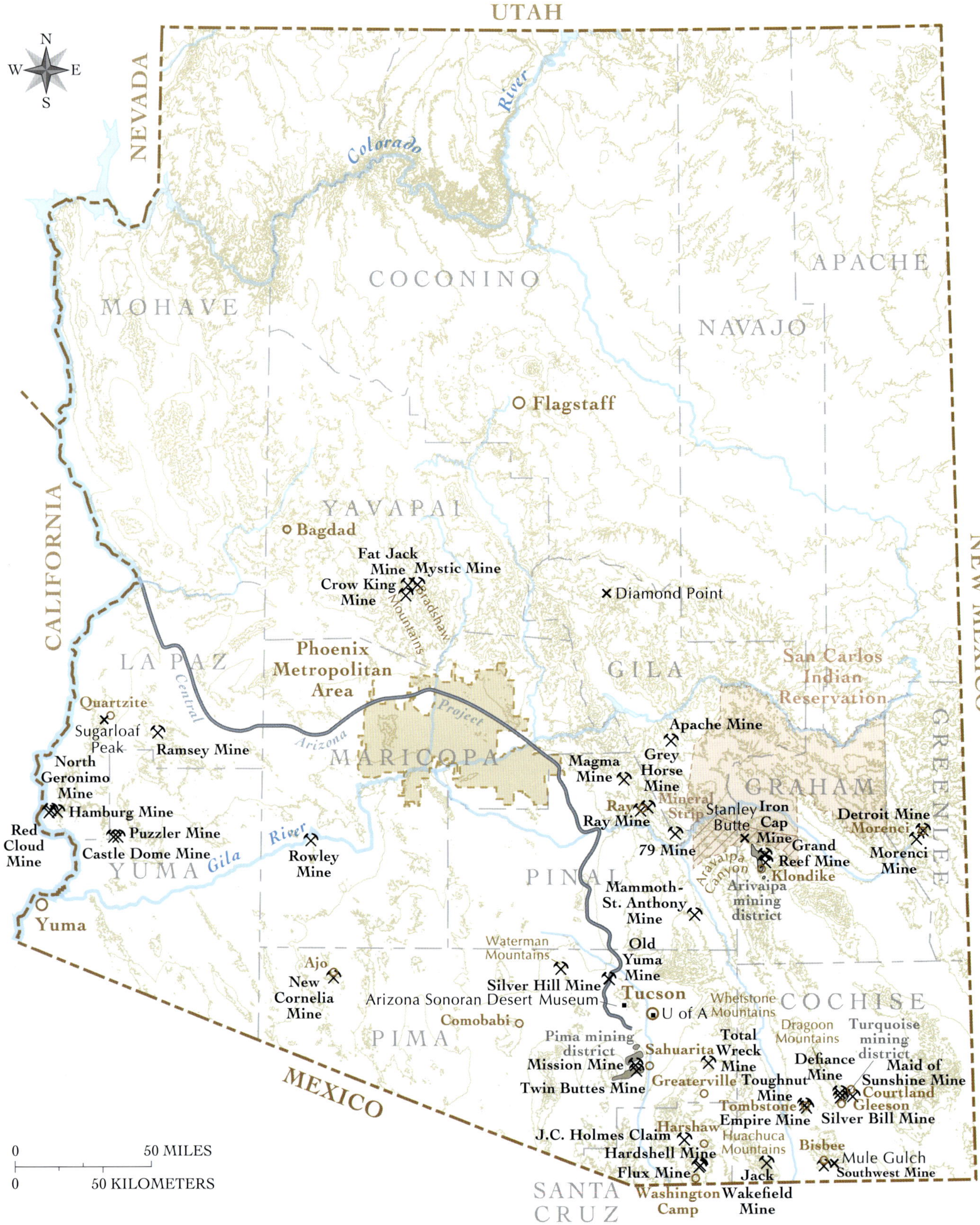

UTAH
NEVADA
CALIFORNIA
NEW MEXICO
MEXICO
Colorado River
MOHAVE
COCONINO
NAVAJO
APACHE
YAVAPAI
LA PAZ
MARICOPA
GILA
GRAHAM
GREENLEE
YUMA
PINAL
PIMA
COCHISE
SANTA CRUZ
Flagstaff
Bagdad
Fat Jack Mine
Mystic Mine
Crow King Mine
Bradshaw Mountains
Diamond Point
Phoenix Metropolitan Area
Central Arizona Project
San Carlos Indian Reservation
Quartzite
Sugarloaf Peak
Ramsey Mine
North Geronimo Mine
Hamburg Mine
Red Cloud Mine
Puzzler Mine
Castle Dome Mine
Gila River
Rowley Mine
Yuma
Apache Mine
Grey Horse Mine
Magma Mine
Ray
Ray Mine
Mineral Strip
Stanley Butte
Iron Cap Mine
79 Mine
Grand Reef Mine
Klondike
Aravaipa Canyon
Arivaipa mining district
Detroit Mine
Morenci
Morenci Mine
Mammoth-St. Anthony Mine
Waterman Mountains
Old Yuma Mine
Silver Hill Mine
Tucson
Ajo
New Cornelia Mine
Arizona Sonoran Desert Museum
U of A
Whetstone Mountains
Comobabi
Pima mining district
Sahuarita
Mission Mine
Twin Buttes Mine
Total Wreck Mine
Greaterville
Dragoon Mountains
Turquoise mining district
Defiance Mine
Maid of Sunshine Mine
Courtland
Gleeson
Toughnut Mine
Tombstone
Empire Mine
Silver Bill Mine
Harshaw
J.C. Holmes Claim
Hardshell Mine
Huachuca Mountains
Bisbee
Mule Gulch
Southwest Mine
Flux Mine
Jack Wakefield Mine
Washington Camp
0 50 MILES
0 50 KILOMETERS
N
E
S
W

The core of the collection was donated by well-known mineralogists and collectors including Richard Bideaux, Richard Graeme, Rukin Jelks, and John Sinkankas. The ASDM holdings include one of three known specimens of graemite (pictured on page 94) from its Bisbee type locality and the world's largest known crystal of matlockite from the Mammoth-St. Anthony Mine (see page 127). The ASDM houses an extensive micromount collection that contains more than 16,000 catalogued specimens. Among the museum's other holdings are the fossilized remains of *Sonorasaurus thompsoni*, a large, Cretaceous Brachiosaurid dinosaur discovered by Richard Thompson in the Sonoran Desert in 1992.

Panczner left the Museum in 1982 and Bob Middleton then Dave Thayer succeeded him. When Thayer left the museum in 1996, the job fell to Anna Domitrovic who curated the mineral collection until she retired in 2005. Domitrovic continues to volunteer in the Earth Science Center and was named the museum's first Curator Emeritus.

Upon John Anthony's retirement in 1984, Dr. Terry C. Wallace became curator of the Mineral Museum at the **University of Arizona**, and for the next 20 years, the museum focused on Arizona and Mexico. The collection grew to include more than 15,000 cataloged specimens. Wallace was an active member of the TGMS and the mineral-collecting community. As a consequence, the museum benefitted from a number of collector patrons including Richard Bideaux, Russ Honea, Rukin Jelks, and Martin Zinn. In early 1997 Dr. Miguel Romero loaned the Mineral Museum several hundred specimens from his mineral collection, which was the finest assemblage of Mexican minerals in history. Romero chose the Mineral Museum largely because of its relationship with the Tucson Gem and Mineral Society show, and indeed the museum became a must-visit for serious mineral collectors. Wallace also initiated the estate gift of the Herbert du Monmonier mineral collection, appraised at $7 million. In June 2011, Wallace's passion for silver minerals and his work developing the museum collection of sulfides and sulphosalts was honored when the new silver sulfosalt mineral terrywallaceite was named for him.

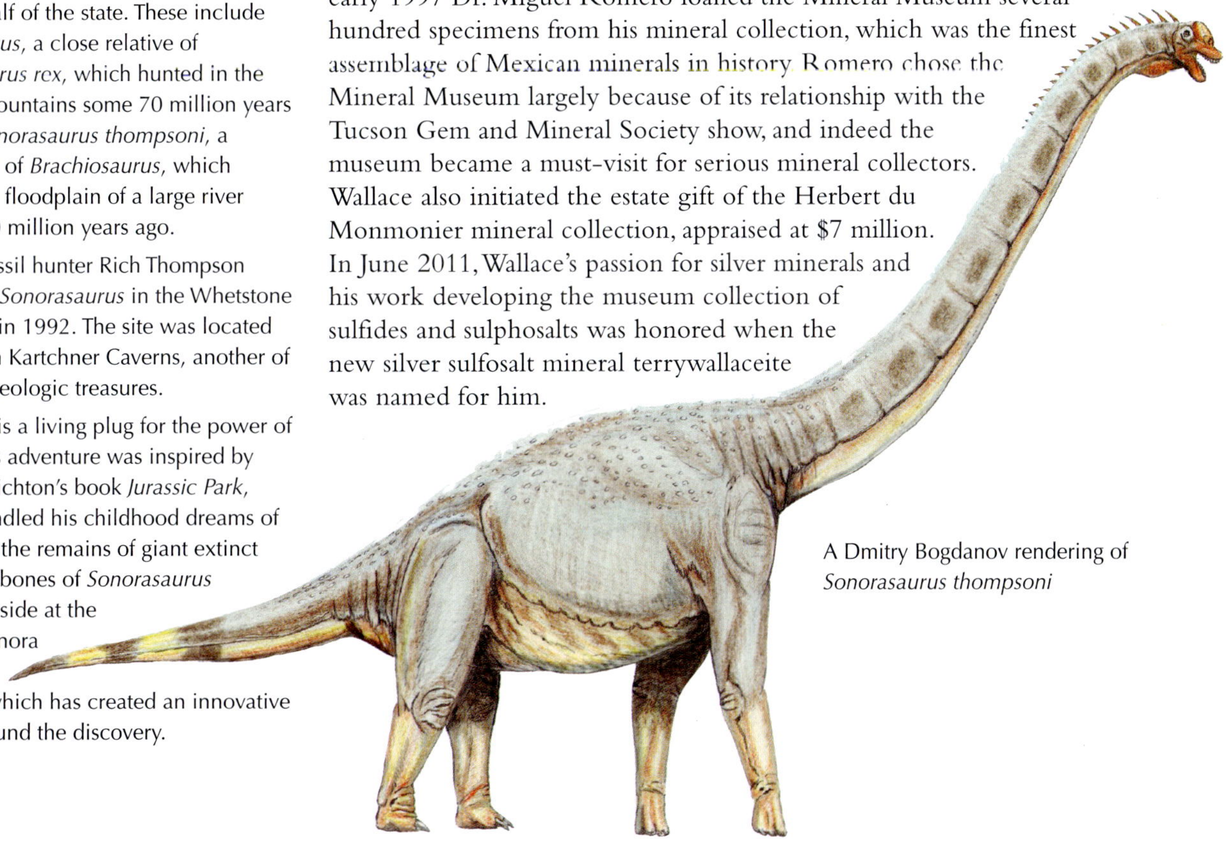

A Dmitry Bogdanov rendering of *Sonorasaurus thompsoni*

Arizona enjoys a wealth not just of minerals but also of fossils, even of dinosaurs. Northern Arizona has long been known for its rainbow-colored petrified logs as well as to a variety of Triassic Period creatures such as the heavily-armored, crocodile-like phytosaurs and their bizarre-looking prey, the also heavily-armored aetosaurs. This region was also the stomping grounds of the spitter from *Jurassic Park*, *Dilophosaurus*. And dinosaurs from the Cretaceous Period have been found in the southern half of the state. These include *Albertosaurus*, a close relative of *Tyrannosaurus rex*, which hunted in the volcanic mountains some 70 million years ago and *Sonorasaurus thompsoni*, a descendant of *Brachiosaurus*, which roamed the floodplain of a large river around 100 million years ago.

Amateur fossil hunter Rich Thompson discovered *Sonorasaurus* in the Whetstone Mountains in 1992. The site was located not far from Kartchner Caverns, another of the state's geologic treasures.

Thompson is a living plug for the power of reading: his adventure was inspired by Michael Crichton's book *Jurassic Park*, which rekindled his childhood dreams of hunting for the remains of giant extinct beasts. The bones of *Sonorasaurus* currently reside at the Arizona-Sonora Desert Museum, which has created an innovative display around the discovery.

Cuprite, 5.4 cm tall
5 level, Southwest Mine
Bisbee, Cochise County
Graeme collection
Jeff Scovil photo

Bisbee

The end of large-scale mining at Bisbee came some 98 years after the chance discovery of mineralization in Mule Gulch. Not all of the copper had been mined—there are more than two years of mineable reserves still in place—but low copper prices, the crushing costs of labor, and new regulations made continued mining too expensive. Phelps-Dodge did maintain a small workforce to mine limited amounts of relatively high-grade gold and silver ore from the **Shattuck** and **Campbell** mines.

The company also engaged Alan Criddle and Charles Stanley, mineralogists at the British Museum, to unravel the puzzling occurrences of both gold and silver in the massive pyrite shell that surrounded the mined out Campbell orebody. The hope was that a better understanding of the mineralogy would aid in prospecting for additional reserves. Criddle and Stanley identified a broad suite of rare or unusual minerals in specimens from the Campbell. Two species new to science and several unknowns were found along with these minerals (Alan Criddle pers. comm.). The first new species to be described was henryite, a copper-silver telluride (Criddle et al. 1983). It was followed by kiddcreekite, a copper-tin-tungsten sulfide for which Bisbee is a co-type locality (Harris et al. 1984). Independently, Sid Williams studied numerous samples from the Campbell orebody as well as a suite of sulfides from the 1300 level of the **Cole Mine**. Interestingly, Williams found many of the same species in the Cole material that had been recognized by Criddle and Stanley in the Campbell ores.

Below: Connellite and cuprite
Field of view 3.5 cm high
5 level, Southwest Mine
Bisbee, Cochise County
Graeme collection
Jeff Scovil photo

During the first dozen or so years after mining stopped, a few former miners continued entering the abandoned mine workings that surround the town. At this time, limited areas of the **Holbrook**, **Gardner**, **Higgins**, and **Southwest** mines were still somewhat accessible. To be sure, entering such areas where the supporting timber was weakened by rot was both foolhardy and risky. Nonetheless these men, like generations of miners before, took the chance, hoping to find something that may have been overlooked.

From the late 1940s numerous aurichalcite specimens were collected in the **Southwest Mine** (Zeitner 1953). Collecting continued in several areas, and some exceptional specimens were recovered. Also during the late 1970s a number of very fine calcite specimens were found in the upper levels of the mine (Graeme 1981). Then, as noted by Wilson (1987), there was the fortuitous discovery of a small number of fine cuprite specimens along with other, often rare species, in this same mine. These cuprite specimens were among the best recovered from Bisbee. One exceptional crystal is 5 centimeters on an edge! Fine examples of atacamite and paratacamite were also found as were a few extraordinary connellite specimens and some impressive examples of claringbullite as tabular crystals to 6 millimeters.

In the early 1990s a few local residents began the dangerous practice of entering partially collapsed workings of the **Holbrook Mine** through the open pit. While interesting calcite specimens were recovered, the liability was simply too much for Phelps-Dodge, which initiated an exhaustive program to seal any and all mine openings. Since that time, the mines have remained sealed and the open pit closely guarded to prevent entry.

Right: Spangolite, atacamite and claringbullite, 4.5 cm wide
5 level, Southwest Mine
Bisbee, Cochise County
Graeme collection
Jeff Scovil photo

Right: This ASTER image uses shortwave infrared bands to highlight in bright pink the altered rocks associated with copper mineralization in the Morenci pit .

The image covers an area of 21 by 16.9 kilometers. It was taken on July 14, 2007, and is centered near 33.1 degrees north latitude, 109.5 degrees west longitude.

NASA/GSFC/METI/ERSDAC/JAROS and the US-Japan ASTER Science Team

Left: Azurite and malachite, 5.5 cm wide
4650 bench, Copper Mountain
Morenci, Greenlee County
Collected in 1986 by Stan Esbenshade and Bob Jackson

Stan Esbenshade collection
Jesse La Plante photo

Clifton-Morenci Mines

with contributions by Stan Esbenshade, Mark Kielbaso and Brian Huntsman

While Bisbee's mines were winding down, Phelps-Dodge's Morenci operation continued to exploit the great copper porphyry that it had been working since the late 1800s. Most of the specimen-rich oxidized zones had been depleted early in the district's history (see page 56). The underground operations were consolidated in the 1920s, closed for a while during the Great Depression, and then reopened as an open-pit operation in 1937. Phelps-Dodge knew in the 1920s that the town of Morenci was sitting on more low-grade porphyry, and it was just a matter of time before the town would have to be relocated, a process that began in the mid-1960s and ended when the last buildings were vacated in the mid-1980s.

The result of a century of extracting copper from the low-grade porphyry was an enormous pit—one that can be seen today from space, but only a modest number of specimens have been recovered, especially given the volume of earth that has been moved. Still, an operation on the scale of Morenci is bound to encounter zones of colorful, crystallized minerals from time to time. Lindgren (1905) published a landmark paper describing many of the less common minerals found in the district. These include coronadite, willemite, hemimorphite, dioptase, morencite (now recognized as nontronite), libethenite, brochantite, spangolite, and gerhardtite, but collectors will always associate Morenci with the world-class azurite and malachite that has been found since the mid-1970s, when Southwest Mineral Associates (SMA), then owned by brothers Wayne and Loren Thompson, was granted a contract to recover specimens and lapidary materials from the pit.

Left: Azurite and malachite, 5 cm tall
4750 bench, Copper Mountain
Morenci, Greenlee County
Collected in 1982 by Brian Hunstman and Jerry Sullivan

Tiziano Bonisoli collection
Jeff Scovil photo

Right: "Electric blue" azurite and malachite, 7.6 cm wide
4700 bench, Copper Mountain
Morenci, Greenlee County
Collected in 1983 by Brian Huntsman and Jerry Sullivan

Dave Bunk Minerals specimen
Jeff Scovil photo

In 1980, SMA hired Brian Huntsman and Jerry Sullivan to work Morenci. Huntsman and Sullivan, both graduates of the New Mexico School of Mines in Socorro, made a good team. Huntsman had a nose for locating good pockets and cutting material, but liked to roam the benches. Sullivan was "The Man" when it came to digging minerals.

Right out of the gate, the newly installed team located sparkling, drusy azurite along with cutting-quality banded azurite and malachite on the 4750 bench of the Copper Mountain section of the orebody. But in those days, the contract collectors had to stay on their toes as management did little to protect their interests from the miners. Lesson learned: the entire zone of this early find of azurite and malachite that was in the pit when they left at dusk was gone when they returned the next day.

While collecting in 1982 Huntsman spotted balls of float azurite and malachite dribbling off of a bench face near an opening into the old underground Detroit Mine workings—the Copper Mountain area of the mine. They placed orange cones around the area, and then Huntsman ascended a ladder and saw a 5-inch-wide crack dropping vertically into the face of the bench. The open crack extended down more than three feet and was covered with chains of malachite and azurite spheres on a darker

oxidized matrix. The spheres were up to 1.5 centimeters wide. The contrast between the dark blue azurite and the grass-green malachite was spectacular. They removed four or five flats of malachite and azurite chains. The chains separated easily from the matrix, so after they gave Wayne Thompson the flats, he spent considerable time piecing the chains back together.

On one memorable day in 1983 on the 4700 level of the Copper Mountain area, Huntsman's heart started pounding when he saw unusually brilliant reflections from azurite in a crack in a large boulder where Sullivan was working. Huntsman fished out a large plate scattered with "electric" blue azurite and grass-green malachite from the crack. The material was unique and beautiful.

It was getting dark, and the agreement with Phelps-Dodge stipulated that they leave the pit before sunset. They sprayed the boulder with orange paint, and Huntsman asked the shovel operator to move it out of the way. They were back early the next morning to find that the boulder had been moved, according to the shovel operator, "to 26 Dump." Huntsman and Sullivan and subsequent collectors searched the dump and gathered what they could—3 flats of top specimens, 10 flats total of the famous find.

For a long time, an abandoned telephone pole marked the site on

Modoc Mountain that became known among collectors as the Crystal Zone. In late 1984 in that area on the 4600 level bench, Huntsman found small pockets of lustrous azurite on gibbsite. But it was windy and cold, and a nearby working shovel was causing the face to constantly collapse. Frustrated, Huntsman left the area. Jerry went back to investigate a few days later when the shovel was being repaired. He found a large rock at the base of the face on the bench and cracked it open to expose a lustrous, 5-centimeter-wide single azurite crystal in a vug of snow-white gibbsite in limonite (pictured left). Huntsman took the specimen to Richard Bideaux, who had become a partner in Southwest Mineral Associates. He bought everyone's share in the specimen for $2,600.

Azurite on gibbsite, 4.5 cm tall
"Crystal Zone," 4600 bench, Modoc Mountain, Morenci, Greenlee County
Collected in 1984 by Brian Huntsman and Jerry Sullivan

Arizona-Sonora Desert Museum collection
Jeff Scovil photo

In October of 1984, Stan Esbenshade started working at Morenci, and shortly thereafter, Bideaux put him in charge of Southwest's specimen and decorator materials recovery operation. In his first few years at the mine, Esbenshade produced plenty of commercial and decorator-grade azurite for SMA but no notable specimens.

Then in 1986 after the Tucson show, Esbenshade began following a layer of color into the bench. It got thicker and better until it opened into a 90-by-50-by-12-centimeter pocket of azurite and malachite stalactites. He had invited collector Bob Jackson to join him at the mine, but Jackson couldn't make the trip. Esbenshade couldn't wait. He opened the pocket to find dozens of long, thin stalactites. The largest was roughly 20 centimeters long and half the diameter of a pencil. Esbenshade pulled a few of the crystals out and then began to enlarge the opening.

When Jackson arrived a day later, he and Esbenshade cleaned out the pocket together. With the exception of a large, shallow, oblong bowl-shaped piece at the bottom of the pocket, they had the pocket contents into flats by afternoon. The bowl in the bottom had a large stalactite on one side and two smaller ones on the other. They removed the "birdbath" and exposed a second chamber just below the first.

The stalactites in the lower pocket were even larger than those in the first, but they were not the smooth botryoidal type like those in the first pocket. Esbenshade and Jackson were obviously excited, but they had to leave the pit for the night. They covered the pocket, and the next day they asked the turquoise contractor to remove 5 feet of the 8 or so feet of overburden so that they could better access the pocket. Hundreds of stalactites and sections littered the bottom, and many more were still attached with columns running from floor to ceiling. Over time, the changing water table had left shelves and rims along the floor of the pocket.

Unfortunately, however, the pocket was only about 6 feet from the bench above and the one in front, so the contents had been severely damaged by heavy blasting. Only a few undamaged specimens were recovered; the remainder would have to be pieced back together. Jackson and Esbenshade spent three days collecting the contents of the lower pocket. They kept everything, made notes, and studied the pocket in anticipation of a reconstruction project.

Back in Phoenix, the sight of Esbenshade's finds in the back of his pickup, spurred Thompson (who had turned his attention to his Indian

Right: Azurite and malachite stalactite repaired, 11.5 cm tall
4650 bench, Copper Mountain
Morenci, Greenlee County
Collected in 1986 by Stan Esbenshade and Bob Jackson

Stan Esbenshade collection
Jesse La Plante photo

Below: Stan Esbenshade loading lapidary material into his pickup at the Morenci Mine in the 1990s

pottery business) to take an active role in SMA. Esbenshade and Thompson reconstructed the pocket over the next three days.

Four months or so later, Esbenshade recovered a smaller pocket of material similar to that from the lower chamber. The area continued to produce good specimens for another 18 months before Phelps-Dodge mined it. A long dry spell followed Esbenshade's 1986 finds as Phelps-Dodge was not active in the specimen-producing zones of the mine.

Tucson-based Zee Haag became a partner in Southwest in 1989. Haag rented an excavator that enabled the SMA team to move much more material in promising areas than they could with hand tools. The newly empowered collectors quickly encountered a zone of high-grade lapidary azurite and malachite—Haag's primary interest. SMA worked this area, which was referred to as the Azurite Pit, off and on for the next dozen years. It produced tens of tons of lapidary, decorator, and landscaping material, as well as some specimens.

In spite of the interest in and the value of good mineral specimens, they

were a minor part of SMA's recovery project. Contrary to what many collectors think, good mineral specimens are not constantly encountered in active mines. Months or even years frequently pass between good finds. It is commonly the lapidary and decorator material that supports the cost of the heavy equipment, fuel, employees, and other expenses incurred by a contract-mining effort.

A number of people participated in SMA specimen and lapidary recovery projects at Morenci over the next 15 years. Many collectors logged significant hours, among them: Terry Glockner, Mike Flaugherty, Mark Kielbaso, Jay Rosenbauer, Wayne Chartier, James Dunn, and Betsy McKintrik. Other collectors, such as Tony Potucek, Les Presmyk, and Ken Roberts, made cameo appearances from time to time.

In 1989 Esbenshade, Thompson, and Haag set their sights on the Crystal Zone. Their target was a small contact zone of hard limonitic rock pocked with oval-shaped vugs on the 4650 bench level just above the spot where Huntsman and Sullivan had their memorable find in 1984. Esbenshade and his crew went to work and found impressive azurite crystals on white matrix. Although they averaged about 2.5 centimeters in length, none matched the size of the one that Bideaux now had in his collection. A number of the crystals were partly or completely pseudomorphed to malachite. The Crystal Zone produced specimens into the 1990s.

In the course of mining, Phelps-Dodge had cut the working face of the

Above: Portions of the "Stalactite pocket" as it was being pieced together in Wayne Thompson's home in Phoenix

Stan Esbenshade photo, Feb. 1986

Facing Page:, upper left: Azurite on gibbsite, 1.9 cm tall
"Crystal Zone," 4600 bench, Modoc Mountain, Morenci, Greenlee County
Collected in the late 1980s

Roland Sherman collection
Jeff Scovil photo

Facing page, upper right: Azurite on gibbsite, 1.6 cm tall
"Crystal Zone," 4600 bench, Modoc Mountain, Morenci, Greenlee County
Collected in the late 1980s

Les and Paula Presmyk collection
Jeff Scovil photo

Facing page, lower: Esbenshade's late dog Fungus poses next to the Crystal Zone vein at Morenci

4650 bench perpendicular to the zone, which as a result could be worked from both above and below. Esbenshade and his crew could not get their heavy equipment into the area as it was blocked by the old telephone pole.

On Father's Day weekend of that year Esbenshade invited Tony Potucek and Les Presmyk to the mine to help excavate a pocket. They recovered a number of nice specimens: attractive arborescent copper specimens covered with azurite; bright foils and blebs of native silver on brown, waxy limonite; and dull red cuprite nodules. Several years later, Phelps-Dodge cut a haul road behind this area and uncovered a few more large azurite crystals on white matrix.

In the late 1980s, Phelps-Dodge started the Northwest Extension project. SMA was anticipating access to fresh ground, but drilling in the Northwest Extension had so far shown no azurite and malachite, only stains of brochantite and chrysocolla. Fortunately, however, the Northwest Extension proved to be far richer than the drilling had indicated, and azurite and malachite were discovered in abundance. Over the next eight years, Esbenshade, Kielbaso, Dunn, Chartier, and others worked several hundred zones and veins in the Northwest Extension.

The Las Terrazas Fault in the Northwest Extension periodically produced collectible material. In December 1991 Mark Kielbaso discovered a thick zone of clay and azurite on the plane of the fault that usually produced specimens and cutting material. This 150-by-150-foot area became

While prolific in terms of lapidary material, the area where the Las Terrazas Fault cuts the Northwest Extension orebody (known as the Pigment Zone), did not produce many specimens. This 11.9 cm tall azurite and malachite stalactite is a remarkable exception.

Les and Paula Presmyk collection
Jeff Scovil photo

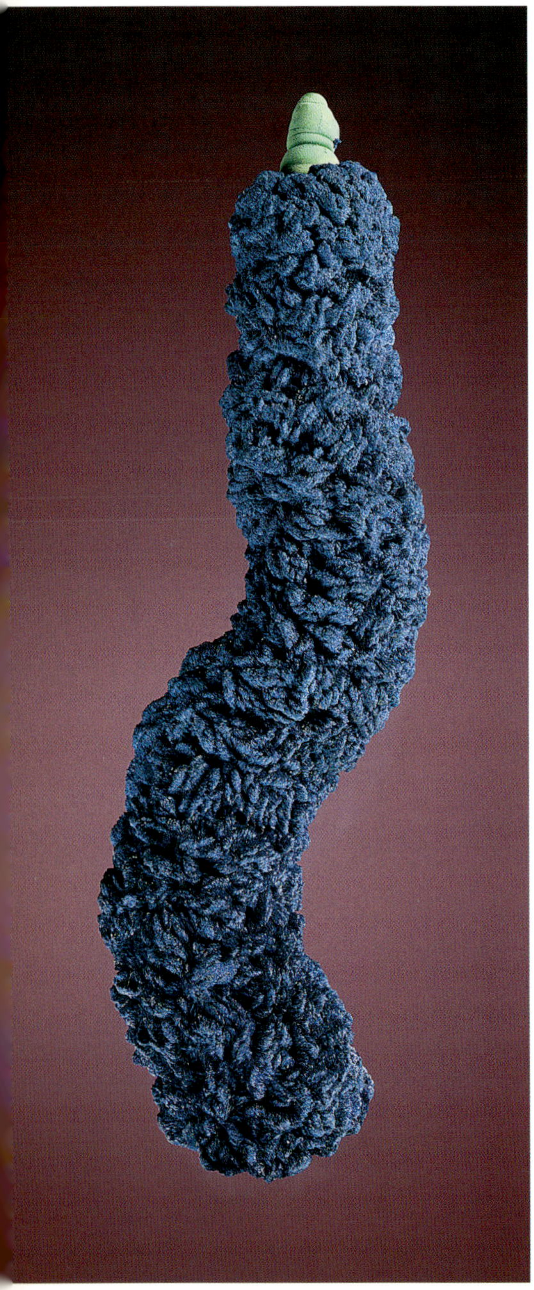

known as the Pigment Zone. SMA worked the exposures in this zone every time the Phelps-Dodge mine plan offered an opportunity. As one bench was mined out, Esbenshade and Kielbaso would work in the floor of the bench until a new blast provided a new exposure. The Pigment Zone produced around 30 tons of pure azurite, mostly in the form of slabs and plates 2.5 to 7.5 centimeters thick. Stalactitic azurite specimens were also found, some with malachite druses.

Kielbaso, James Dunn, and Betsy McKintrik fanned out across the bench after a blast to see whether they could find any color. Dunn was the first to find anything. He picked up a chunk of clay-covered rock that resembled a brick, but a nick in one corner showed blue. Dunn called McKintrick and Kielbaso. They broke the chunk in half, and it was solid azurite—a two-pound chunk of solid azurite! They had all heard or read stories of veins of solid azurite at Morenci, but that was a hundred years ago. Now here they were holding a huge chunk!

They started breaking brown lumps and checking the heft of pieces (heft is a key indicator as to whether a chunk is azurite or just rock). Within 30 minutes, they had accumulated close to five hundred pounds of solid azurite and had narrowed the debris field. Kielbaso went to the bench above to retrieve "Kitty," SMA's Caterpillar 25-ton excavator.

He fired up the excavator and headed to the azurite zone, where he cut parallel trenches into the bench to locate the source of the azurite. Within about 20 minutes, the bucket was raining blue. There must have been 200 pounds of solid azurite in the bucket. Kielbaso pulled dipper load after dipper load of azurite out of the trench. Meanwhile Dunn and McKintrick filled two 55-gallon drums that they had on the truck, dumped them at the SMA yard, and filled them up again.

Dunn returned with the empty drums, and they had begun to load the truck when the shift foreman for the Northwest Extension came by. He told them that would have to vacate the area as the shovel on the bench below was going to be digging in the area just below them. Having just hit the mother lode, they were not all that excited about moving on. Furthermore, according to Kielbaso, they "were already contemplating the new Harley-Davidsons that we would be riding with our SMA bonuses."

Kielbaso cut a shallow ramp into the azurite-producing hole. He put the excavator into the hole, out of sight. Dunn went to the front of the bench and watched for the Phelps-Dodge shovel below, as they were somewhat concerned that "Kitty" might be scooped into a haul truck. Every few minutes, the arm of the excavator popped over the bank to drop a load of azurite. This went on for about forty-five minutes until the sun began to set and they had to start winding down for the evening. They had pulled almost five tons of azurite from the bench in two or three hours, and there was no end in sight. Azurite plates up to two feet square and four inches thick lined both the hanging wall and footwall of the vein, with up to six inches of red mud in between.

After the shift foreman agreed to flag the area so that it would not be mined and hauled off, Kielbaso, Dunn, and McKintrick packed up and headed for the mine gate. When they returned in the morning, they found

Azurite on malachite, 8.5 cm wide
4850 bench, Northwest Extension,
Morenci, Greenlee County

Les and Paula Presmyk collection
Jeff Scovil photo

that the azurite vein had been mined overnight. As disappointed as they were, they continued to collect what was left of the "Blue Hole." When all was said and done, they had collected close to ten tons of solid azurite over three days. And for the next two weeks, they continued collecting the zone into the floor of the adjacent bench.

SMA thought that the azurite would make excellent pigment material for the paint industry, but it was tenorite-rich and created too dark a color. Most of the Blue Hole material instead wound up instead at various cutting factories in China where it was turned into carvings.

Another terrific find occurred in November 1992; Esbenshade found azurite crystals in a boulder on the Northwest Extension. He identified their source on the nearby 4850 bench. Esbenshade located Kielbaso, and they began to work on the fracture, which produced well-defined large bladed azurites as singles and clusters. Some of the azurites were altered to darker green, lustrous, crystallized malachite. There were many loose crystals and clusters, some of which were attached to matrix. Because the bench was not a working face, Esbenshade and Kielbaso took a luxurious two days to carefully collect the pocket—a welcome departure from the grab-what-you-can method they so often had to employ. They mostly collected plates of malachite after azurite with overgrowths of azurite. The largest crystals were 4 centimeters with plates up to 25 centimeters across. They were the best azurite and malachite-after-azurite specimens to come from the Northwest Extension orebody.

Postmodern-Era Collectors

Dedicated collectors such as Dick Morris, Mark Hay, Neil Prenn, Brad Archer, Pete Knudsen, Tony Potucek, Tom Hughes, George Godas, Les Presmyk, Graham Sutton, Stan Esbenshade, Andy Clark, and David Shannon, carried on in the tradition of Dick Bideaux, Gene Schlepp, and Dick Jones by fanning out across the state in search of mineral specimens. It is fair to say that these Arizona "institutions" visited every mine in the state during this period. They worked long abandoned mines such as the Grey Horse (page 298), used metal detectors to make discoveries such as of the Sunshine #8 Claim in the Bradshaw Mountains, and found erratic boulders that yielded one-of-a-kind specimens!

The stories from this period are unique in that they are fresh in the memories of the collectors, most of whom are still active in the hobby. Their reminiscences are powerful reminders that mineral collecting is an adventure! It is difficult to find a collecting story that does not mention an encounter with Arizona's silent sentinels—the diamondback rattlesnake.

Thirteen of the thirty-six known rattlesnake species are found in Arizona, which is home to more of those species than any other state. The Western Diamond-backed Rattlesnake (*Crotalus atrox*) pictured **above** is the largest rattlesnake in the western United States. The Western Diamond-backed bites more humans than any other rattlesnake in the nation. But collectors can take comfort in the knowledge that only about 1 percent of rattlesnake bites result in death.

Dean Hueber photo

Along with passionate "amateurs," a cadre of professional collectors such as John Mediz, Wayne Thompson, and Jimmy Vacek, acquired contracts to collect minerals and lapidary materials. Often these deals were struck with owners of the state's operating mines such as Ajo, Ray (pages 205 and 295), and Morenci (page 353). But contract collectors also worked idle properties. Localities such as the Grand Reef Mine (pages 182 and 264); the 79 Mine (pages 193 and 300); and the Apache Mine (page 302) heard drill compressors or bulldozers in the pursuit of linarite, vanadinite, or wulfenite. The largest and most ambitious of these projects—an open pit at the Red Cloud—crossed into the twenty-first century.

It is difficult to do this period justice; many people visited many mines. We visit a few of the highlights, beginning here with remote Graham County, and travel through 10 more of Arizona's 15 counties to give a glimpse of this era, its finds, and its collectors.

Graham County

with contributions by Jimmy Vacek, Bobby Lane, Andy Clark, and George Godas

The Aravaipa mining district, south of Stanley Butte in western Graham County, covers the eastern entrance to the scenic Aravaipa Canyon. The remoteness of the district has long hindered collectors targeting the area's significant and unusual mineral deposits.

For a short time in the late 1920s, the district's Grand Reef Mine was Arizona's second largest lead producer and the most important mine in the district. The **Iron Cap** was second; in fact, these two were the only mines in the district rich enough to support sustained mining. The Iron Cap worked a lead-zinc skarn deposit in Paleozoic limestone (Reiter 1981).

Arizona's mines generally do not produce good crystallized sulfide minerals, but the Iron Cap mine is an exception. In the winter of 1986-1987, Jimmy and Joyce Vacek accompanied their friend Doug Brown to the Iron Cap Mine to collect quartz. When they arrived, Brown scurried under the stress-cracked 12-by-12-inch beam. Jimmy cautiously followed while

Sphalerite and johannsenite
6.2 cm wide
Iron Cap Mine, Graham County
Part of Bobby Lane and Bob Johnson's first 1987 find
Mark Hay collection
Jeff Scovil photo

Joyce elected to remain outside. Brown immediately began whaling on a wall containing pockets of pagoda-type quartz crystals. This was arduous labor and other drifts beckoned to Vacek, who soon stumbled upon a labyrinthine stope that had droplets of water coming down from the ceiling. Vacek sat on a large boulder on the edge of which he saw facet-grade, green sphalerite crystals and modified galena cubes! Vacek was stoked!

He sent Bobby Lane and Bob Johnson back to the mine for a more serious look around. On their first trip in, Johnson and Lane had the sobering realization that the block that Vacek had first found had fallen from the back of the stope. The mine was obviously dangerous. Cave-ins were everywhere. The collectors had to climb through broken timbers and rock falls, and when they reached the main stope, they could see that numerous large boulders, now on the floor, had obviously come from the roof of the stope. Nevertheless, they collected many flats including some very nice specimens from small pockets in many of these boulders.

On one trip, Lane and Johnson began digging in the muck at the base of a boulder about the size of a pickup truck. They followed a trail of specimens under the boulder. About three feet back, a large galena crystal came out, then another and another. They were huge crystals. One of the specimens that Johnson kept later won the Best Arizona Specimen plaque at the Tucson Gem and Mineral Show.

Collecting began in earnest after Vacek's discovery, and a number of excursions into the mine netted exceptional specimens. George Godas was with Bill Hawes when Hawes collected what is probably the best galena pocket from the Iron Cap. Hawes had been working on a promising boulder

Galena and sphalerite, 5.4 cm tall
Iron Cap Mine, Graham County
Les and Paula Presmyk collection
Jeff Scovil photo

Les Presmyk working a ledge in the Grand Reef "glory hole" in August 1991
Tony Potucek photo

all day, and he was getting frustrated. As they were packing to leave, Hawes decided to take one last whack at it. He hit it and knocked a chunk off to expose a large pocket. A fine miniature of galena and sphalerite slid out of the pocket and (luckily) onto a sandy cushion on the floor.

On one trip, Godas and Graham Sutton were working on the surface and set off a blast above the stope. Sutton planned to set off an entire 50-pound bag of ANFO (ammonium nitrate and fuel oil), and Godas moved his truck in anticipation of the blast. The blast sent a grapefruit-sized rock into an arc and through the closed window of Godas's truck.

Many of the specimens from the Iron Cap are coated with minor quartz or other less attractive minerals. The sphalerite generally occurs on a matrix encrusted with gray johannsenite crystals to 3 millimeters. The smallest crystals are generally transparent and green, but as crystals increase in size, they appear progressively darker. Three-centimeter-wide crystals are black and opaque. Galena crystals up to 2 centimeters across occur in association with sphalerite or on quartz. They generally have a matte luster although the very best exhibit smooth faces and medium luster.

The Iron Cap has a unique assemblage of manganese minerals for Arizona: axinite-(Mn), manganbabingtonite, johannsenite, neotocite, and kutnohorite. Quartz, calcite, and nekoite are also associated with galena and sphalerite from the mine. The quartz forms white crystals to 4 centimeters long that frequently have stepped faces reminiscent of pagodas. Quartz occurs on johannsenite and may be found with white calcite crystals in the 3 to 6 centimeter range.

Also in the Aravaipa mining district, the **Grand Reef Mine** (also see page 182) was popular with collectors for much of the last quarter of the twentieth century. That popularity was rooted in the mine's potential to yield great linarite specimens. The fact that collectors could work on the surface without risk of entering decaying underground workings was a bonus. But the mine had better potential than production, as the matrix that hosts the specimens is, as previously mentioned, very tough.

Wayne Thompson's stories of the linarite crystals that got away are cautionary tales. In the late 1970s he took several collectors to the mine and spent a couple weeks of mining there. The best single specimen they found was a crystal nearly 5 centimeters long that they discovered as they were peeling rocks off of an exposure in the silicified breccias at the top of the glory hole. Of course this crystal stayed on the rock face. Thompson left the group with the order that they were not to touch the piece until he returned with glue and other materials to protect the crystal and keep it in the matrix while they removed it from the wall. As Thompson walked back up the hill, one of the other collectors met him with the crystal in his hand, not only was it off the matrix, but it was in two pieces.

On a later visit Thompson made a discovery that elevated the mineralogical importance of the Grand Reef Mine. While looking for linarite, he encountered a vug that contained tuffs of 1.5-centimeter-long, white, translucent, acicular crystals of a mineral that he did not recognize. Thompson showed the specimen to Les Presmyk who bought the specimen and took it to the University of Arizona for identification, but the X-ray

Right: Linarite on quartz
3 cm wide
Grand Reef Mine, Graham County
Gail and Jim Spann collection
ex Ernie Schlichter
Jeff Scovil photo

Above: Laurelite
Field of view 5 cm high
Grand Reef Mine, Graham County
Les and Paula Presmyk collection
Jeff Scovil photo

analysis only proved it was an "unknown and likely new mineral." In the meantime, collector Bill Besse acquired another piece of the same vug. Besse, who lived in California, took the specimen to Tony Kampf at the Los Angeles County Museum of Natural History for identification. In 1989 Kampf described four new lead fluoride minerals from that one vug from the Grand Reef: laurelite ($Pb(F,Cl,OH)_2$), grandreefite ($Pb_2(SO_4)F_2$), pseudograndreefite ($Pb_6(SO_4)F_{10}$), and aravaipaite ($Pb_3AlF_9 \cdot H_2O$). These minerals all occur exclusively at the Grand Reef Mine and are among the most difficult Arizona species to obtain.

In 1994 Graham Sutton approached Presmyk about starting a linarite-mining operation. Sutton negotiated a lease with owner George Badger of Salt Lake City, Utah. They worked for a weekend; then Presmyk went back to his day job leaving Sutton to continue. Two days later, Sutton had hit a small pocket of laurelite. Presmyk returned to the mine the following weekend with Tony Potucek, and they continued drilling and blasting. They collected several flats of laurelite and three or four aravaipaite specimens, but they did not find any grandreefite or pseudograndreefite. When later asked where they might find grandreefite and pseudograndreefite, Kampf observed that the rarer minerals might be found if one removed all of the laurelite from a pocket. None of the collectors was ever that curious.

The Grand Reef has gained a reputation for offering some harrowing collecting. In the 1980s, Sutton along with Bill Hawes and "Ranger" Rick Bernard were benching an area above the opening to the large underground stope on the west side of the vertical reef. Hawes was working above Sutton

when a large boulder came loose, hit Sutton, and sent him over the edge onto the boulders below. He was medivacked off the reef with numerous broken bones. Sutton's rehabilitation included a metal contraption anchored to his foot and lower leg that made friends cringe every time they saw it.

Many other collectors tried their hand at the Grand Reef; some had modest success. Others such as Dave Riley, Dave Penney, Ray Grant, and Dave Shannon spent considerable time at the locality, but even those experienced collectors struggled to recover unscathed linarite crystals.

Graham and Presmyk invited a small group of collectors including John Lucking and Dick Morris to join them on their last effort at the Grand Reef in 1996. They had a twofold plan: to tackle the cerussite knob in the floor of the glory hole and to try to reach the azurite pod in the mine that was visible but out of reach. Sutton widened the adit so that he could get his bulldozer into the stope to push some muck out of the way and expose the azurite pod. Unfortunately when they reached the pod, they found that the intense heat and smoke of a 1967 fire had destroyed the azurite.

Their efforts on the cerussite knob were at least somewhat successful. They drilled holes and filled them with an explosive mixture that broke some of the breccia apart. Presmyk thinks they probably destroyed as many fine cerussite specimens as they recovered, but they did come away with some very good crystals, even a few on matrix. All told, the find produced about 20 flats of lustrous cerussite crystals up to 2.5 centimeters long. Some were tinted blue. And Sutton had filled an order for one ton of galena. The group also collected a flat of ktenasite and some yellow mimetite crystals.

Facing page, lower left: Chrysocolla pseudomorph after cerussite
the crystal is 1 cm wide
Grand Reef Mine, Graham County

Arizona-Sonora Desert Museum collection
Jeff Scovil photo

Facing page, upper left: Cerussite on matrix, 6. cm tall
Grand Reef Mine, Graham County

John Lucking collection
Jeff Scovil photo

Facing page, upper right: Caledonite
The crystal is 5 mm long.
Grand Reef Mine, Graham County

Previously in the collection of the late Dave Shannon
Jeff Scovil photo

Facing page, middle right: Graham Sutton strikes an offensive pose as he passes a Gila Monster on the path into the Grand Reef Mine

Les Presmyk photo

Facing page, lower right: Named for collector David Shannon, shannonite is one of the mineral species for which the Grand Reef Mine is the type locality. The field of view of the photo of this example from the Mesa Southwest Museum collection is 1.1 cm wide.

Jeff Scovil photo

Right: Leadhillite, 2.2 cm wide
Grand Reef Mine, Graham County

University of Arizona collection
Acquired in trade from Dick Bideaux
Jeff Scovil photo

Right: Andradite, 5.7 cm wide
Stanley Butte, Graham County
Evan Jones collection
Jeff Scovil photo

Above: Quartz and andradite
10.5 cm wide
Stanley Butte, Graham County
Evan Jones collection
Jeff Scovil photo

North of the Aravaipa mining district and in the southwest quarter of the San Carlos Apache Reservation, **Stanley Butte** is a 7029-foot landmark of the "Mineral Strip." Since 1990, entry to the Mineral Strip by non-tribal members has been restricted. San Carlos Game & Fish rangers say the remote area was closed because of trouble with poachers, rustlers, and mineral collectors. Parts of the 15½-mile dirt access road have always required four-wheel-drive. By 2006 the washed-out track had sprouted boulders and pinon pines. The area remains largely unexplored.

A couple of miles northeast of Stanley Butte and very much within the boundaries of the Mineral Strip, a skarn holds many varieties of crystallized quartz and garnet. It is the specimens from these inclined beds of limestone and garnet that are said to be from Stanley Butte. Reservation signs and maps warn that this is now a "closed area." Crystal treasure lies beyond the signs, beneath the brush, and up the slopes.

All across the four-mile swath of skarn, float weathers from surface pockets. Icy-clear, frosty, or amethystine quartz and black, brown, or green andradite garnets can be chased uphill to open pockets. North of the Stanley town site, abundant float led collectors to the pockets of sharp-edged garnets at the Grey Throne. Lustrous, almost black, root-beer-colored garnets sparkle on the slopes and in the gullies of Black Crystal Canyon. Miners followed clusters of amethyst to the Billingsley prospect, and for decades, amethyst rough, and loose crystals of quartz and garnet have been collected from the pockets of Crystal Peak.

In 1871 by presidential executive order, the San Carlos Reservation was created to be the new home of the San Carlos Apache people. In 1896, realizing that great mineral and timber wealth had been given away, the government re-opened 232,000 acres of the reservation to all Americans for "…occupation, location, and purchase." Land status of this "Mineral Strip" was further clouded in 1908 when borders of the new Crook National Forest were drawn over a portion of the reservation. In 1963 ownership of the Mineral Strip was restored to the San Carlos Tribe. Non-Indian homestead owners and ranchers were to be bought out and forced to leave. By February of 1983, all ranches had been abandoned. The Forest Service, however, held that mining claims were still valid in the overlap of National Forest. In the late 1980s a federal court ruled that the Mineral Strip would be administered by the Department of the Interior (Bureau of Indian Affairs). Mining claims were declared null and void.

In October of 1990, President Bush signed the San Carlos Mineral Strip Act, which transferred all rights to the tribe. As of January 2011, the Mineral Strip was still closed. Like San Carlos olivine (variety peridot), the resources of Crystal Peak belong to future generations of the Apache people. Although a few attempts have been made to continue to explore the area, or access it from the Klondyke and Aravaipa side of the mountains, very little collecting has been done in the Stanley Butte area since 1990.

Quartz and andradite
6.1 cm wide
Stanley Butte, Graham County
Paul Harter collection
Jeff Scovil photo

Right: Wulfenite on quartz, 10.7 cm tall
Toughnut Mine, Tombstone
Cochise County
Collected in 1988 by George Godas and David Riley

Carole Lee collection
Jeff Scovil photo

Above: Shootout on the streets of modern-day Tombstone

Dwight Hanson photo

Cochise County

with contributions by Peter Megaw, Mark Hay, Ryan DeWitt, and George Godas

In 1881 the Arizona Territorial Legislature created Cochise County by slicing off the eastern third of Pima County. The move was made because of the rapid increase in population caused by the great silver rush to Tombstone, which was the original county seat. The county is named for the famous Apache chief who made his home in the Dragoon Mountains, a rugged range near the town of Wilcox; in fact, before 1878 the area that would become Cochise county was essentially off limits to prospectors and ranchers. This all changed when Ed Scheffelin discovered the vein that would become the Toughnut Mine and started the Tombstone stampede. In 1929 the county seat was moved to the king of mining camps, Bisbee.

The mines of **Tombstone** (see page 39) lay dormant for decades, and tourism became the town's primary industry. Boot Hill Cemetery and many of the old buildings were rebuilt. The town has became a kind of living museum with stagecoach tours and shootouts staged in the streets.

Mining had all but vanished until 1987 when Peter Megaw was contracted to map and sample the underground workings of the **Empire Mine**. Megaw is a collector, so he was on the lookout for minerals while underground. He and friend Jim Walker recovered a number of rare species, including dark green, 3-millimeter balls of dugganite; khinite crystals; yafsoanite crystals to 1 millimeter; girdite; and the world's best utahite—crystals to 4 millimeters!

Later that year Megaw was underground again, this time with Duncan Reisemeier. With dozens of mines, many of them interconnected, the Tombstone underground is a maze of workings. Megaw was exploring the workings when he found himself face-to-face with a 12-foot-high wall that was completely covered with wulfenite crystals! Megaw called Walker, and the two spent a week and collected 80 flats of wulfenite and mimetite. The find consisted of lustrous, square wulfenite crystals up to 2 centimeters wide that are bright honey-yellow to orange, some verging on red. The mimetite formed as green to red spheres to 5 millimeters and as powdery yellow dustings on matrix.

Megaw and Walker also collected about 8 flats of pale gray-green to purple fluorite cubes to 2 centimeters across. The edges and corners of the crystals are rounded, and some are associated with microscopic yellow-orange wulfenite crystals.

A year or two later, Dave Riley and George Godas found the stope and began harvesting wulfenite. Mark Hay joined them on a subsequent trip after which Godas, Hay, and collector Dick Morris made numerous nighttime mining trips and collected 30 or 40 flats of wulfenite.

The wulfenite area is on about the 450-foot level of the **Toughnut Mine** and getting down there was quite an ordeal. Hay remembers their

In January 1989, 450 feet below the streets of Tombstone, (from left to right) Dick Morris, George Godas, Mark Hay, and Tony Potucek explored the underground workings of the Toughnut Mine in Tombstone.

going "down the decline, across tunnels, down vertical shafts and through stopes to access other shafts that accessed other tunnels. Then at the end, there was a 20- or 30-foot vertical shaft that bottomed out in a rabbit warren of crawl holes and twisting tunnels that was the wulfenite area." He is still amazed that Godas and Riley found the place. In the tunnel at the top of this shaft there was no sign of wulfenite. But at the bottom, in the tangle of tunnels and crawl spaces, wulfenite was everywhere.

During their Tombstone excursions, the group made a number of interesting finds. At one point Godas and Riley discovered a 4-by-6-foot, carrot-shaped mass of fine-grained, etched, grey-white quartz with brilliant, orange-yellow wulfenite blades scattered across it. On another trip, Hay got lucky and uncovered an old, partly buried pocket that extended back into the wall about 10 centimeters. The sides were carpeted with green-brown velvety mimetite spheres to 3 millimeters in diameter with brilliant, yellow to root beer-color wulfenite blades to about 1 centimeter. They were among the most aesthetic pieces that the group collected; unfortunately they only recovered a half-dozen specimens.

Some collectors speculate that these specimens came from the Empire rather than the Toughnut Mine and that Hay, Godas, and Morris had crossed from one mine into the other somewhere underground. They know, however, that the shaft they went down was the Toughnut, they therefore label their specimens "Toughnut Mine" as well.

Hay recalls one visit when he and Godas were "taking five" in the tunnel and noticed wulfenite crystals in a pile of "backfill" that was in a short tunnel connecting two stopes. The fill consisted of fist size cobbles of rock in piles to almost face height on both sides of the tunnel. Piles of cobbles are not unusual in old mines, as the miners didn't want to move waste rock any farther than they had to.

Wulfenite and mimetite, 5.7 cm wide
Toughnut Mine, Tombstone
Cochise County
Collected in December 1988 by Mark Hay

Mark Hay collection
Jeff Scovil photo

Quartz, 4.5 cm tall
Jack Wakefield Mine, Cave Canyon
Huachuca Mountains, Cochise County
Mark Hay collection
Jeff Scovil photo

On closer inspection, however, they noticed colorless crystals in the voids between the rocks. To their shock, they found that the crystals were attached to the stones and in many cases, spanned two or more stones. The wulfenite crystals were in place! The cobbles were not backfill but an unusual natural formation through which the tunnel had been driven.

As in many old mining towns around Arizona, the "Town Too Tough to Die" has entered an age that is for whatever reason averse to allowing mines to remain accessible. The openings have thus been plugged or secured with bat gates, and collecting is not allowed.

The bulk of Tombstone's water comes from springs in the **Huachuca Mountains**, some 25 miles to the southwest. The mountains are also a source for a variety of quartz crystals, including clear, Japan-law twins associated with calcite. Collector Andy Anderson hit a bonanza pocket in the Cave Canyon area of these mountains in 1979. The **Jack Wakefield Mine** produces clear, lustrous groups of quartz crystals. Numerous collectors have worked this mine and found crystals exceeding 6 centimeters, some of which are doubly terminated. Word of the finds began to disseminate so that by the 1990s, a number of active field collectors were tapping the remote canyons. Ryan Dewitt, who specializes in quartz, was among them.

Quartz pockets are not common in the Huachucas, but when they are found, they can be large, producing more than one hundred points up to 25 centimeters long, some smoky and some clear.

Shortly after 2000 Dewitt began collecting in the southern canyons, Miller Canyon in particular. He has since observed quartz pockets as large as 6 feet across, but they are normally closer to a foot in diameter. The pockets are commonly collapsed and contain light brown clay, which keeps the crystals from contacting one another; hence, many specimens are recovered undamaged. Dewitt's best finds in the Huachucas are of lustrous, clear smoky singles and groups up to 30 centimeters.

Like those in Tombstone to the southwest, the mines in the Turquoise mining district around the ghost towns of Courtland, Pearce, and Gleeson had not seen commercial mining activity in decades. The Defiance (see page 170) and Silver Bill (see facing page) near Gleeson were regular collector haunts, but the **Maid of Sunshine Mine** near Courtland was virtually unknown. Then in January 1981 while searching for azurite in the area, Gary Novak and Neal Pfaff found cyanotrichite in the "Maid." Pfaff figured that they were the first to collect in the old mine because the material was relatively untouched, but other collectors such as Dick Morris later reported having poked around in the mine in the 1970s.

Dave Shannon worked the Maid of Sunshine for specimens for a couple of years in the early 1980s. He brought in a Cobra drill and explosives and recovered a small lot of calcite crystals in addition to a large number of flats of cyanotrichite. The rock is hard and the specimens of dark blue crystals up to one centimeter long that Shannon collected tended to be small, generally ranging from 2 to 5 centimeters across. Novak and Pfaff continued to make periodic trips to the mine, coming away with cyanotrichite and the occasional exceptional azurite specimen.

Right: Smithsonite and rosasite, 5 cm wide
Silver Bill Mine
Near Gleeson, Cochise County
Evan Jones collection
Jeff Scovil photo

Facing page, upper: Azurite, 6 cm tall
Maid of Sunshine Mine
Courtland, Cochise County
Les and Paula Presmyk collection
Jeff Scovil photo

Facing page, lower: Cyanotrichite
8.8 cm wide
Maid of Sunshine Mine
Courtland, Cochise County
Les and Paula Presmyk collection
Jeff Scovil photo

In the 1970s Richard Bideaux prompted several collectors to explore the mines near Gleeson in Cochise County. A handful of localities are in close proximity to one another, but the **Silver Bill** Mine was the primary target. Tucson-based collector Peter Megaw made his first trip to the Silver Bill with a few friends in the late 1970s. On one of his first probes, he buried the bar to the hilt in a blind pocket. Megaw recovered several flats of isolated, clear yellow and zoned wulfenite crystals on wad from the pocket. Working from Dick Jones's hand-drawn map, Stan Esbenshade and Neal Pfaff had equally good luck collecting a long tube lined with wulfenite on thin plates and in soft wad.

While wulfenite was the mainstay at the Silver Bill, collectors also recovered magnificent rosasite and smithsonite specimens from the mine's Copper stope in the mid- to late 1970s. Accessed by a series of vertical ladders, the Copper stope belled out from a slot driven perpendicular to the manway. The walls of the stope walls were composed primarily of soft, dark wad, with patches of rosasite, malachite, and smithsonite showing here and there. Silvery blue aurichalcite crystals up to 2 centimeters were intermixed with the wad.

The Copper stope attracted collectors as word of its potential and of the specimens recovered began to circulate. In May 1979 Tom Hughes and Pete Knudsen made a trip to the Silver Bill. Knudsen and his partner Rob Jeffreys had been working an area near the decline for silver ore, which at the Silver Bill was found in the form of cerussite. Hughes began working a smithsonite and rosasite zone. He opened a seam in the wad and exposed an

undulating carpet of mammillary rosasite. The blue-green specimens to 16 centimeters across are the best for the species that the mine produced.

In June 1979 while a graduate student at the University of Arizona, Will Wilkinson spent some time collecting in the Copper stope with his brother Jeff. They uncovered a small pocket of malachite after azurite specimens—the best the Silver Bill ever produced—which were lightly attached to the shiny, black, botryoidal manganese and iron oxide that was lining the vug. The Wilkinson brothers collected about a flat of specimens including several nice crystal groups, many of which were thinly coated with aurichalcite. The crystals reach lengths of 2.5 centimeters and the groups are up to 10 centimeters wide.

The Silver Bill and nearby Defiance Mine (see page 170) are privately owned. They were closed in the early 1980s, and mineral collecting is no longer allowed. Although in the early 1990s, mine owner Scott Lewis mounted a commercial operation to recover wulfenite from the Defiance. Over an 18-month period, a solution cavity that he had exposed yielded numerous, thin bladed wulfenite crystals.

Facing page, upper: Rosasite, 8.2 cm wide
Silver Bill Mine
Near Gleeson, Cochise County

Evan Jones collection
Jeff Scovil photo

Facing page, lower left: Neil Prenn in the Silver Bill in the late 1970s
Pete Knudsen photo

Facing page, lower right: Wulfenite with smithsonite, 2.5 cm wide
Copper Stope, Silver Bill Mine
Near Gleeson, Cochise County
Collected by Tom Hughes

Tom Hughes collection
Jesse La Plante photo

Santa Cruz County

with contributions by Mike Shannon, Mike New, John Lucking, Malcolm Alter and Stan Esbenshade

The oldest mining region in Arizona continued to produce mineral specimens in the last quarter of the twentieth century. Collectors combed the many mines in Santa Cruz County, and went home with a variety of interesting, and in some cases fine-quality specimens, but the most notable producers in this period were the Flux Mine in the northern Patagonia Mountains, Washington Camp-Duquesne in the southern Patagonias, and the J.C. Holmes Claim in the Santa Rita Mountains.

Dozens of collectors found good cerussite specimens at the **Flux Mine**, (see also page 212) but *the* great pocket was discovered shortly after the Tucson show in 1980 by the father and son team of David and Mike Shannon. The Shannons had gone to the Flux to work on an anglesite discovery they had made the year before. But the mine had other plans: an eight-foot high wall of backfill had collapsed in their absence and buried their anglesite find ... and their hopes.

They sat down on the pile to regroup and suddenly realized that they were not alone. Somewhere in the mine, someone was banging on rock. Curious as to who it might be, they followed the sound through the underground maze and finally into a 20-foot-wide room. The floor of the lucky collector's stope was littered with snow-white jackstraw cerussite specimens—not one foot of space was unoccupied.

They did not know the collector, a Mexican-American. He explained in broken English that his grandfather had worked in the Flux and had brought him to the mine to collect specimens in years past. As it turns out, he was collector Eddie Granillo's father Theodore (who worked for Susie Davis).

More amazed by the potential he saw in the walls than by the specimens that Granillo had already dug, the elder Shannon immediately began negotiating to buy the stope, the specimens—everything. They eventually

Cerussite on wad, 10.1 cm wide
Flux Mine, Santa Cruz County
Collected in 1980

Dick Morris collection
Jeff Scovil photo

came to an agreement: Shannon gave Granillo a check for $200 and a $100 bill, and Granillo left the room to the Shannons.

The Shannons spent the next 24 hours digging, wrapping, and covering their tracks. They had moved Granillo's pocket back another six feet, cleaned out every exposed crystal, and brushed away the obvious path out of the stope, out of the mine, and all the way back to camp—they had made the trip 21 times. Ready to leave, they felt secure that no one, save Granillo, would find the stope before they returned. And they would not return immediately: it was just six weeks before the 1981 Tucson show, and they had a lot of work ahead of them.

They had pretty much kept their secret by show time. When the entire contents of the pocket had been trimmed and boxed, they had 191 flats of Flux Mine cerussite. With a new find of great material, they had high hopes for the show. But it was a bust. Visitors to the show seemed to feel the material was just too delicate. Their mail-order customers, however, were buying the snowy crystals, and by summer's end, they only had 75 flats left, so the Shannons returned to the mine in the fall to replenish their stock.

They did some housekeeping on the stope and began advancing the trench. They drove wide, flat-tipped bars into the soft limonite and moved six feet of rock in a few hours. Small cerussite crystals started showing up almost immediately, and the Shannons were into a small pocket after only a few minutes. Within about three hours Mike had opened a black "limonite" stalactite vug with snow-white jackstraw crystals to several centimeters.

Most were well attached; nonetheless, each of his efforts to extract them brought a downpour of white needles onto David, who was digging along the base of the seam.

Mike asked David to bring him a flat and David returned with a flat and a mucking shovel—not the tool one associates with professionally collecting a pocket of fragile cerussite crystals. David pushed the shovel into the face and began shoveling out scoops of nearly pure, loose cerussite crystals. Mike was horrified,

> *I slid down into the trench to block the shovel. In doing so, I put all my weight on to the pry bar I had driven in to the face, and it and a pile of rock came tumbling down. When the dust cleared, I saw why he had been shoveling! The 3-by-5-foot opening I had created formed a window in to a huge, open, crystal-lined pocket! He had just begun opening it from the bottom and had wanted to be the first to peer in; ergo the shovel!*

The 6-by-6-by-12-foot pocket was studded with cerussite crystals to 20 centimeters! Every flat that they had brought in with them was full in an hour, so they began hauling flats out of the mine. They carried ten loads out of the mine each day for 4 days. But they were unable to cover their tracks this time. No huge crystal-covered plates were exposed, but they were leaving an obvious, interconnected series of cerussite pods. They would have to come back as soon as possible, but another Tucson show was already looming, and they had plenty of specimens to prepare.

The largest specimen they collected was 70 centimeters wide; it took both of them to carry it out. The longest crystal was nearly 30 centimeters long, but it broke in half on the way down the "road" to Patagonia.

Before Mike and David had prepared the specimens, dealers Mike New and Wayne Thompson stopped by the Shannons' house, presumably to buy boxes. They saw the cerussite haul and bought a good number of specimens. The Shannons prepared the remainder and had a very good Tucson show.

Sometime after the show, they heard rumors that New had been arrested while camped at the Flux Mine. Law enforcement had contacted ASARCO, which owned the mine, and New's operation had been shut down. When the Shannons returned to the mine the following summer, they found that the open pit at the top of the mountain had been bulldozed, and dozens of entrances had been backfilled. They made their way to the pocket area and found that a tunnel had been driven 30 feet into the rock face. There was no cerussite in sight. It looked as though the pocket had never existed.

Months earlier after seeing the Shannons' cerussite bonanza, New and dealer Wayne Thompson considered a cerussite recovery effort. Thompson and New brought a crew out to the mine to assess the occurrence. They spotted the pocket that the Shannons had been working. New described the prolific zone as the oxidation of a large pod of galena. The former lead sulfide pod had altered to four layers of cerussite-rich ground. New's crew began mining about four to six feet below the level where the Shannons had been working. They drove a foot or so into the face and New recalls, "The rock just opened up. There was cerussite above and on both sides of the pods—lots of cerussite."

Thompson and New left the crew at the Flux with $478 for food and supplies and went back to Phoenix, returning a few days later with tools and more food and supplies. They returned to find their miners out of both food and money, which as it turns out had not been spent on supplies. The crew had instead gone to Tucson where they hired prostitutes. New and Thompson gathered the 25 or so flats of cerussite that their team had collected and went back to Phoenix. On their way out of town, they stopped at the general store in Patagonia, and the owner asked them to pay the $130 bill that their miners had run up buying groceries. Needless to say, Thompson and New were not impressed with their team. (They were even less impressed later when they learned that their miners had sold the best material to explosives expert Bill Bowman who has a ranch near Patagonia.)

New and Thompson were back at the Flux a week later. This time the miners had gathered about 70 flats of good material and another 15 flats of top-quality specimens. But the operation was short-lived. The Patagonia Town Marshal and a deputy sheriff from Santa Cruz County had arrived at the mine and wanted to speak to the "person in charge." U.S. Border Patrol had noticed seismic activity after the miners' blasts, and they called the marshal who, in turn, had called ASARCO. The company asked that the culprits be arrested. They hauled New to the Patagonia City Jail. The parties ultimately agreed that New would give ASARCO some of the specimens that they had collected, clean up the site, and stop their mining operation. No charges were filed.

Shortly after Thompson and New abandoned the Flux Mine, Gene Schlepp, Stan Esbenshade, and Andy Anderson decided to give it a go. They knew that New's crew had collected a lot of cerussite, but they were also aware of their troubles with law enforcement and ASARCO. They decided to go nonetheless. Schlepp, Esbenshade, and Anderson followed a trail of debris through the underground labyrinth, found the spot, decided that it looked promising, and made plans to return.

Dealer Herb Obodda joined Schlepp and Esbenshade on their follow-up trip. They collected for a day and a half when Schlepp broke into a hole while probing what appeared to be the end of the tube. The hole opened into a 5-by-2-foot tunnel with cerussite needles, 10 to 12 centimeters long, in a spectacular heap toward the back of the 15-foot deep hole.

The smallest of the group, Esbenshade got into the tube and began handing the fabulous specimens out to Obodda and Schlepp. Most of the specimens had been removed from the tube two hours later, when the collectors were interrupted by a pair of ASARCO geologists. The collectors had 50 or 60 boxes of specimens packed and ready to go, and Esbenshade was afraid that they might lose it all, so he hid in the cave while Schlepp and Obodda talked to the geologists. The collectors wound up giving the company some of their specimens, and the geologists suggested that they finish what they were doing and not return. They complied, as they felt that the pocket had probably played out anyhow.

ASARCO sealed the mine entirely not long after. The great continuous Flux cerussite pocket surely produced tens of thousands of specimens from 1979 to 1982. There have been no significant finds at the Flux since.

In 1976 Fleetwood Koutz noticed colorful lead mineralization while researching his PhD thesis in the Hardshell Mine. He told fellow students Frank Sousa, Ty Schuling, and Bob Bodner about his observations, and the trio went to investigate.

A pillar on the 375 level of the mine grabbed their attention as exposed in it were small vertically linked pockets of apple-green pyromorphite crystals on mottramite/descloizite. Another stage of green pyromorphite formed on curled clay chips. The clay was deposited between the descloizite and the pyromorphite.

The crystals formed curved prisms similar to those seen in the campylite variety of mimetite. The generally less than 5-mm long crystals commonly formed spherical groups to 7.5 cm in diameter. The spheres had nucleated on yellow pyromorphite and, along with single crystals, were scattered across clay chips or quartz.

Sousa, Schuling, and Bodner took turns collecting plates and crystals in the limited space. They collected yellow pyromorphite on shattered quartz from a small stope above the pillar while they waited for their turns at the pillar. The upper stope seemed unstable, so they limited their time there. They did, however, manage to recover one cm-long pyromorphite crystal from the stope.

Pyromorphite on quartz, 7.6 cm tall
Hardshell Mine, Santa Cruz County
Collected in 1976 by Frank Sousa, Ty Schuling, and Bob Bodner

Les and Paula Presmyk collection
Jeff Scovil photo

About 2 miles south of the Flux Mine, Harshaw (now a ghost town) was settled in the 1870s and was named for prospector David Harshaw. The town was the hub of one of Arizona's must productive later nineteenth century mining districts. The largest mine in the district was the Hermosa, which produced a large amount of silver. Although mining has long since ceased, the area is now a target of exploration and public land use controversy. Two of the mines near Harshaw, the **Hardshell** and the **Aztec** saw some collecting in the late 1970s. In October 1976 acting on a tip from a fellow student, Frank Sousa, Ty Schuling, and Bob Bodner collected the most interesting mineralization from the Hardshell: botryoidal crusts on quartz of the best pyromorphite known from the state (photo above).

North of the town of Patagonia, the Santa Rita Mountains are home to the copious Glove Mine (see page 223). These mountains are also home to several modest localities, the **J.C. Holmes Claim** among them. The J.C. Holmes is a series of trenches and diggings in a cliff face along one side of the dry Temporal Wash; in the 1950s Dick Jones mentioned the J.C. Holmes as a vanadinite locality, although he never produced any specimens of note.

In the mid-1980s the "Mineral Minions," an informal group of collectors, made a fieldtrip to the locality. The group of experienced collectors including George Godas, Mark Hay, Dick Morris, Bob Jones, Evan Jones, Bill Panczner, Tom McKee, and Gary Coleman met at the site on November 23, 1985. The last to arrive, John Lucking and his son Charlie hammered on the wall for a while before the senior Lucking asked Hay for advice. "Look for clay seams and dig." Lucking found a red streak and went to work. It took just a few minutes for him to dig about a foot into the rock face. The seam opened up, and the he spent the next several hours carefully lifting fragments of silicified rock covered with caramel-colored vanadinite crystals out of the pocket.

Above: Vanadinite, 2.6 cm tall
J.C. Holmes Claim, Santa Cruz County
Collected November 25, 1985 by John Lucking
John Lucking collection
Jeff Scovil photo

One of the best vanadinite discoveries at the J.C. Holmes, the largest specimen was a 12-by-10-centimeter plate covered with crystals as long as 1 centimeter. A few smaller specimens had crystals up to 2 centimeters long. The mud in the bottom of the pocket contained many loose single crystals up to 1.5 centimeters in length. As John dug, Panczner and Charlie wrapped specimens and loaded them into beer flats. By late afternoon, they had six flats of specimens recovered, wrapped, and lugged to the truck. McKee and Coleman camped at the Holmes that night. The next day they were rewarded with a second major pocket not far from Lucking's.

The **Washington Camp-Duquesne** area (see also pages 219 and 355) remained a favorite collecting locality, and many collectors came away with prized Japan-law twin quartz crystals. The largest of the Washington Camp Japan-law twin pockets was literally stumbled upon around 1986 by a husband and wife team—the Parnells—who were new to the area.

The pocket produced several hundred specimens. The Parnells had no mineral collecting experience when they found their pocket. Collector Malcolm Alter asked them how they had found the pocket, they said they "noticed a few crystals on the ground and just started digging."

The pocket produced large plates with numerous twins up to 15 centimeters long on an edge. They sold plates measuring 3 feet across for several thousand dollars—in the 1980s that was a large sum for a quartz specimen. One of the large plates had more than 40 twins on it. Alter also remembers a small cabinet specimen sporting two perfect 10-centimeter tall twins, which had grown parallel to one another on the matrix.

The Parnells buried the discovery site, but were convinced that they had not collected all of the crystals. Malcolm Alter, Glen Miller and Roger Parnell spent several days mucking out the 6-foot wide, 18 foot deep pocket, which unfortunately bottomed out. There was however, a 3-foot-deep section in the bottom of the pocket that was filled with broken and partly rehealed crystals, some of which were twins. The largest of the exhumed crystals was about 14 centimeters on an edge.

Pima County

with contributions by Malcolm Alter, Bill Williams and Stan Esbenshade

When Arizona became a territory in 1864 the first territorial legislature carved the lands into four counties Mohave, Yuma, Yavapai and Pima. The latter defined as the area south of the Gila River, east of longitude 113°20' and stretching to the border with New Mexico. The original Pima County encompassed the area covered today by Pima, Cochise, Graham, and Santa Cruz counties. As such Pima was home to tremendous mineral wealth. Eventually the other counties were carved off, but modern Pima County is still huge (more than 9,100 square miles) and stretches from Ajo in the west to the minor mines of the Empire District in the east.

Although the mining heritage of Pima is rich both in history and mineral specimens, by the last quarter of the twentieth century, only the large commercial copper mines south of Tucson were active. However, mineral collectors were visiting the county's many closed and abandoned mines, and some very fine specimens were found during this period.

In 1985 the grandson of a wildcat miner who had worked in **Greaterville** in the 1950s showed up at Jimmy Vacek's house with a small box of exquisitely crystallized gold specimens: clusters of clean, bright, parallel crystal groups. Vacek called Ray Grant and Bobby Johnson to look at the lot. Ultimately Johnson purchased the best specimen and Grant and the late Harold Michel took the rest of the specimens.

The specimens are among the best Arizona has ever produced and each of these sought after specimens has gained interesting provenance since they surfaced. The miner was Dewy Keith. Working alone and on faith, he had driven a 100-foot drift in the Yuba Mine. According to his grandson, he was running out of money and was ready to give up when he had a dream that

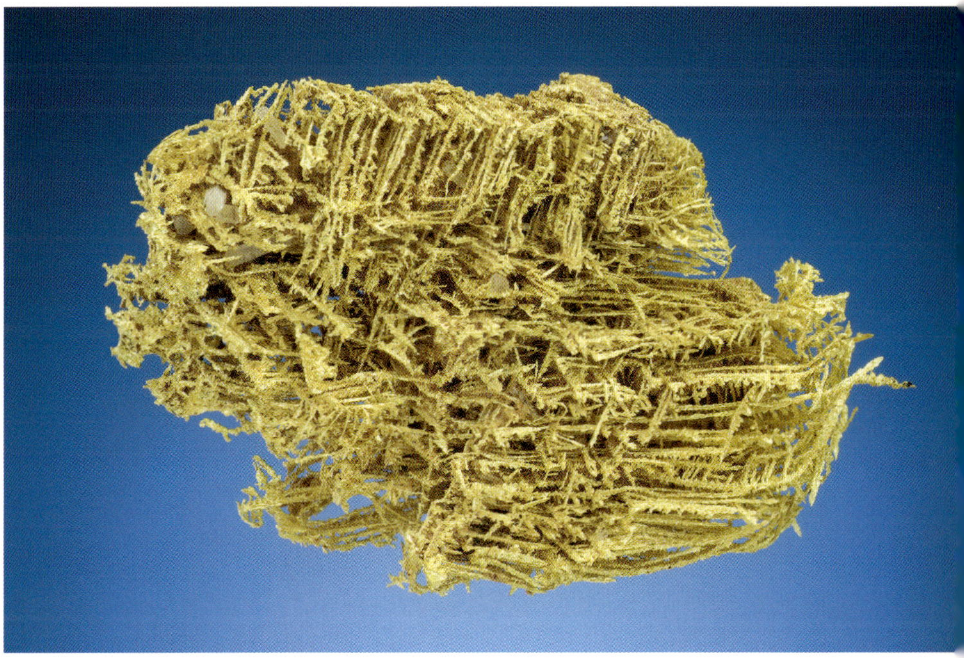

Gold, 4.6 cm wide
Yuba Mine, Greaterville, Pima County
Collected in the 1950s by Dewy Keith

Les and Paula Presmyk collection
ex Harold Michel
Jeff Scovil photo

Gold on quartz, 4 cm tall
Greaterville area, Pima County
Found with a metal detector by Bob Melzer

Evan Jones collection
Jeff Scovil photo

he would hit pay dirt with his next blast. Classic! The next day he drilled a round, fired it, and as he cleared the rubble, he saw that his nocturnal premonition had been correct. The blast had exposed a pocket containing nearly 1,500 grams of intricately crystallized wire gold. The best of the specimens formed 3-centimeter-wide patches of interconnected wires loosely resting on a matrix of brown quartz.

Keith put a dozen or so of the specimens in a box where they sat for 30 years. The total weight of the gold in the specimens offered to Vacek is about 100 grams, so 1,400 grams of the original find were probably sold as bullion, although someone along the way may have preserved one or more of the other beautiful specimens.

From time-to-time gold specimens with textures similar to those from Keith's find turn up. Whether or not they are part of the 1955 pocket is not known, but they almost always bring frenzy to the community of Arizona collectors. Greaterville remains a favorite target of prospectors with metal detectors, but none has been known to have recovered another of the fine crystallized gold specimens.

Due east of the Santa Rita Mountains and Greaterville, a 7-mile-long, north-south-trending range of hills and ridges called the Empire Mountains are home to a series of lead-silver mines organized into the Empire mining district. **The Total Wreck Mine** was a popular collecting site in the 1980s. According to Malcolm Alter,

> *The best pocket ever was found by Kevin Cole and Tim Sutter, both graduate students at the University of Arizona. They found the wulfenite in a watercourse high up in a stope just below the caved "glory hole" stope. Cole and Sutter collected a dozen flats of lustrous thick crystals, of good orange color, and up to 2.5 cm individual crystals.*

The potential of the **Waterman Mountains** has long teased collectors. Dave Shannon found the rare lead-copper sulfate osarizawaite in the **Silver Hill Mine** (see also page 160), but the most important discovery from the mine in this period was of smithsonite and rosasite.

In 1984 Graham Sutton and Bobby Lane took a collecting trip to the Silver Hill Mine. They entered the mine and worked their way back to a large stope. They did not realize that they were standing on a false floor of rocks and dirt covering a jumble of cribbing and stulls—the "ceiling" above another stope. They encountered rice-grain-size white smithsonite scattered on limonite plates and associated with rosasite spheres to 3 millimeters and in some cases laths of sky blue aurichalcite. Their discovery brought them back to the mine repeatedly over the next few months. Bill Hawes was along on one trip when they discovered several pockets of blue rosasite spheres scattered about smithsonite grains and occasionally small lath-like aurichalcite crystals. The find yielded 200-300 flats of specimens.

While Sutton, Lane, and Hawes were working the smithsonite and rosasite deposit, Eddie Granillo and his partners were digging azurite in a hot, cramped area jumbled with broken mine timbers—the stope below Sutton's. The deposit that Granillo was collecting was an occurrence of medium blue drusy masses made up of azurite crystals to 7 millimeters. The walls of this stope also produced frothy, colorful masses of green malachite.

The **Pima mining district** south of Tucson covers the eastern side of the Sierrita Mountains and stretches for over 20 miles from Green Valley to the San Xavier del Bac mission. The district included a number of large open pit copper mines including (from south to north): **Sierrita**, **Esperanza**, **Twin Buttes**, **Mission**, **Pima**, **Eisenhower**, **San Xavier South**, and **San Xavier North**. From 1950 through 2000, this district was one of the largest copper producers in the United States. All of the mines target porphyries that were emplaced between 64 and 55 million years ago. Ultimately, the Pima, Mission, Eisenhower, and San Xavier South pits became one huge open pit. The present Mission complex covers over 15,000 acres. Since 1987 ASARCO has controlled the consolidated operation.

Considering the scope of material moved in all these mine operations, there have been a relatively minor number of specimens collected. According to Neil Prenn, the mine had no formal mineral collecting policy simply because there were so few crystallized mineral specimens found on their property. ASARCO has a policy of no mineral collecting by its

Right: Rosasite, 10.3 cm wide
Silver Hill Mine, Pima County

Les and Paula Presmyk collection
Jeff Scovil photo

Above: Smithsonite with malachite
6 cm tall
Silver Hill Mine, Pima County

Dave Bunk Minerals specimen
ex Jim and Dawn Minette
Jesse La Plante photo

Lower right: Calcite on wulfenite
7.4 cm wide
Total Wreck Mine, Pima County

Evan Jones collection
Jeff Scovil photo

employees, but salaried personnel were able to collect without problem. Of course, hourly employees collected and took home specimens in their lunch boxes. For many years, this occurred simply because operating rules at the time required them to exit their truck cabs and wait on the ground away from their truck as the loader or shovel loaded it, thus enabling them to look around for specimens. With the advent of larger trucks, this operating rule was changed and the drivers remained in the truck while being loaded, so their opportunity to collect was limited.

In the late 1970s several mine geologists began to focus on collecting minerals. There were never great minerals recovered, but Bill Williams had a hand in nearly every discovery,

> *As mine planning engineer, I spent a lot of time in the field studying locations of haul roads and utilities, as well as examining ore trends. Since I had a driver, most of the mineral specimens that I found were located by driving along mined ore faces or on the base of the bench. However, since an engineer in a wheelchair could not easily access the shot muck, the other collectors were generous in sharing some of their better finds with me.*

The Mission Mine is primarily known for one type of specimen: exceedingly rare, finely crystalline native copper suspended in clear gypsum. Found on the Golden West property of the Eisenhower Mining Company, these specimens were mined from the 2970-G bench and the 2930-H benches in 1979. Gypsum included with materials such as chalcocite crystals, chlorite, unidentified "grunge," and wall rock were also found. Specimens varied in size and thickness often depending upon how the inclusions crossed through the cleavage planes of the gypsum. Specimens exceeding 30 centimeters wide are known but are extremely difficult to come by. Finding undamaged specimens was difficult simply because of the softness of the gypsum, which scratches easily.

Copper in gypsum, 10.2 cm wide
Mission Mine, Sahuarita, Pima County
Les and Paula Presmyk collection
Jeff Scovil photo

Copper, 38 cm wide
New Cornelia Mine, Ajo, Pima County
Freeport-McMoRan Copper & Gold collection
Jeff Scovil photo

The **New Cornelia** pit at **Ajo** (see also page 148) closed in the 1980s, but the ancient property gave up a few surprises while in its death throes. Phelps-Dodge transferred Richard Graeme from Bisbee to Ajo after the mines at Bisbee had closed. He found the change "agreeable climate at Bisbee to the searing heat of Ajo" the most difficult part of relocating. By mid-1976, Graeme had been promoted to the number two position at the mine, which required and enabled him to move about the pit to inspect the operating areas. During his time at Ajo, Graeme never saw mineral specimens en masse as he had seen in the Lavender pit at Bisbee; still, he made several memorable discoveries.

In August of 1976, for example, while inspecting one of the operating areas, Graeme saw a mass of what was obviously native copper sticking out of the bench face some 40 feet above him. As Ajo was justifiably famous for its copper, he wanted a closer look:

> *The large electric shovel working just a few yards away was temporarily idle while waiting for a train to load with the waste he was digging. I told the shovel operator of my desire to look at the copper and he said he had seen it and had tried to "tease it out" but with no success. At my*

request he lifted me up in the shovel bucket to the copper, which apparently continued back into the bank for some distance and could not be dislodged. I advised the shovel operator that I would go to the level above and, between trains, would guide his digging to free the copper.

It took a while, but we exposed a calcite-coated flat mass of copper that was nine feet long before it broke under its own weight. Using a rope and the huge shovel bucket, we loaded the mass into my company pickup, which groaned under it weight. It was far more difficult to unload as the whole mass of copper and calcite weighed most of 1,500 pounds.

At the geology office, a makeshift tank was constructed of wood and black plastic and numerous gallons of chemically pure hydrochloric acid purloined from the assay office were used to remove the calcite. Several days of etching revealed a beautiful, immense tabular mass of copper with countless thumb-sized crystals. Now what to do with a 1,000 plus pound copper specimen? Consulting the General Superintendent solicited the response that in as much as I had already wasted a good deal of company resources on this foolish effort, it should be displayed at the company office.

The mass broke as it was being moved, and the smaller portion, which was some four feet long, was displayed in the window of the closed Ajo Theater. The larger and much heavier piece went into the yard of the General Manager's residence.

Malachite pseudomorphs after azurite
4.6 cm tall
New Cornelia Mine, Ajo, Pima County
Evan Jones collection
Jeff Scovil photo

At the request of Stan Esbenshade, Southwest Mineral Associates (SMA) reactivated their contract at Ajo in 1986. Esbenshade talked with Mike New who had been part of Thompson's crew when SMA worked it in the 1970s. New indicated that the mineralized fractures were confined to an area in the northwest corner of the pit and that the zone ran across 4 or 5 benches.

Gene Schlepp accompanied Esbenshade on his first trip to the mine. They spotted the fracture zone from the other side of the pit. When they got over to it, Esbenshade began pulling rocks out of an opening and a few pseudomorphs rolled out of the crack. Looking up in the hole, he could see a jumble of rock fragments and malachite. More poking brought down a pile of malachite and rocks. They collected 3 or 4 flats of specimens before they had to leave, which they were glad to do given that they had underestimated how hot it would be and did not have nearly enough water with them.

Esbenshade made a few more trips and then abandoned Ajo until October. Esbenshade continued to work the pseudomorph zone but regularly checked the other benches for mineralization. The benches above and below the one he was working also yielded pseudomorphs of different habits. Two benches below the malachite after azurite pseudomorph zone, he also found small metallic-gray, cuprite crystals. But most of their (other collectors including Ted Parkevich, Ken Roberts, Mark Kielbaso, Bob Jackson, and Tim Smith periodically worked the mine for specimens) attention was focused on the pseudomorph zone which consistently produced specimens. He collected the pseudomorphs on and off until the spring of 1988, when the fracture was not longer producing as readily.

Pinal County

with contributions by Garry Alexander, Mark Hay, Bob Jones, and Andy Clark

Pinal County was formed on February 1, 1875 from parts of Maricopa and Pima counties. The Dripping Springs Mountain Range sits in the eastern extreme of the county and is home to the Arizona Silver Belt and more than 250 mines. The range is a 35-mile-long, northwest-trending uplift of Paleozoic sediments (limestones, shales, and quartzite), which were intruded by Cretaceous-Tertiary porphyries to create one of the most mineralized zones in Arizona. Mining operations and collectors continued to expose crystal-lined pockets in the last quarter of the twentieth century, but not at the same rate or to the same degree as in decades past.

More than 700 million tons of ore were removed from the underground workings of the **San Manuel** mine during the last 25 years of the twentieth century. It was a period of expansion and eventual closure of the large mining operation. Unfortunately for mineral collectors, block cave mining is not conducive to preserving or even finding minerals, and this method was employed at San Manuel for more than 50 years.

The two 500-foot-tall smoke stacks at the San Manuel smelter south of Mammoth were brought down on January 17, 2007. The mine closed in 2003 and BHP, the property owner is making reclamation efforts.

Photo courtesy of the *Tucson Citizen*

Collector Garry Alexander, a long-time San Manuel employee, was fortunate to be able to preserve some of the specimens encountered during mining. Over its considerable life, the San Manuel Mine produced collectible species including pyrite, copper, calcite, chrysocolla, azurite, malachite, copper, and some barite. Unfortunately, rising costs, escalating environmental requirements, and decreasing grades caused many open pit copper mines to permanently shut down in the 1990s. BHP Billiton ceased mining and smelting operations at San Manuel in June 1999 and later announced that it would close the operation entirely in November 2003. The dumps were reclaimed and the signature headframes were scrapped. Today little remains of the once great San Manuel Mine.

The mineral specimens that the **Magma Mine** in Superior (see also page 201) produced were atypical for Arizona: calcite, pyrite, and barite. In the mid-1970s the mine hired a series of professionals—mining engineers, geologists, and samplers—who happened to have a strong interest in mineral collecting. While the competition between them was intense, the upside was many eyes were on the lookout for specimens. Reg Barnes and Bob Falletta were among the first of these collecting professionals to work at the mine.

Barnes and Falletta found rich pockets and seams of calcite specimens in their first two years at the mine. Les Presmyk was hired two years later in 1976. After six months at the mine, Presmyk too encountered a rich pocket of calcite. Everyone learned quickly that collecting in an active mine comes with challenges. The pockets are generally exposed by the mining process and are thus in "public" view, so other potential collectors—competitors for crystals—quickly learn of specimen strikes. Supervisors, whose concerns range from safety to how their employees are spending company time, also observe a collector/employee's activities. And even when permitted, collecting in active stopes was frequently hurried, as mining marches on.

The mechanized "continuous miner" uncovered one calcite pocket on the 3512 level of the 3500-2C stope early in Presmyk's career at Magma.

Above: In the late 1970s, Garry Alexander found four specimens of well-crystallized pyrite on top of a muck pile in the San Manuel Mine. This crystal from the chance find is 10.4 cm wide.

Garry Alexander collection
Jeff Scovil photo

Right: Barite, field of view 5.4 cm wide
3440 level, Kalamozoo orebody
San Manuel Mine, Pinal County

Mark Hay collection
Jeff Scovil photo

Right: Calcite on hematite, 7.8 cm tall
Magma Mine, Superior, Pinal County
Collected in 1980

Mark Hay collection
Jeff Scovil photo

Above: Calcite, 5.6 cm tall
Magma Mine, Superior, Pinal County

Paul Harter collection
Jeff Scovil photo

Lower right: Barite, 3.2 cm wide
Magma Mine, Superior, Pinal County
Collected on Nov. 18, 1981, by Mark Hay

Mark Hay collection
Jeff Scovil photo

While Presmyk was able to squirrel away a few boxfuls of calcite and roll one large boulder out of the path of the continuous miner, the bulk of the pocket was destroyed in the mining process. The stope miners who had worked at Magma for some time recalled hitting crystals on the level above, and in fact, miners encountered the watercourse in which calcite crystals occurred three more times over the next two years.

While the calcite crystals on each level of the mine are different enough to be distinguished from one another, many exhibit a wheat-sheaf appearance. The tips of the crystals from the 3512 level appear slightly pinched. One pocket on the 3440 level has very sharp, pointed tips, and another pocket just 10 feet away contained crystals that resembled bulbs.

About two weeks before Thanksgiving in 1980, engineer Ted Miller emerged from the graveyard shift and showed Presmyk the nicest barite Presmyk had seen since he had started working at the mine. Miller also had several nice crystallized gypsum specimens. Miller and the level supervisor collected barite from the 3620 level for a few nights before word got out.

Geologist (and mineral collector) Mark Hay had been added to the geology staff a few years earlier. He learned about the barite from Presmyk, who had also said that the pocket had been worked out. But the mineral collecting community at Magma was in Hay's words, "cutthroat." Each member had to "bleed badly" to obtain specimens from one another over the years. Hay was thus not inclined to take Presmyk entirely at his word. He asked about the discovery and learned that the pocket had been in the footwall of the D-bed in one of his stopes. Being a junior geologist, Hay was not required to attend planning meetings, so while the others were tied up he went down to the pocket level and asked a miner to lead him to the area that had been producing the barite. The miner pointed to the space between two 10-by-10-inch wood support posts. Mining had encountered the lower boundary of the orebody—the footwall—and about 60 percent of the working face in the area was in barren "waste" limestone. They had stopped working the heading, and no one would be in there until they were ready to backfill it. For the moment, he had the place to himself!

Hay directed his mining light into a fist-sized hole near the floor,

> *I couldn't see much, so I took my gloves off and reached in with my hand. It was a pretty good-sized hole—I couldn't reach the end of it. All I felt were broken rock fragments and mud, nothing that felt like a barite. I began taking handfuls of loose rock and mud out of the bottom. Going through the mud in the light of my cap lamp, I realized that there were barite crystals in the loose material! There were no two-inch crystals, but there were probably a dozen in total. They were fat little diamond-shaped, mud-covered crystals with sharp edges. "Pay dirt!"*

He got lucky with the first rock he moved. It was large and being mostly hematite, was pretty heavy. But he dislodged it and rather than letting it fall to the ground, he caught it in his arms and rolled it over on its back. To Hay's delight, there was a small hole on the backside of the rock and in that hole was—still attached—a perfect barite rose (pictured on page left).

Hay carefully wrapped his prizes and showed them to Barnes and fellow

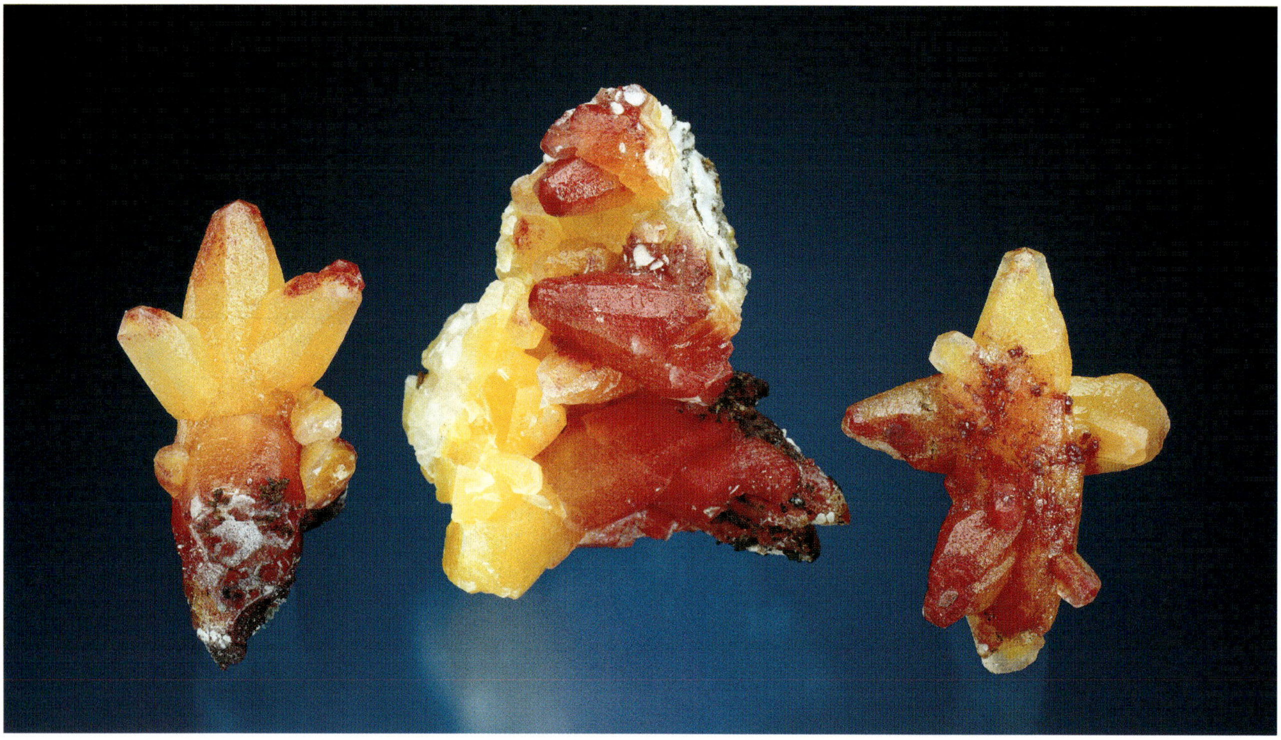

Calcite with cuprite and copper inclusions
The center crystal is 6 cm tall
2100 level, Silicate orebody
Pearl Handle pit, Ray, Pinal County
Collected in 1975 by Dan Fountain
Robbie McCarty, Paul Harter and Les Presmyk collections
Jeff Scovil photo

geologist Don Atkinson. The next morning the three of them descended to the 3620 level. Everything was just as Hay had left it. They enlarged the opening so that Atkinson could crawl through. It was a tight fit, but he was able to get into the untouched part of the pocket. The best piece they discovered was a fine group of 2.5-centimeter crystals on matrix.

The ambient rock temperature at that depth in Magma was over 140 degrees Fahrenheit; it was hot in that hole! Atkinson emerged from the pocket after 5 or 10 minutes of collecting. He said the pocket continued, twisting and turning, and he did not see the end, and he said that there was barite still in there. He had handed out the loose ones that he could reach, but there were single crystals and clusters still attached to the walls.

The three began to discuss their options. They were on company time and it was day shift, the busiest possible time at the mine—their options were limited. They were discussing coming back when they would have time to open up the pocket when a light shone done from the level above. One of the mining engineers had overheard their conversation and instructed them to clear out.

Hay, Atkinson, and Barnes headed back to the surface where they were invited to a closed-door conference with Chief Geologist Russell Webster. Although no one lost his jobs, they were not able to collect the remaining barite crystals. Shortly after the panel was backfilled. They later learned that the mining engineer who had caught them collecting was Presmyk, who was shift foreman that day.

The Magma Mine shut down in June 1984. It reopened a few years later and all of the replacement bed ores were mined out. No crystals were produced in this last period of mining although there is reasonable certainty that some pockets were encountered.

About six miles southeast of the Magma Mine, the **Ray Mine** (see also page 205) is still in production and specimens are now and then recovered from the Pearl Handle open pit. The old underground workings are periodically encountered through the open pit and wood timbers containing visible native copper are recovered. Although the copper formed as a result of mining, the specimens are still quite popular among collectors.

Calcite is rare at Ray but late one night in 1975 on the 2100 level of the Silicate orebody, shovel operator Dan Fountain found bright yellow calcite crystals included with rich red to violet-red cuprite needles. There is no telling how many of the dogtooth crystals were destroyed before he saw the reflections of the calcite crystals in his headlights. Crystals as long as 6 centimeters occurred as singles or as small subparallel clusters. When one of these rare specimens comes available, there is usually strong competition to acquire it as few were saved.

In its early days, the underground Ray Mine, Pinal County, required significant timbers to hold the ground open. Copper-laden water moved through the openings and found receptive environments, in this case a 19.8 cm tall section of timber, in which to crystalize. The grain of a second timber is also preserved in copper as is the space where an iron spike had been.

Flagg Mineral Foundation collection
Jeff Scovil photo

Chrysocolla, 14.4 cm wide
Pearl Handle pit, Ray Mine, Pinal County
Scott Rudolph collection
Jeff Scovil photo

In September 1976 Southwest Mineral Associates (SMA) had contracted with Kennecott Copper Corporation to collect minerals in the pit at Ray. Thompson's crew consisted of several seasoned field collectors including Gary Fleck, Andy Clark, Jeff and Brad Archer, Ed Anderson, Wayne Johnson, Don McCoy, Jack Lowell, and Pat Garrett.

Chrysocolla is ubiquitous at Ray and is the major ore mined from the Silicate orebody, east of the Emperor fault. It occurs mostly as crusts and seams and on occasion is coated by drusy quartz crystals. Chrysocolla specimens have been collected throughout the mine's history, but the well-documented great finds began during the latter part of the 1970s. It was not too long into their contract before SMA had discovered chrysocolla pseudomorphs after gypsum (or possibly azurite) on the 2300 level of the Silicate orebody on the eastern side of the mine. Then in September, Fleck, Thompson, and Clark found chrysocolla near the quartzite contact. Those lower in the zone were friable and tan rather than the more desirable blue. Two interconnected pockets produced about 100 significant specimens. The pseudomorphs occurred in sprays up to 4 centimeters long and as clusters.

Gorgeous sky-blue pseudomorphs were found 25 feet or so above the contact in the Whitetail Conglomerate. The crystals were brittle and had been damaged by blasts in the pit. Some of the larger matrix pieces contained as many as 30 pseudomorphs. Most of the specimens consist of randomly divergent clusters several centimeters across. In total the find

Upper left: Cuprite, crystal is 1.5 cm wide
Pearl Handle pit, Ray, Pinal County
Evan Jones collection
Jeff Scovil photo

Lower left: Chrysocolla, malachite and quartz, field of view 2 cm wide
1100 bench, Hercules area
Ray, Pinal County
Mark Hay collection
Jeff Scovil photo

Right: Chrysocolla after gypsum (or possibly azurite), 14 cm tall
Pearl Handle pit, Ray, Pinal County
Flagg Mineral Foundation collection
Jeff Scovil photo

produced several hundred specimens (Jones 1983).

As their collecting was limited to the pseudomorphs, a few seams of chrysocolla, and some copper, SMA exited the property to pursue more specimen-rich areas in the Phelps-Dodge family of mines. And in 1980 John Mediz obtained the mineral mining contract for the Ray pit.

The pit produced a constant stream of native copper and cuprite for Mediz. It was not uncommon for him to collect a pickup full of native copper in a single day. The copper was generally found as three-dimensional, loosely consolidated masses more than 25 centimeters across.

In early 1981, Mediz discovered and dug more than 90 flats of mostly cuprite crystals on matrix. The largest crystals reached about 2.5 centimeters across. Many of the lustrous, well-developed crystals were scattered on dark brown to charcoal-colored limonite. The discovery was made along the Emperor fault at the north end of the pit. Arizona State University has a crystal from this find that shows narrow dodecahedral modifications.

Right: Vanadinite and calcite, 5.4 cm wide
Grey Horse Mine, Pinal County
Collected by Andy Clark and Gary Fleck in 1977

Les and Paula Presmyk collection
Jeff Scovil photo

Above: Andy Clark at the Grey Horse Mine, Pinal County, circa 1977

Following a tip by Curt Van Scriver and Mike New, Andy Clark and Gary Fleck decided to explore the **Grey Horse**, an old vanadium mine between the Ray and 79 mines, not far from Kearny. New and Van Scriver had found a large vanadinite crystal on the mine dump, but they could not locate its source.

On their first trip, Clark explored the upper levels of the mine while Fleck used ropes to lower himself to an intermediate level. Clark hadn't found much but Fleck had found vanadinite crystals more than 1.5 centimeters long in an abandoned, open pocket on the 2810 level. Vanadinite crystals covered one vertical surface of a crack in the limestone. They could see that the pocket stretched more than 3 feet into the hanging wall and continued upward at least 18 inches into the back.

They used hammers and bars to collect the specimens on blocks of limestone, and later (out of the mine), they used a diamond saw to reduce the many blocks to a mosaic of thin vanadinite-covered plates. Wayne Thompson and Bud Standley arranged for all of the pocket pieces to go to the Arizona-Sonora Desert Museum. Under the direction of curator Bill Panczner, crews were reconstructing wulfenite and selenite pockets for a new underground mine exhibit. The large colorful pocket of Grey Horse vanadinite was to be included.

In June of 1977 Clark and Fleck were back at the Grey Horse, this time with an air compressor. They opened another pocket on the 2810 level and recovered considerably more vanadinite, including a cabinet-sized specimen that was awarded the "Best of Species" trophy at the 1978 Tucson Gem and Mineral Show.

Clark and Fleck continued collecting on 2810 level until the pocket vein pinched out. Calcite commonly coated the vanadinite and served to protect the crystals while the limestone blocks were removed. After the specimens were trimmed, any excess calcite could be removed by bathing the crystals in a solution of Calgon and phosphoric acid.

The **Vekol** mining district is about 15 miles south of Interstate 8 and located within the Tohono O'odham Reservation. The area is remote, isolated, and starkly beautiful. The small Vekol Mountain Range rises up from the flat saguaro-studded desert. The mining claims are patented, privately owned "islands" in the midst of the tribal nation.

Below: The Vekol area is a popular spot for amateur astronomers to set up their telescopes and see the constellations. This Dwight Hanson photo of Rho Ophiuchus cloud complex was taken from the Vekol Road stargazer locality. The remote area offers dark skies, good visibility, and comfortable nighttime temperatures, although being remote and quite close to the Mexican border, it is presently not as safe as it might be.

The Vekol got its start as a silver mine. The named reportedly means *grandmother* in the Tohono language (Murbarger 1964). Legend has it that the Tohono Indians told local trader John Walker about the mineralization (Shannon 1983). On February 5, 1880, the claim was recorded, and the following year, Walker's brother acquired a third interest in the Vekol.

The Vekol Mine is high and dry. Typical of older mines, the stopes are irregular and because the old miners followed leads until they petered out, the workings can appear haphazard. Hand-gobbed walls are common.

In 1976, Ray Grant and Dan Helm were taken to an area in the Vekol mine by an Arizona State University geology graduate student to see the "siderite" pockets. They made many trips to the pocket area and collected 20 flats on one trip. They were accused of blasting the area, but the work was all done by hammer and chisel. Later, David Shannon collected there and determined that the "siderite" was iron-rich dolomite. Thin crusts of calcite commonly cover portions of the crystals, which range up to about 2.5 centimeters, although the average crystal is considerably smaller. Glen Miller did as well at the Vekol Mine as any collector. But he noted "even the better specimens were not fast sellers."

Right: Dolomite and calcite
Field of view about 4 cm wide
Vekol Mine, Pinal County
Jeff Scovil collection and photo

Right: Smithsonite, 4.8 cm tall
79 Mine, Gila County

Marty Zinn collection
Jeff Scovil photo

Upper left: Dick Morris hauling flats full of rocks up the raise in the 79 Mine
Tony Potucek photo, August 1991

Lower left: Wulfenite
The "red dot" crystal is 1.8 cm wide
79 Mine, Gila County

Dick Morris collection
Jeff Scovil photo

Gila County

with contributions by John Mediz, George Godas, and Malcolm Alter

Pieces were shaved off of Maricopa and Pinal counties in 1881 to form Gila County, which was intended to serve the rich mining area around Globe, an important mining hub. The county is home to two of the most important mining districts in the state: the Globe-Miami district in the Pinal and Apache mountains and the Banner district in the eastern Dripping Springs Mountains.

In the Dripping Springs district, John Mediz acquired a lease on the **79 Mine** (see page 193) in 1976. He and partner Jim Burke spent months cleaning up the rubble-filled stopes and drifts left in the wake of years of neglect and rabble collecting. Then Mediz and Burke went to work on the aurichalcite stope on the fourth level, which turned out to be a veritable treasure chest for them. It took Mediz and Burke six years to recover all of the aurichalcite from the level. The stope had produced tens of thousands of specimens, most of which were sold on the European market.

Mediz was more reluctant to work the wulfenite stope because rubble from the fault was constantly falling into the area. And to be quite honest, the wulfenite deposit was not as lucrative as the aurichalcite vein. Still, in 1982 Mediz hit a pocket of wulfenite. Atypical for the mine, the crystals resembled those from the San Francisco Mine in Sonora, Mexico. The yellow crystals were up to 1.2 centimeters wide and were scattered across a yellow mimetite-covered matrix.

Mediz encountered his best pocket in the late 1980s. Absentmindedly, he pried a large slab out of the wall and opened a large area of sooty, gray-black wad. Digging by hand, he found fist-sized specimens covered with red and yellow wulfenite crystals. He spent five days clearing out the pocket.

A couple of months later, Mediz packed stulls into the fault area in the wulfenite room in an effort to shore up the ceiling. He returned a few days later to find that the stope had collapsed, and the stulls and his tools were buried under 10 feet of muck and boulders.

Mediz found an unending supply of hemimorphite on the third level of the mine. He also did well in the chrysocolla stope (470 level), which produced pseudomorphs of chrysocolla after hemimorphite and cerussite. One notable pocket produced what initially appeared to be coarsely crystallized aurichalcite coating knobs and chunks of matrix and, in some cases, wulfenite crystals. The coarse "aurichalcite" turned out to be pseudomorphs of chrysocolla after aurichalcite, some of which were enhanced by a scattering of small cerussite crystals.

George Godas at the 79 Mine
Tony Potucek photo, Augusr 1991

The chrysocolla stope also produced smithsonite ranging from gray to various shades of green. But the aurichalcite stope yielded Mediz's best smithsonite specimen—a deep lustrous botryoidal green specimen that is on display at the Arizona-Sonora Desert Museum.

In 1991 Mediz lost the lease on the 79 Mine. To mark the end of his term at the 79, Mediz invited friends and local collectors to the mine for a "Bon Voyage to the Shithouse" gala at which the outhouse was blown to smithereens. For a time, the mine was once again open to collectors.

In the 1980s, Coeur d'Alene Mines had acquired the 79 Mine when it purchased Callahan Mining Company. Collector Mike Jaworski was working for Coeur d'Alene Mines in 1996 when he learned the company was planning to dispose of the patented claim covering the 79 Mine. Jaworski was interested in keeping mineral collectors in the 79 Mine, and he worked to broker a deal. Scott Lewis and George Godas ultimately agreed to buy the claim, including the historical mine and production information, for a lump sum payment of $2,500.

Lewis and Godas owned the 79 Mine together for six years. During that time they tried to empty the wulfenite stope. The only memorable specimen that they found was a wulfenite from a small pocket on the west side of the stope. Godas recalls how he wound up with the specimen:

> *I had made two piles, one with everything that we had gotten that month except for the wulfenite, which made up the other pile. Scott went first. To my relief, he picked the large pile and I wound up with the wulfenite. At the time, the pile he picked was worth more, but that wulfenite is still my favorite from the 79. The crystal is 1.7 centimeters on an edge, transparent, and golden yellow in color. It sits on a sliver of matrix and has good luster. The beveling is exquisite, starting from a thick square and thinning out in three stages to razor sharp edges.*

The **Christmas Mine** is located on the eastern slopes of the Dripping Springs Mountains. It overlooks the Gila River and Dripping Springs. Around 1880 Dennis O'Brien and William Tweed staked a claim on the mineralization that is today known as the Christmas Mine. Phelps-Dodge later acquired the property, which U.S. officials then determined was on the San Carlos Indian Reservation, and the claims were declared null and void. On December 22, 1902, in a one-sentence executive order, President Theodore Roosevelt had the western boundary of the reservation redrawn so that the contested mineral lands would fall outside of the boundary of the reservation. On Christmas Eve of that year, G.B. Chittenden relocated the claims, "jumping" Phelps-Dodge's ownership. The Christmas Mine eventually produced more than 350 million pounds of copper.

In 1975 Jimmy and Joyce Vacek successfully negotiated a contract to collect minerals at the Christmas. They set up a joint venture for the specific purpose of extracting non-crucial crystal and lapidary materials from the open pit. From 1975 to early 1976, the Vaceks and Grant Richards collected dark blue kinoite in one area of pit and dioptase associated with cuprite and apophyllite in another.

Perhaps their most significant find was a pocket of rosasite on hemimorphite attached to a quartzite matrix. Rosasite "fingers" on one specimen reached lengths of 8 millimeters. This more than 70-centimeter-long pocket was exposed in a quartzite boulder that Richards had peeled from the wall of the pit. The boulder was lined with forest-green rosasite spheres stacked on water-clear hemimorphite blades up to 4 centimeters.

The **Apache Mine**, located in the Globe Hills mining district in Gila County was an early favorite of mineral collectors for the cherry-red vanadinite crystals to about 1 centimeter that were scattered quite aesthetically about a darker matrix. While Apache Mine vanadinite has since been surpassed by global finds, particularly those in Morocco, Arizona mineral collectors still prize specimens from this mine. The Apache exploits an almost vertical, north-south trending vein cutting through a hard, fine-grained quartzite. The vein is composed of galena with minor copper and zinc sulfides. Vanadinite occurs in the upper, oxidized portion of the vein.

According to Keithley (1945), vanadinite and descloizite were best developed in large fractures in the quartzite west of the vein. He further reported that while removing material from a sidewall, he

> *... accidently opened a narrow fissure. From it poured a stream of detached vanadinite crystals, ranging in color from light to dark red, from apprx. ⅛" to ¼" in diameter. Their modification was so great they hardly resembled vanadinite. Their forms might well be described as fantastic. Due to their flower-like "arms" or extensions they were quite fragile.*

Local collectors such as Wayne Thompson, Sr., began collecting the simple, red hexagonal crystals in earnest in the 1950s. In 1957 Richard Thomssen reported finding mottramite, descloizite, calcite (small clear to white rhombs), galena, cerussite, anglesite, and willemite in addition to vanadinite. Thomssen also reported rarer occurrences of wulfenite, boleite, brochantite, malachite, chrysocolla, and covellite. Richard Bideaux later

Above: Vanadinite, 7.2 cm wide
Apache Mine, Gila County
Evan Jones collection
Jeff Scovil photo

Facing page: Kinoite and fluorapophyllite
field of view 1.7 cm tall
Christmas Mine, Gila County
Marty Zinn collection
Jeff Scovl photo

added matlockite to the list of rare species that occur in the Apache Mine.

Around 1976 John Mediz acquired the lease to the Apache. Through all of the years the Apache was collected, especially in the 1970s, collectors noticed soil and roots in the fractures of the quartzite. While some desert vegetation will send roots down over 20 feet, collectors figured it was only a matter of time before someone working the underground would break through to the surface. That never happened. But John had ideas about working the mine from the surface. In 1982 with the assistance of Andy Clark and Gary Fleck, he brought a bulldozer to the mine. Clark recalls that in the last days of underground collecting, they would saw roots and soil in cracks in the stope. During the rainiest of days, the water would get into the underground collecting areas soon after the storms. Once Mediz and the boys got the surface cut open, they found that the uppermost end of the stope was within 10 feet from the surface.

From 1976 to the end of 1983, Mediz collected large quantities of commercial-grade vanadinite specimens from the surface. In January 1984, he drilled fifty 14-foot holes above mine and loaded them up. He wanted to remove the remaining overburden and open up what he believed to be better vanadinite targets. The blast opened up the upper workings and the popular vanadinite chimney area, which many collectors had worked. John mucked out the tons of rock and debris that hoards of collectors had dug,

Above left: Bob Jones and Tom McKee collected this 4.8 cm tall vanadinite from a 30-foot chimney in the upper stope of the Apache Mine, Gila County

Mark Hay collection, ex Tom McKee
Jeff Scovil photo

Above right: Tom Garrett (left) and Casey Jones at the Apache Mine in the summer of 1969

Jones Family photo

Right: Vanadinite, 6.3 cm wide
Apache Mine, Gila County

Paul Harter collection
Jeff Scovil photo

Above: The Apache Mine
Les Presmyk photo, September 1972

Right: On March 2, 1975, Reg Barnes and Sam Nasser collected this 5.4 cm tall vanadinite from a tunnel on the 50 foot level of the Apache Mine, Gila County

Mark Hay collection, ex Reg Barnes
Jeff Scovil photo

which had buried the bottom part of the stope. In doing so, he uncovered a vein containing massive calcite encrusting larger vanadinite crystals. The calcite had served to protect the more delicate vanadinite by encasing the crystals. The vein material was loose. Using his hands and a minimum amount of hammer and chisel work, Mediz took out flat after flat of vanadinite specimens, mostly on matrix. He had the assistance of Richard Hauck, a New Jersey collector, for part of the time. One pocket in the zone contained the greatest known hit of Apache vanadinite. Numerous crystals up to 3 centimeters long grew into the calcite. Yellow-orange to orange-red crystals were found as skeletal groups and solid crystals associated with the white calcite, and there were matrix pieces up to 20 centimeters across. Many of the largest crystals were hoppered and were found loose in the pocket. In all, Mediz removed several hundred beer flats of Apache vanadinite. Mitchell Dale purchased Mediz's claim around 1990 and the mine has neither been accessible nor has it produced much vanadinite since.

Right: Last light on the Superstition Mountains in Maricopa County

Dean Hueber photo, January 16, 2011

Above: Gold wires on quartz, Mystic Mine, Maricopa County

Evan Jones collection

Jeff Scovil photo

Maricopa County

with contributions by Ed Huskinson and George Godas

More than half of the population of Arizona lives in Maricopa County, which encompasses the Phoenix metropolitan area. One of the original counties in the territory, it was mostly known for the agriculture made possible by the Salt and Verde rivers. Although the county was never a great producer of mineral wealth, a number of localities fall within its borders.

Maricopa is the supposed home to the famous Lost Dutchman Mine (see page 29). In 1891 while on his deathbed, Jacob Waltz claimed that his secret gold mine was located in the Superstition Mountains, the imposing mountain range east of Phoenix. That claim has since drawn prospectors of all ilk to the steep cliff faces of the Superstitions in search of gold.

While little has been found to substantiate Waltz's claim, collector and economic geologist Ed Huskinson did help to disburse an unusual find of native gold recovered between December 1991 and August 1992, from the **Mystic Mine** near, of all places, the Phoenix metropolitan area! The story of the specimens' discovery bears retelling.

Al Harding and his buddy Leith Pike retired with their wives to Sun City, Arizona. They started prospecting in their spare time, although their efforts were restricted to Wednesdays by the women (Wednesday was Ladies Day at the Sun City Golf Course). The men were free to prospect as long as they did not stray too far. They found mineralization at the southern end of the Hieroglyphic Range, only 9 miles distant. They staked 12 claims around a series of small prospect pits and leased another claim east of the old workings. Their surface rock chip samples assayed barely above background for gold, but they decided to drill it nonetheless.

Harding and Pike hired a Bucyrus Erie churn drill (better suited for drilling for water) and set to work. At a depth of slightly over 100 feet, gold could be seen glittering in the cuttings. They used core drills and hammer drills thereafter and bored almost 100 holes. They wanted to develop the mine themselves, so they hired several tramp miners. But after sinking ten feet of a shaft and receiving their first paychecks, the miners over-celebrated and failed to return to work.

Discouraged, the partners realized they would have to bring in a mining company to advance the project. Harding and Pike approached several companies and in 1980, signed a deal with Ranchers Exploration and Development Company of Albuquerque, New Mexico.

Ranchers greatly expanded the claim block and then combined forces with ACNC (American Copper and Nickel Company). Times turned tough, and ACNC bowed out in 1982. Still, with several multi-ounce intercepts of gold, Ranchers stuck with it. But in June 1983, Maxie Anderson, the CEO of Ranchers, died in a balloon crash. Without his vision and drive, Ranchers lost its way, and the company sold its interest in the property to Goldsil.

Fischer Watt Gold Company signed a lease with Harding and Pike in 1987 and working in partnership with Mining, Drilling, and Milling Company, began sinking the shaft that Al Harding had begun in 1978. They went after the deposit by driving a spiral ramp into the orebody. Production started in 1991, and in 1992 they shipped their first concentrate.

Mineralogy of the Mystic Mine

The mineralogy of the Mystic system is simple. Gold-bearing solutions passed up the high-angle footwall fault, and then spread out along horizontal to sub-horizontal shear (ladder) zones between the footwall and hanging wall faults. The schist is structurally perturbed; it is so altered that only crushed unrecognizable bits of material are left. The hydrothermal solutions that then passed through the schist also altered it intensely. The solutions carried quartz, then a pulse of gold, then more quartz, with a final overprint of iron- and manganese-oxides. The last (and uppermost) stage was carbonate rich, depositing calcite, which is stained black by included manganese oxides. There is a dusting of gold at or near the schist interface.

The gold at first appears crystalline, but this is simply a reflection of the gold having filled interstices in the crushed quartz, yielding a myriad of shiny faces in select specimens. A few small vugs hosted wire gold.

Huskinson first heard of the Mystic Mine in 1981 from a drilling company crew. They told him about the job they were working near Sun City. "The gold is so rich that you can pick up sections of the core and wiggle it, like it is glued together with rubber cement."

Huskinson was impressed,

> *Later, when the project was being mined, it was possible for me to visit the "Gentlemen's" bars in north Phoenix and buy specimens from the pole dancers, who the miners had tipped with gold nuggets and other specimens. Both of the geologists for Fischer Watt approached me about selling their gold. I brokered the specimens for Fischer Watt and carried thumbnail, toenail, and miniature-sized specimens to the 1992 Tucson Gem and Mineral show. The larger ones had either been mined outright and thus saved from the crusher or had been picked up on the 200 level when the morning shift arrived to find every horizontal planar surface in the heading covered with high-grade gold.*

The geologists told Huskinson that one morning they found two miners from the previous shift buried beneath a large slab; their feet sticking out into the stope. The slab had come loose and fallen on the high-graders, breaking their backs and legs. They survived and then sued the company for "unsafe mining conditions," despite the fact that they themselves had undermined the back and had neglected to bar it down, all while high-grading gold on company time! Huskinson thought that their case was thrown out of court—dismissed outright.

Fischer Watt curtailed its activities in 1992 when the mill at Congress, Arizona, closed. About 65 miniature to toenail specimens were sold at the 1992 Tucson Gem and Mineral Show. The rest of the specimens were returned to Fischer Watt's president Perry Durning.

The property made the last lessees rich. The Mystic Mine passed into the hands of a land developer in a land exchange with the BLM for some property near Tucson. The mine was closed and reclaimed. This exciting property is now one of the holes on a Sun City golf course.

Wulfenite on mimetite, 4.5 cm tall
Rowley Mine, Maricopa County
Dick Morris collection
Jeff Scovil photo

Dreams of glassy red-orange wulfenite in combination with yellow-orange mimetite (as pictured above and on cover) drew many collectors to the **Rowley Mine** through the 1970s and into the 1980s. And it even generously yielded specimens into the twenty-first century (see page 351).

Various combinations of the collecting team of Neal Pfaff, Gary Novak, and Stan Esbenshade were among those drawn to the mine. Together the trio extracted more than 100 flats of wulfenite from the mine. On one trip in 1979 Esbenshade and Pfaff hit groups of bright yellow wulfenite crystals to 1.25 centimeters. Brown-orange vanadinite after wulfenite pseudomorphs occurred on some of their specimens.

George Godas made regular trips to the mine from his home in Tempe. In the fall of 1986 Godas found a pocket of groups and isolated wulfenite crystals exceeding 2.5 centimeters on edge in a side stope just above the main level of the Rowley. When he got to the mine, there were three geology students from New Mexico at the top of the #1 shaft. They had been there for several days and had set off some small charges at the bottom of the shaft, near the first shoot (through which one could go into the Jobes stope). Nothing significant had shown up in the area while they were working. After they left, Godas went down the shaft and began to explore their zone. He began digging in the middle of the zone and collected a steady stream of very small wulfenite crystals on pieces of barite. He collected about four flats that day, and returned twice more to collect what turned out to be a seam with four compartments (pockets) for a total of about 15 flats.

George continued making periodic trips to the Rowley and in January 1991, hit another fine pocket of brilliant orange wulfenite crystals up to 2.5 centimeters (1 inch). Many of the smaller crystals were transparent and occurred with yellow mimetite on barite plates up to 7.5-by-20 centimeters. Godas recalls noticing,

> *...some promising mimetite fibers on the outer face of the northern flank of the inclined (#1) shaft. Some of the fibers were fairly long (2 cm) and some others, the shorter ones, were gemmy. I tried digging with chisel and hammer, but the barite there was tight and hard. I decided that the time had come to try the Cobra drill.*

About two months later, he returned with Darren Vernier. He drilled one hole, set off a shot, and came face to face with a breccia zone so large and loose that he had to hold it back with his left hand while plucking with the right. He went back to the zone 8 times before someone got wind of it and traffic increased considerably. One day there were five people digging in the same zone (including two snowbirds in their seventies who Godas had previously met on the surface and who had said then that they would never venture underground). One of the very best specimens, if not the best, was a small cabinet of four flawless crystals of brilliant yellow wulfenite crystals up to about 2.2 centimeters on a barite matrix.

Above: Stan Esbenshade and Neal Pfaff, just in from the mine, January 1978

Right: Gary Novak in the main incline of the Rowley Mine

Stan Esbenshade photo, 1980

Yavapai County

with contributions by Jeff Scovil, Les Wagner, Bruce Barlow and George Godas

Yavapai County stretches from the northernmost reaches of the Phoenix urban corridor to Sedona, and encompasses the Jerome, Prescott, and Bagdad as well as the gold mines near Wickenburg. Yavapai was one of the four original counties in Arizona. It was named for the Yavapai people who were the primary inhabitants of the area when the original county lines were drawn. It was carved into five units: Yavapai, Apache, Coconino, Maricopa, and Navajo counties in 1891.

The **Fat Jack Mine** is Arizona's premier amethyst locality. Located about 40 miles south of Prescott in the Bradshaw Mountains, the mine is best known for sceptered amethyst crystals but it also produces dull yellow-brown stolzite crystals up to 4 centimeters! The Fat Jack is another Arizona locality in which rare osarizawaite is found. The lead copper sulfate occurs as massive, pale green pocket-filling.

Ernest "Fat Jack" Campbell and A. M. Heath located the initial claim on June 18, 1896. Campbell's partner was a man named Major "Slim Jim" Strong. Their nicknames are not indicative of the statures of these two men: Fat Jack was reportedly tall and slender, while Slim Jim was short and squat. The claim owners reportedly packed out a considerable amount of high-grade gold ore in the early days of the property. The last assessment with Campbell's name on it was 1919.

Upper: George Godas stands up in the hole where he is working at the Fat Jack Mine, Yavapai County

Tony Potucek photo, December 1987

Lower: Quartz, 4.6 cm wide
Fat Jack Mine, Yavapai County

Les and Paula Presmyk collection
Jeff Scovil photo

An exhibit of Fat Jack quartz at the 1984 Arizona State Fair piqued collector interest in the locality. Jimmy Vacek was offered an early lot of Fat Jack quartz that he partnered with Dave Shannon. Later in the summer of 1986, Shannon collected a large pocket (Scovil and Wagner 1991), containing hundreds of amethyst scepters and "regular" quartz crystals. Shannon's pocket also contained Japan-law twin crystals—the first from the locality. As a bonus, the pocket contained an 8.4-centimeter-long doubly terminated quartz scepter with amethyst heads—an icon of the locality (pictured on the facing page). Commercial collecting interests took over the Fat Jack in the fall of 1986, and free-lance collecting was stopped. Nothing is known about the quality of material during this time, but the property re-opened in the summer of 1987 and the free-for-all resumed.

Jeff Scovil and Les Wagner did some collecting in the latter half of 1987 after the Fat Jack reopened. Their best pocket was found to the left of the adit. Wagner recalls that the pocket was a mud-filled, collapsed cavity approximately 70 centimeters wide and 50 centimeters deep. They carefully removed the scepters one-at-a-time using only their bare fingers and a couple of nylon probes from Wagner's electronics repair box. The pocket yielded about 100 pieces of all lengths and shapes.

The Scovil-Wagner amethyst scepters are uniquely gemmy, and their milk-white shafts are more slender than on other crystals from the locality. Individual crystals from the find are as long as 7.5 centimeters. The pocket included a few larger pieces that appeared to have been intergrown, broken apart, and later re-healed with another growth of quartz.

Moreover, the pocket contained loose quartz crystals that were

Collector Harold Michel bought this unique crystal from David Shannon, who dug it in 1986. Michel was a fixture in the Phoenix mineral community. He built one of the finest collections of Arizona minerals in the area. He proudly and frequently shared his collection, displayed in cases encircling his living room, with school groups and other collectors. Michel died in May 1997. He was 74 years old.

Quartz, 8.3 cm long
Fat Jack Mine, Yavapai County

Evan Jones collection
ex Harold Michel
Jeff Scovil photo

overgrown with somewhat zoned amethyst. In some cases, this overgrowth almost entirely enclosed the earlier crystal. This is evident on a few of the larger pieces that have doubly terminated amethyst overgrowths.

There was only one "matrix" specimen in the find, but the matrix was actually a growth of small, intergrown, doubly-terminated quartz crystals, with a couple of scepters interlocked in this growth. The mud and caliche was very thick in the bottom of the pocket, and that is probably what protected the specimens from further separation or damage.

Wagner remarked that the stolzite crystals that he and Scovil found are reminiscent of those from France, although they are darker in color, which Wagner described as a muddy brown-yellow. A few of the crystals were in excess of 3 centimeters on edge. Ray Grant found one or two comparably

Quartz, 6.3 cm wide
Fat Jack Mine, Yavapai County
Les and Paula Presmyk collection
Jeff Scovil photo

sized stolzite crystals. Grant's are dark gray and very etched. It seems that the larger they grew, the further from the classic yellow that they became. The larger crystals also tended not to show much fluorescence.

Collecting continued to be very productive throughout the latter 1980s and into the 1990s. George Godas and Howard Hyman leased the property in 1992-93 and in 1994 Godas held the lease with David Shannon. Godas said there were quartz crystals everywhere, at least in the beginning. One Saturday in June, George went up to the Fat Jack with three wine "flats." He filled them and their lids, piled crystals two layers deep in the back of the Isuzu. Godas filled his digging boots with quartz groups and removed his pants and shirt and rolled some of the better crystals in them as well. Unfortunately, there is no known photograph of Godas driving back to Tempe *sans* clothes!

As long as the mine was under the ownership of Fred Lorette, it was leased to or available to collectors one way or another; in fact, Lorette encouraged collecting. However, new owners acquired the property later in the 1990s. They made the Fat Jack a fee-to-collect site. It has since seen little specimen production.

The **Bagdad** is an active open pit copper mine that exploits a modest sized porphyry stock. Copper mineralization was discovered on the Bagdad group of claims in 1882, but there was no mining activity until 1906. The Giroux syndicate was unsuccessful in developing the property, and a rapid succession of owners followed. Finally in 1911, the Bagdad Copper Company acquired the claims.

Bagdad was steadily developed into the 1970s, and although there were not many minerals produced during the first 50 years of mining, there were hints that Bagdad might one day produce specimens. Wayne Thompson recalls, as a boy, going with a local rock club to the open pit. The company allowed club members to collect whatever they could carry. Wayne remembers pit benches with large boulders of chrysocolla sitting on them; some contained drusy quartz pockets. Many of the boulders were layered with green malachite and made for attractive combinations. When cut they produced beautiful slabs.

Richard Bideaux wrote about groups of lustrous blue-black azurite crystals up to 10 centimeters that a miner had collected around 1960 (Bideaux 1973). And, further hinting at future finds, in the late 1960s, an area was encountered that produced pseudomorphs of malachite after azurite and of chrysocolla after malachite after azurite in sprays up to 12 centimeters long. While the malachite pseudomorphs were stable, the chrysocolla specimens were not: they became dehydrated, cracked, and fell apart. Collectors preserved the specimens by covering them with shellac or lacquer, both of which yellowed in time. Jimmy and Joyce Vacek were

Chrysocolla, 4.5 cm wide
Bagdad Mine, Bagdad, Yavapai County
Tony Potucek collection
Jesse La Plante photo

involved in the area off and on for years. Bagdad mineral dealer and mine employee Stan Pike was a friend of theirs, and he sold them large quantities of drusy quartz on chrysocolla. The mine also produced some of North America's best conichalcite.

About 1986, dealers Bruce and Debbie Barlow secured the collecting rights for the pit from Cyprus-Bagdad (in 1973 Bagdad Copper Corporation had merged with Cyprus Mines Corporation). The most colorful oxide minerals were found in the Hawkeye Fault in the Bagdad pit. In 1989 the Barlows made their first significant strike, finding chrysocolla, malachite, and drusy silica over both the malachite and chrysocolla.

They did not have another significant find until 1992, but it was big. Mining on the 2900 to 3250 benches of Giroux Point had exposed significant amounts of chrysocolla and malachite along with drusy quartz. Barlow brought in Guy Myers and Graham Sutton to help with the find. The trio collected over 40 tons of chrysocolla from the area. One 15-centimeter vug had several 1.5-centimeter balls of grass-green malachite on drusy quartz on chrysocolla. It was cut out of a 300-pound boulder.

Another locality higher up in the open pit but beneath Giroux Point produced a boulder that started as a 5,500-pound chunk measuring 7-by-5½-feet. The mass was reduced to a 400-pound chunk of gem chrysocolla, which later sold at auction for $70,000.

In the early 1990s while looking for agate in Yavapai County near the **Agua Fria River**, collectors Al and Mary Rose found a small odd-looking, pinkish crystal in the sand. Following the float, Al went to work on a small knob and found a pocket containing a few blocky quartz pseudomorphs; they did not know at the time what mineral the quartz had replaced.

The couple returned a year later with their infant son Erik and a few friends. After searching for several hours, they noticed weathered domes in the reddish rhyolite in a contact between two types of volcanic rock, not far from a quartz ("chalcedony") vein that they had worked before. Al went to the top of the formation, broke a large rock loose and opened a pocket of salmon-pink crystals, a few more than 5 centimeters long. Some of the crystals were attached to the walls of the cavity while others were loose in the 3-foot-long pocket. The bottom of the pocket was covered with chalcedony such that the crystals in the bottom were "frozen" in black agate.

Conichalcite (bright green) on malachite (dark green) on chrysocolla (blue-green), 6.4 cm tall

Paul Harter collection
Jeff Scovil photo

The Roses had the material analyzed and learned that they were pseudomorphs of quartz after anhydrite. The crystals on most of the specimens were lying flat on the matrix, although a few specimens had crystals standing up on the matrix in various directions. The average crystal was 1.5 centimeters long.

The Roses returned annually to the spot for a few years without finding another pseudomorph pocket. They stopped going back until 2000 when Al had a dream in which he saw a hump under a bush. Compelled to go back, they returned to the locality, and Al said, "the hump from my dream was there! I broke into the rock and found another cavity!"

The Roses recovered approximately 40 specimens, ranging in size from 2 to 7 centimeters, from this second pocket. Roughly half of them were similar to those discovered six years prior. Some of the blocky crystals were

Singular Finds from Yavapai County

Above: Quartz after anhydrite
12.7 cm wide, "Agua Fria River"
Collected in 1994 by Al and Mary Rose
Les and Paula Presmyk collection

Facing page, upper left: Calcite
6.1 cm tall
Old Dick Mine, Bagdad
Les and Paula Presmyk collection

Facing page upper right: Vanadinite
7.7 cm tall, Lincoln Claim, near Humboldt
Evan Jones collection

Facing page, lower left: Gold, 5.6 cm tall
Sunshine #8 Claim, Bradshaw Mountains
Les and Paula Presmyk collection

Facing page, lower right: Cerussite and barite, 8.6 cm wide
Amethyst Hill
Flagg Mineral Foundation collection

All five photos are by Jeff Scovil.

partly or completely coated with translucent, gray to black chalcedony or hyalite opal. The exact location of the site remains a family secret.

North of Wickenburg and just in Yavapai County, **Amethyst Hill** is covered with pits and shallow shafts, but quartz is not the target of prospectors and miners working this knob; the mineral of note is cerussite. In 1985 an old prospector told Stan Celestian and Ken Sprik that cerussite was found at Amethyst Hill. The area was small, but the digging easy. The cerussite was found associated with barite in a watercourse. Two of the best reticulated specimens were sold at the Tucson Gem and Mineral Society show in 1986 for $150 and $200. In 2009 Stan and Dick Zimmerman claimed the locality and continued their efforts. They uncovered a 3-foot wide, barite-lined pocket that contained more than 130 cerussite crystals. The vast majority were in layers on the floor of the pocket with only a few still clinging to barite.

Yavapai County also hosts a number of "singular" occurrences. The **Old Dick** underground mine near the Bagdad Mine produced exceptional calcite specimens in the 1960s and 70s; metal-detecting "gold bugs" located the **Sunshine #8 Claim** to protect an interesting find of crystalline gold in the Bradshaw Mountains; and back in the late nineteenth century the **Lincoln Claim**, west of Humboldt, produced some of the earliest known vanadinite specimens from Arizona.

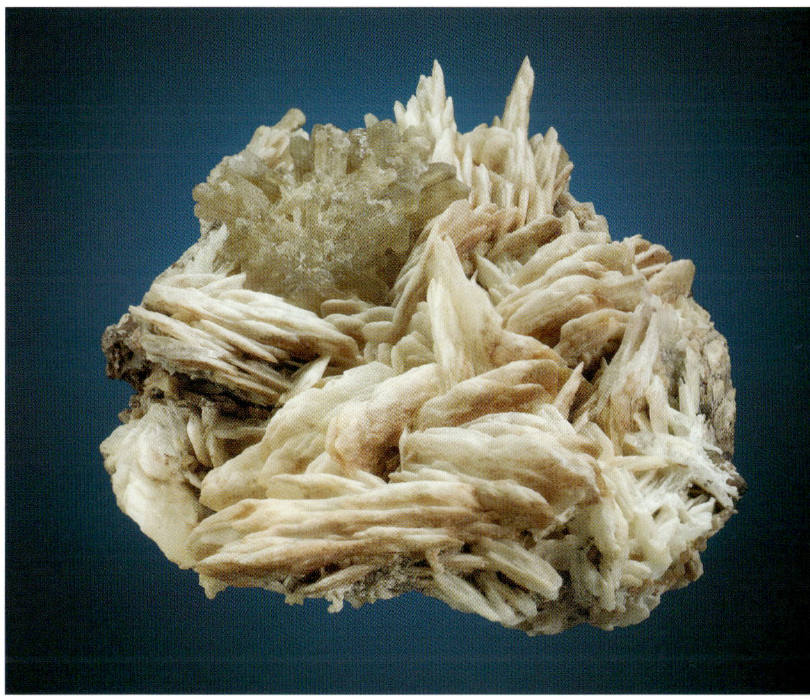

Mohave County

with contributions by Mark Hay, George Godas, and Andy Clark

Mohave County was one of the original Arizona Territory counties, and it spanned the northwestern extent of the state. Today Mohave County is a strip bounded by the Colorado River and extending from the Utah border to Lake Havasu City in the south. Lake Havasu is a large reservoir on the Colorado River, and the starting point for the Central Arizona Project canals delivering water to Phoenix and Tucson. The county seat for Mohave County is Kingman, the heart of Arizona gold country.

Mineralization, indeed crystallization, in the Cerbat Mountains north of Kingman has been known for better than a century. Hodge (1877) reported, "In the Poland mine, are many beautiful specimens of white crystallized sulfate of lead (anglesite), a rare mineral in all mining countries." He also noted the occurrence of native silver in the veins of the Keystone and the Lone Star mines, both of which are near the town of **Mineral Park**. Many collectors remember the exotic curled specimen of native silver from the collection of the late mineral collector Dave Eidahl that is now in the Los Angeles County Museum of Natural History collection. The specimen originated in one of the mines in the vicinity of Mineral Park and dates to the late-1800s. It came from the Dohrmann collection, which was auctioned in December 1886.

From a collector's point of view, the most noteworthy mine in the Cerbat Mountains, north of Kingman, is the small, inactive **Western Union Mine**. "Ore" production from the mine was a trivial 12 tons (vanadinite), and workings consisted of a 130-foot-long adit with a 40-foot deep winze.

Andy Clark, Brad Archer, and Jeff Archer were in the Kingman area in

1974 picking up turquoise for resale when Clark ran into Jack Harrison, owner of the Western Union Mine. Harrison encouraged them to come to the mine. Over the course of several visits in 1975, the group collected four to five flats of the brown to red-brown vanadinite from the back of the main adit above the winze and down the winze. The vanadinite occurred as isolated brown, blocky simple hexagonal crystals to about 1 centimeter. It was scattered on an altered gray-white earthy granitic matrix. They sold some specimens from a room in the Desert Inn at the 1976 Tucson show. They specimens that they gave to Harrison for allowing them to collect were ultimately donated to the Arizona State mineral collection (Hay 1997). The specimen that Brad Archer had kept (and later sold to Les Presmyk) piqued the interest of Mark Hay and Dick Morris.

Ready to take on another vanadinite locality, the pair made their first trip to the Western Union Mine in December 1988. Unable to secure permission to cross private property on the east side of the canyon, Hay and Morris went up the west side of Cerbat Canyon. They found the mine in excellent shape.

They took a quick look around and determined that the best potential for vanadinite was down the winze. The mineralization was richer and the color more red than brown toward the back of the short drift at the plugged bottom of the winze. They found a crack, which was completely filled with groups of tightly intergrown, slender vanadinite crystals in various shades of orange and brown. Within several hours they collected six flats of specimens. The best cabinet specimen went to the Arizona-Sonora Desert Museum.

In May 1991 Mark Hay and George Godas got the best pocket the mine has ever produced. Hay hit a 5-centimeter high, 25-centimeter-wide pocket that produced about a dozen specimens. The vanadinite is up to 1.6 centimeters, brick red, and is geometrically bizarre, with sceptered growths, simple barrels, and skeletal features.

Godas was working above and a few feet away from Hay when he hit that best pocket—a vertical gash 30 centimeters tall and 10 centimeters wide. In classic Western Union style, the whole zone was a mish-mash of intergrown vanadinite crystals. It took several hours for them to extract the loose pieces. When they reached the end of the 45-centimeter-deep pocket, they had collected 150 specimens; the largest specimen was more than 13 centimeters across. In the mid-1990s Jack Harrison allowed the claim to lapse, and George Godas filed a claim on the mine. It is still in his possession.

Facing page, upper: Vanadinite, 4 cm tall
Western Union Mine, Mohave County
Collected in Oct. 1989 by George Godas

Mark Hay collection
Jeff Scovil photo

Facing page, lower: Vanadinite
7.3 cm wide
Western Union Mine, Mohave County
Collected in May 1991 by George Godas and Mark Hay

Dick Morris collection
Jeff Scovil photo

The **Portland Mine** is on the western flank of the Black Range. It was discovered late—around 1920. In the 1970s several mining companies explored the area for precious metal potential. At that time the Portland Mine consisted of shafts and underground workings with some open cuts on the surface that had been developed in the 1930s and 40s. Western States Mining Company picked up the property in the early 1980s and developed the small orebody. The deposit was exhausted by 1988, and reclamation efforts commenced.

While working for Western States, collector Tom Hughes visited the Portland Mine when the orebody was nearing depletion. He found battered,

clear to white, modified calcite scalenohedra on one of the leach dumps. Hughes contacted Tony Potucek about his observations and suggested that Potucek visit the mine and try to determine the source of the crystals.

In the spring of 1989, Potucek traveled to the then abandoned mine with his brother-in-law. They walked the benches searching for the source of the calcite. On their second day looking, they found the deposit: a high angle vein cutting across two benches. The vein was truncated above and below the benches by other faults exposed in the wall of the pit. He noted three habits of calcite in the occurrence: modified, clear to white, scalenohedrons; flattened barrel-shaped crystals with truncated tops; and clear platy, parallel growths on dacite. He also collected a very few "butterfly" twins to 2.5 centimeters that had formed amidst some of the smaller scalenohedrons. The largest calcite crystals are 6 centimeters long, and Potucek collected crystal groups as large as 25 centimeters.

Potucek later enlisted the help of fellow Arizona collector Dick Morris to continue collecting the vein, which was productive along its entire length. They and other collectors including Mark Hay, George Godas, John Callahan, Benjy Kuehling, and Paul Harter intermittently collected the Portland Mine calcite for about 18 months from the late 1980s into the early 1990s. All told, the vein produced about 150 flats of calcite specimens.

George Godas was a frequent visitor to the mine. He "really liked to camp out at that place—stars, big horn sheep, burros, and a green lake (in the bottom of the pit)."

Right: Fluorite with wulfenite, 6.2 cm tall
Hull Mine, Yuma County

Evan Jones collection
Jeff Scovil photo

Facing page, upper: Calcite, 5.7 cm tall
Portland Mine, Mohave County
Collected in March 2003 by George Godas

Mark Hay collection
Jeff Scovil photo

Facing page, lower: Calcite, 12.8 cm wide
Portland Mine, Mohave County

Tony Potucek collection
Jeff Scovil photo

Yuma County

with contributions by Jack Crowley, George Godas, Mark Hay, and Ray Grant

Yuma County is in the southwestern corner of the state. It was one of the original counties, but it lost more than half of its area when La Paz County was split out of it in 1983. Only a few mining districts are now within its boundaries.

The mines in the **Castle Dome** mining district first began producing specimens in the 1800s. Mineralogists such as William P. Blake became a mining company president in the district and collected specimens from the company mines in Castle Dome. But specimen collecting is not recorded from the mines from 1900 through the 1950s.

In 1976 Jack Crowley and Gary Moss made a trip to the Castle Dome mining district. Bill Sanborn had mentored Jack for years, and Sanborn suggested that the district offered strong specimen potential. Crowley and Moss found the **Puzzler Mine** covered over with wood and a small tramcar was sitting on that. They removed some of the boards and dropped down the shaft to the second level and found gray-green vanadinite crystals exposed along the walls in the northern stopes and next to the shaft. The crystals were tapered and up to about 7 millimeters with generous crusts of crystals scattered on matrix. The find produced around 100 flats of material.

During the mid-1970s Wolfgang Mueller gained access to the **Hull Mine** (formerly the Rialto Mine) and found orange wulfenite crystals from an opening near the shaft on the 300 level. The wulfenite was associated with altered galena crystals (cerussite/anglesite coatings) and transparent blue-green fluorite. The wulfenite crystals reached about 1.2 centimeters on edge. Years later in 1990, Mueller and George Godas accessed the lower workings by the mine's escapeway and found small specimens of lemon-yellow wulfenite associated with bright yellow-green balls of mimetite. The area held promise for better production but unfortunately the escapeway timbering burned; the access and ladders were destroyed in the fire.

Upper: Dick Morris in the "Winze" pocket, 100 foot level Puzzler Mine, Yuma County Mark Hay photo, June 1988

Lower: Vanadinite from the "Winze" pocket, 100 foot level Puzzler Mine Castle Dome district, Yuma County Mark Hay photo, June 1988

Some exploration and development work in the 1970s had opened up the mine making it accessible via a circular decline large enough to accommodate a pickup truck. In the mid-1980s Graham Sutton and Bobby Lane began collecting in the Hull Mine. The decline made almost two loops on its way to the bottom, which is equivalent to the 300 level. Graham and Bobby went down to about the 250 level and encountered a galena vein bearing crystallized cerussite and unusual wulfenite. They collected cerussite crystals in the 5 to 12 millimeter range, some with small 3 to 4 millimeter blocky orange wulfenite crystals perched on the cerussite. The cerussite formed lustrous, gray-white aggregates on corroded galena and on a propylitized altered volcanic rock. The largest aggregates reached 1.5 centimeters. Over a dozen or so trips, they collected about 20 flats of material. The best of the find went to Harold Michel. At about the same time, Ray Grant collected very fine cerussite crystals from the same area. No one is sure who found the area first.

In early 1987 Mark Hay, Dick Morris, and George Godas visited the district. They found cerussite in the Hull Mine, and investigated a stope on the 300 level where they found a crack. In the sandy material in the bottom of the crack they found several thick, yellow, tabular wulfenite crystals to 1.2 centimeters. The crystals were found as loose singles; there were no groups, and they were never on matrix. The large crack twisted up and out of sight. It appeared barren except for the wulfenite. Morris also pulled out a 3.8-centimeter thick reticulated and twinned cerussite crystal.

In July of 1987 Hay, Godas, and Morris turned their attention toward the Puzzler Mine that Crowley and Moss had rediscovered in 1976. The weathered ore car was still chained to the wooden mineshaft cover. They moved the ore car slightly and accessed the shaft. Brown vanadinite crystals seemed to cover everything, but they were of little interest.

On New Year's Day, Jan. 1, 1988, Hay and Morris returned to the Puzzler. They felt confident that if they invested some time and hard work, they would hit a good pocket. They found a seam that opened into a pocket of lustrous gray-green vanadinite, about 45 centimeters high by 60 centimeters deep and up to 10 centimeters wide. The lustrous crystals curved and were up to 5 millimeters long, although they appeared larger having formed in parallel groups on matrix. The largest specimen that they collected was about 18 centimeters across. White tabular barite crystals up to 6 millimeters are associated with the vanadinite.

Above: Vanadinite, 6.1 cm wide
"Winze" pocket, Puzzler Mine
Yuma County
Collected on May 14, 1988 by Mark Hay

Mark Hay collection
Jeff Scovil photo

Right: Vanadinite, 5 cm wide
Green pocket, 180 level
Puzzler Mine, Yuma County
Collected on Jan. 1 1988 by Dick Morris and Mark Hay

Mark Hay collection
Jeff Scovil photo

Quartz and hematite, 5.8 cm wide
Veta Grande Claim, Sugarloaf Mountain,
La Paz County
Collected in 1987 by Bob Lane
Evan Jones collection
Jeff Scovil photo

La Paz County

with contributions by Mike Sprunger, Jim Walker, Graham Sutton, Bobby Lane Mary-Fong Walker, Stan Esbenshade, and Mark Hay

Yuma County was divided roughly along the Gila River into a northern (La Paz) and southern (Yuma) section in 1983. The heart of La Paz is the town of Parker, which is located on the Colorado River and is very much a tourist destination for desert water sports. Parker Dam was built in 1928 and created the placid Lake Moovalya. When La Paz spun off from Yuma, it took many of Yuma's most famous mineral localities with it, which has caused some confusion when looking at mineral specimen labels that predate the 1983 county reorganization.

La Paz County is home to the town of Quartzsite, which in turn hosts the remarkable annual phenomenon "Quartzsite," in which waves of RVs descend on the town in January to participate in an open market of rocks, rough, and fossils. In 2010, more than 2,000 vendors (the year-long population of Quartzsite is 1,900) participated in the event.

Quartzsite was supposed to have been named after the metamorphic rock, quartzite, when the post office was re-established in 1897. But the postal service misspelled the name, adding an "s" to it.

Local lore, however, has the name originating from the nearby **Crystal Hill**, known for large, clear and chlorite-included quartz crystals and clusters, which became a popular collecting site in the 1950s. Three collectors uncovered a notable pocket containing three 25-centimeter plus crystals. One of the specimens is buried with its owner in the Quartzsite cemetery, the whereabouts of the second one is unknown, and the third specimen is part of the Flagg Mineral Foundation collection.

A few miles southwest of Quartzsite on Sugarloaf Mountain, Henry Hubbard and a partner recorded the **Big Bertha Claim** in 1949. Hubbard drove a tunnel along a fissure "vein" and encountered a large cave-like opening approximately 12 feet long, 6 feet high, and more than 3 feet wide. Hubbard offered a group of hematite specimens to Ward's Natural Science Establishment, Inc. Ward's kept three of the specimens, priced at $15, $10, and $1, and returned the balance. Although Hubbard continued mining, not much is known about his subsequent finds.

In time the locality fell idle, but a Big Bertha hematite specimen that Mike and Sandra Sprunger saw in Tom McKee's collection inspired the couple, in 1974, to try their luck there. They worked the vein and found the best specimens Big Bertha had ever produced.

Pyrophyllite and quartz
Field of view 4.8 cm long

ex Fred Pough collection
Jeff Scovil photo

Veta Grande Claim, Sugarloaf Mountain, La Paz County
Collected in 1978 by Dick Jones, Neal Pfaff and Gary Novak

The Srungers headed to Quartzsite to sort through their fabulous find and reconstruct the crystal groups. While they were working, Mike mentioned that it would be nice if they found some "iron roses" to complete their hematite suite, so when the pocket was complete, they hopped on their motorcycles and headed up the side of Crystal Hill. They found a seam that Mike believes old-time prospector and mineral collector Joe Livingston had been working.

They worked the dump, and Mike dug what was left of the seam. He found a section containing palygorskite "mountain leather" and iron roses. They collected more than 300 hematite roses which, when cleaned, were mirror bright. More than 100 of the crystals exceeded 1.2 centimeters; the biggest exceeded 3.2. Almost all of them were floaters, but a few crystals were found on quartz plates. The three best specimens came from the vein. According to Mike, one end of the vein had matrix roses, which were mostly weathered loose. The other end produced the doubly terminated "floater" crystals often associated with needle quartz. One of the three best went to Leonard Bedale, who said he sold it to Jim and Dawn Minette. Sprunger and his wife had recovered the best hematite not only from Big Bertha but also from Crystal Hill!

In the course of collecting at both localities, the Sprungers also found a total of 15 to 20 "Faden" quartz crystals. Mike described the largest crystal from Crystal Hill as being bright, clear, and about 6 centimeters long. Later while in Casa Grande enjoying one of Helen Jones's pot roast dinners, the Sprungers gave the large crystal to Dick Jones.

Jones later filed a mining claim on Big Bertha, renaming it **Veta Grande**. He along with Neal Pfaff and Gary Novak worked the locality in October and November of 1978. They hit a 4-foot pocket containing tabular, brilliant hematite crystals as large as 2.5 centimeters wide on a white quartz matrix. The quartz crystals were up to 9 centimeters long. The pocket also contained pale green mats of pyrophyllite crystals. Jones sold the specimens at the November 1978 Pasadena show.

In 1984 California-based collectors Jim Walker and Gary Moss were exploring the area around Veta Grande when they found an 12-by-9-by-9-foot erratic boulder that had some hematite crystals showing. Not known for their delicacy, Walker and Moss drilled the boulder, loaded it with dynamite, dubbed it the **Electric Meatball,** and blew it apart. While the shattered boulder did not fulfill their hematite dreams, embedded in the pieces were interesting, pale, triangular crystals up to 2 centimeters on edge—some of the best zunyite every found. Never a dull moment, Jim recalls wrapping up their specimens,

> *We were busy loading up chunks of the more promising looking material when a voice from the sidelines shattered the desert still. "Whatchu boys doin here?" "YOU AIN'T TAKIN' OUT ANY GOLD, ARE YOU? Our finely tuned survival skills, honed by two lifetimes of mineral pillaging and occasional surprise visits by actual claim owners, kicked in. Usually, if you live through the first few seconds of these encounters, you're home free. Such was the case here. In the ensuing minutes, we convinced the codger connected to the voice that we were severely retarded; we did this by showing him what we had been feverishly loading into all those milk crates. "That ain't worth nothin, boy!" trickled from his lips like gas from a dyspeptic cow's rear. We fervently agreed with him and replied that, lowly as it was, this humble material from his fabulous gold claim was nevertheless beautiful to us. Could we take some home? All the while we were simultaneously planning as hasty a retreat from the vicinity as we could muster without appearing nefarious.*

Zunyite, 17.5 cm wide

"Electric Meatball" near the Veta Grande Claim, La Paz County

Collected in 1984 by Jim Walker and Gary Moss

Mary Fong- and Jim Walker collection

Jeff Scovil photo

The following year, Wayne Thompson sent a crew headed by Bill Hawes to the site with a bulldozer, but money ran out before any pockets were found. Graham Sutton and Bob Lane filed a new claim on the property in 1990. They renamed it the **Purple Cow Ledge** and put considerable energy and effort into developing the property, but the hard quartz zone only gave up a few pockets. This time it was not the hematite but the quartz crystals—chlorite-included groups and Fadens—that paid the bills. The property has been all but inactive since Sutton and Lane stopped working the claim.

A line noting a "spectacular abundance" of pyromorphite and mimetite on the 352 foot level of the Ramsey Mine in a 1962 Still and Still Consulting Mining Engineers and Geologists report caught the attention of Dick Morris in 1990 while he was researching potential collecting localities. Consulting engineers almost never write in extravagant terms about a mineral occurrence in a mine. This occurence *had* to be investigated!

Morris organized a trip to the mine with George Godas and Mark Hay in August 1990. Their initial impressions were bleak: the upper 300 feet of workings were barren. But true to form, Godas made the first find. On the 350 foot level he found a crack near the main incline that contained moderately lustrous curved greenish vanadinite crystals to 0.5 centimeters. Hay explored further down the inclined shaft and discovered vanadinite on about the 400 foot level. These crystals were far superior to those on the 350 level. The crystals ranged in color from mahogany-orange to yellow-brown, some with greenish overtones. All had brilliant luster. The matrix is invariably sprinkled with brilliant black micro-crystals of descloizite. Morris, Godas, Hay and fellow collector Ken Algier, collected the locality for the next three or four years before it was closed by the owners.

Above: Les Presmyk spent some time collecting at the Purple Cow Ledge (aka Veta Grande or Big Bertha) while Graham Sutton and Bob Lane were working the property in the early 1990s. Presmyk collected an exceptional hematite and quartz (in fact it won Presmyk the 1991 Prospector's Trophy at the Denver Show). Presmyk also took this photo of Graham Sutton working the vein.

Right: Vanadinite, 3.5 cm wide
400 foot level
Ramsey Mine, La Paz County

Collected by Ken Algier

Mark Hay collection
Jeff Scovil photo

"Sméagol" Fluorite, 15.5 cm wide
Heson Mine, Mystery Hill mine group
La Paz County
Collected by Stan Esbenshade
Stan Esbenshade collection
Jesse La Plante photo

The Buckskin Mountains are in a remote region of northern La Paz County. In those mountains, the Mystery Hill mine group in the Alamo mining district on the south side of the Bill Williams River has produced some notable fluorite.

In the 1960s collector Bob Bartsch dug in the **Mystery Hill** mine group. Although he met only moderate success, he always felt the fluorite locality, frequently referred to as the **Heson Mine**, held potential In the late 1970s Neal Pfaff and Stan Esbenshade visited the mines on Bartsch's recommendation.

Esbenshade recalls that an adit provided access to a small room-and-pillar type mine, which was located in a 4- to 5-foot-thick, flat-lying silicified fault in schist. The fault contained fluorite and galena mineralization, which was thin and tight in the southern stopes, but seemed to open up in the northern workings. The lower part of the limestone bed had many thin bands of quartz with openings between them that contained small, green to purple fluorite octahedra. The upper 12 to 18 inches of the bed was massive quartz, which contained pockets of larger, color-zoned octahedrons and the infrequent and unusual pale green, lustrous and transparent crystal.

The upper band of fluorite was in very hard rock, so it was difficult to remove undamaged specimens. Additionally, most of the material was ugly and as a result didn't sell well. They continued to occasionally go to the Heson nevertheless.

On one trip after Pfaff and Esbenshade had been at the mine for two days and had retired to their camp. Pfaff decided to turn in, but Esbenshade wasn't ready. They had shot their drill hole before leaving that evening so Esbenshade went back into the mine to see what the blast had uncovered. He looked through the rubble then turned his attention to the face that they had been working. Not seeing much, he started barring some pieces from the thick quartz layer near the top of the bed. After a few rocks were barred down, he saw a small hole that had opened up. Looking in the hole, Esbenshade could see a heap of loose, green fluorite crystals and rocks at the bottom of what appeared to be a tubular pocket. He could also see a larger, greener, crystal toward the back of the tube. He wanted to get at the large crystal right away, but it was out of his reach, so he began removing the loose specimens from the front of the pocket. Although these were the best specimens he had ever seen from the mine, Esbenshade was focused on that large crystal, which was becoming more interesting to him as he got closer to it.

The crystal was obviously loose but was pinned in by the pocket rubble. A little careful work freed it. He pulled it out slowly, so as to not bump it on the pocket walls. The specimen (pictured facing page) is a 17.5-centimeter group of green crystals with purple edges. Esbenshade remembers sitting there and looking at it for a long time,

> *It was 100 times better than anything we'd ever seen from that mine. There was nothing more in the pocket so I put the others in some flats. I wrapped the good one well and stuffed it under the seat of the jeep before going to bed. Neal had already seen the other specimens by the time I got up and thought they were pretty good for the locality. I decided not to show him the good one because I figured he would want to get out of there as the hope of finding anything better probably didn't exist in his mind. We worked the rest of the day and went beyond the pocket and got nothing. So, we loaded up the jeep and left. As we were getting ready to turn onto U.S. 95 to Yuma, I told Neal to pull over. I pulled the specimen from under the seat and handed it to him. He was as stunned as I was the first time I saw it. He agreed that the motivation to keep beating on that hard crap would have evaporated after seeing the piece. We named it named the specimen "Sméagol" after the character in the* Lord of the Rings*, who murdered for the One Ring.*

Esbenshade and others eventually mined that rib through to another drift, but no one ever found anything even close to the secondary specimens, let alone "Sméagol."

One of only a handful of hardcore female field-collectors, Mary Fong was turned on to the wonders of the underground by then boyfriend Jim Walker. He took her first to the **Hamburg Mine** in La Paz County (when it was still Yuma). This is Mary's story:

> *It's a sunny spring day in the early 1970s and I am reminded about the question that Jim asked me when we started dating, "Do you like rocks?" I wasn't so sure after eating a pile of dust as we drove down a dirt track that seemed more roller coaster than a road. Jim had targeted the Hamburg Mine as the site for my first underground mine experience.*
>
> *We arrive at the mine, which to my novice eyes was literally in the middle of the back of beyond. (It's actually north of Yuma and east of the famous Red Cloud Mine, in an area between the Colorado River and the Yuma Proving Ground.) We got out and surveyed the terrain. Since the mine dates to the 1880s, there was the typical debris from an old mining operation: pieces of mine equipment and an open mine entrance that went to black. No one had obviously been there for a long time.*
>
> *We strapped on mine lights and head on down. I'd never been underground before, so I approached this foray with more than a little trepidation. The mine entrance is a decline that goes down for about 30 feet before it levels off. It takes a while for our eyes to adjust to the dark, but as we head down the slope we notice a small alcove off to our left. As it is, I'm not too sure about this whole adventure: it's dark, it's dusty, I'm uncomfortable because I'm sliding down this slope and it's hard to get my footing—what have I gotten myself into?*
>
> *Meanwhile, Jim is examining the mine walls—looking for glints of red on the walls that would indicate vanadinite. Although the walls are also covered by dust, there are some obvious fissures in the back wall of the room and Jim begins to poke around while I just try and keep from falling down. He begins to find bright, prismatic crystals of vanadinite. He calls me over to show me what we are looking for, and I gingerly get closer to the wall rock. But once I spot the brilliant red vanadinite, my unease gets erased and I am hooked by the excitement of discovery: Hey this is fun! And a field collector is born.*
>
> *We spent the next few hours moving rock, revealing specimens covered with small colorful red prisms. We carefully wrapped everything in toilet paper (the field collectors wrapping material of choice). The pièce de résistance came, when in the back of the fissure, Jim spotted unusually large crystals for the mine—6 to 7 millimeters in length. Because they were literally perched on a piece of rock that was wedged into the back of the fissure, it took quite awhile to remove the surrounding rock without dislodging and breaking the specimen. But time stood still while Jim patiently worked the specimen out. I had completely forgotten about the discomfort of being underground, stooped in a cramped position with my gloved hands held below the specimen to prevent it from hitting the floor*

Facing page: Vanadinite, 8.7 cm wide
Pure Potential (North Geronimo) Mine
La Paz County
Dick Morris Collection
Jeff Scovil photo

Right: Wulfenite, 4.2 cm wide
Pure Potential (North Geronimo) Mine
La Paz County

Evan Jones Collection
Jeff Scovil photo

Above: Vanadinite, 3.5 cm tall
Pure Potential (North Geronimo) Mine
La Paz County

Dick Morris Collection
Jeff Scovil photo

of the mine should it fall unexpectedly. And we were greeted by success: that Hamburg Vanadinite specimen still resides in our collection today and, in 35 years, we have never seen larger crystals from that locality.

Mary Fong and Jim Walker had married and were a veteran collecting team in the early 1980s when they were prospecting near the Red Cloud Mine and Jim observed "a curious remnant of the original fault/ore zone, no more than ten feet by eight by six feet high-an odd-looking little mesa."

As it turns out, this was the up-dip remnant of the Geronimo ore zone immediately west of the underground workings (see below). They discovered wulfenite in the quartz veins and, when extraction techniques proved fruitless, Jim applied his working knowledge of "instant cave" (dynamite). He accidentally overloaded the hole, but one large wulfenite specimen on quartz survived the blast. The ground zero specimen consists of a square tabular bright orange wulfenite crystal 2.5 centimeters on edge associated with a "flower" of quartz.

For decades after their limited production stints in the late 1800s, the **Geronimo** mines sat idle. Dr. Ray Grant was one of the earliest mineral collectors to access the North Geronimo Mine and recognize its potential for specimens. Tom McKee had always held that he had found the "lost" Romaldo Pacheco Mine and he had a sweet 2.5-centimeter single cherry-red wufenite as evidence. In the late 1980's he led a group of local collectors to see if he could find the mine again. That group included Casey Jones, George Godas, Dick Morris, and Mark Hay. They found the mine Tom was looking for and Casey Jones found a large hoppered red vanadinite on the mine dump. A short time later Godas filed a claim on it. He renamed it the **Pure Potential**. He contracted Graham Sutton to empty the mine of muck and stabilize the shaft. Sutton installed culverts and ladders for access down the inclined shaft. He excavated a 60-foot-long vertical shaft and installed a headframe and hoist to remove waste material and allow them to lower a small diesel-powered front-end loader into the mine. Drilling, blasting, and mucking advanced the drifts both north and south. Sutton's knowledge of mining and equipment was key to Godas's success at the North Geronimo.

Godas encountered good vanadinite and even the occasional wulfenite pocket over the next two years. Both species from the Pure Potential have the same intense red color. Then in October 1996 George pulled back a slab to reveal a 70-by-70-centimeter pocket that was lined with vanadinite. The resulting specimens are some of the best of the species from Arizona.

By the end of 1997, Sutton had purchased Godas's share in the Pure Potential. Sutton sold it to The Collector's Edge, Inc. Sutton extended the headings and continued mining, but the project only broke even. The operation never producing anything to equal the pockets that Godas had dug. There have been no reported finds from the mine since 2005.

George Godas invited Mark Hay and Dick Morris to the Pure Potential (North Geronimo) Mine to witness and assist with the recovery of Godas's historic pocket.

Mark Hay photo, October 4, 1996

Left: Garth Bricker was one of those drawn to the Red Cloud in the 1980s and 90s when the historic mine was open to collectors. He was also part of the Thompson-Horner team for its first season working the vein through the open pit, but Bricker collected both this specimen and the rough for the cut stone from a vug that he tapped into in the 1980s.

Wulfenite specimen, 5.2 cm tall, and 5.61-carat faceted stone
Red Cloud Mine, La Paz County

Garth Bricker collection
Photo Robert Weldon © 2008 GIA

Below: Wulfenite, 4.6 cm wide
Red Cloud Mine, La Paz County
Collected by George Godas in 1985

Dick Morris collection, Jeff Scovil photo

With the meteoric rise in silver prices in 1979, Don McDaniel, owner of the famous **Red Cloud Mine** (see also pages 47 and 113), initiated a mining operation, which crashed along with silver prices in 1980. With mining halted, the stopes were empty for the first time in decades, and collectors descended on the remote La Paz County mine. Bob Lane, Bruce Barlow, Dave Shannon, Dick Morris, and George Godas were among the many collectors to come away with striking finds.

The death of McDaniels and the eventual sheriff's sale of the property put an end to unfettered collector access to what had become hallowed ground for nearly a generation of collectors. But the sale, to dealer Wayne Thompson, in partnership with collector and composer James Horner, was certain to bring more and better specimens to the international market than the independent miners and weekend collectors possibly could.

Thompson and Horner purchased the property in 1995. Les Presmyk and McDaniels' daughter also held stakes in the operation. They hired Graham Sutton as the mine foreman. Don McCoy, Bud Standley, Bob Johnson, and George Stevens came on board to work as miners/collectors. Presmyk developed the mining plan.

The old stopes of the silver mine followed the lead and silver-rich mineralized stringers. The old timers had provided a map, if you will, of where to look for wulfenite. Presmyk's plan was to expose the vein through an open pit, which would allow them to use large equipment and to work in the sunlight.

Right: Wulfenite on quartz, 4.3 cm tall
Red Cloud Mine, La Paz County
Collected in 1997

Les and Paula Presmyk collection
Jeff Scovil photo

Above: The headframe of the Red Cloud Mine, La Paz County

Les Presmyk photo, circa 1987

Facing page: Wulfenite, 7 cm tall
Red Cloud Mine, La Paz County
Collected in March 1997

Private collection, Jeff Scovil photo

They began stripping overburden in the fall of 1995. For the first month or so as they exposed about 20 feet of the vein, the only material they moved was waste. Then the collecting began. As is typical for the Red Cloud, pockets were scarce, and most were generally less than 10 centimeters wide. After four months of mining, they had about a flat of decent specimens to sell at the 1997 Tucson Show.

In March 1997 they hit an unusually large pocket; in fact, it was the largest that they would encounter. Sutton was operating the excavator and opened up a large crack, which contained wulfenite crystals up to 1" scattered across a dark, almost black, matrix. When it was over, the initial wulfenite-lined pocket measured more than 6 feet long, 10 inches wide, and more than 3 feet high. The pocket yielded specimens as large as 30 centimeters across and crystals up to 3.5 centimeters on edge. The zone would ultimately be traced for almost 40 feet as it opened and closed. Large, lustrous, orange-red, undamaged Red Cloud wulfenite specimens were available in the local and international mineral market as never before.

Left: Wulfenite, 10.6 cm wide
Red Cloud Mine, La Paz County
Private collection
Jeff Scovil photo

Right: Wulfenite, 3.1 cm tall
Red Cloud Mine, La Paz County
Collected in 2002 by The Collector's Edge
Paul Harter collection
Jeff Scovil photo

Mining progressed to the middle 150 of the vein, which proved relatively barren of wulfenite and any other mineralization. Even moderate strikes were scarce. A few good specimens were recovered in 1998, but the operation was not paying off. The following year saw no major pockets, so at the end of 1999, Horner and Thompson sold the property to The Collector's Edge, Inc.—Bryan Lees of Sweet Home Mine (Alma, Colorado) fame was to have a go at the famous and beloved locality.

From November 2001 to mid-May 2002, Sutton and Collector's Edge continued open pitting the Red Cloud vein. They moved below the Thompson pocket and into the legendary "Forest" area of the south stopes. They opened some pockets and in fact, one beautiful specimen with a 3.5-cm crystal perched aesthetically on matrix reportedly brought $125,000. Sutton followed the vein south until it pinched to a mere trace.

In March 2003, Collector's Edge decided that it was time to shut down the mine. Sutton blasted one more time on the north end toward the last of the stopes that were still visible in the pit and opened one last pocket, aptly named the "Red Gem Pocket." It was the last gasp of the open pit.

CHAPTER 12

The Centennial State

The dawn of the twenty-first century bears witness to the legend of untamed Arizona deserts and a wilderness barely touched by human hands juxtaposed with the reality of a rapidly growing population that strains the state's natural resources. On the one-hundredth anniversary of Arizona's statehood, the state's population will exceed 6.5 million, and more than 90 percent of those people live within 30 miles of the ribbons of freeway that traverse the state. Arizona's natural wonders are still magnets for tourists—the Arizona Office of Tourism estimates that more than 100,000 people visit the state every day, the majority coming to visit national parks, monuments, or public lands. But the tension between those with a desire to preserve the "wild" nature of the state and those favoring development frames nearly every political discussion and decision.

Above: Unveiled in 2011, this new first class "Forever" stamp commemorating Arizona's Centennial will be released on February 14, 2012. The stamp depicts Cathedral Rock in Sedona as painted by Arizona artist Ed Mell.

Left: Beryl, repaired and restored
7.5 cm wide
Santa Teresa Mountains, Graham County
Collected in 2005 by John Crowley

Evan Jones collection
Jeff Scovil photo

Mining remains a pillar of Arizona's economy. The state is still the nation's premiere copper producer, with more than 1.5 billion pounds of copper produced in 2010 (about two-thirds of the total national output). Morenci is the largest; the mines of the Sierrita and Mission districts south of Tucson are also major contributors. The worldwide demand for copper is growing and the mapped reserves in Arizona suggest that mining will be important to the state for the next 50 years. However, the proposed opening of new mines in Arizona precipitates bruising battles over the use of public lands and the perception that mining is "bad" for the environment. On Arizona's hundredth birthday, three proposed copper mining projects in the southern part of the state are pending approval: the Resolution Copper Project, which targets the deep porphyry beneath the Magma Mine near Superior; the Rosemont Copper Open-Pit project south of Tucson in the northern Santa Rita mountains; and the Florence Copper Project, a large *in situ* project designed to recover 2.8 billion pounds of copper. All three of these proposed mines face considerable opposition ostensibly on environmental grounds.

Mineral collecting in Arizona has also changed in the twenty-first century. Access to localities even for casual collecting activities is much more restricted. Many of the abandoned specimen-producing mines have been permanently sealed. In the last two decades of the twentieth century, a national consensus emerged that abandoned mines were an issue both from a safety and an environmental standpoint. In 1977 Arizona's legendary Congressman Morris Udall introduced the Surface Mining Control and

Reclamation Act, which required that coal-mined land be remediated. The act also provided federal funding to help address safety issues and environmental problems at other abandoned mines.

In Arizona a number of well-publicized accidents shaped public opinion and brought safety issues to the fore. In the late 1970s partiers in Tombstone and Phoenix disappeared in abandoned shafts, and in the mid-1980s a teenager went missing while exploring an adit near Globe with a group of friends; his body was found at the bottom of a vertical shaft 200 feet below the adit. The state of Arizona reacted by passing a law that substantially increased the penalties for owners of mines (or of property that contains mines) who do not fence all adits and restrict access to mine workings.

The shaft of the legendary Glove Mine in the Santa Rita Mountains, Santa Cruz County was accidentally and ironically set ablaze by a welder installing a bat-friendly, safety gate over the mine opening in 2011. The gate was being installed as part of a federally funded program to close open mine adits. When the fire started, the shaft acted as a chimney, and smoke could reportedly be seen for miles.

John Cesar photo

These laws changed the landscape for collectors, who could no longer visit old mine sites to look for minerals: the openings were being sealed, and it became illegal to circumvent the measures. In 1992 the Arizona State Mine Inspector entered into an agreement with the Bureau of Land Management to survey federally managed lands and inventory abandoned and inactive mines. By the year 2000, approximately 10,000 abandoned mines in Arizona had been visited, and today underground access to more than 1,500 of them has been permanently closed.

The impact of the internet as a marketplace has also been profound. There are conservatively more than 2,000 websites that sell specimens online This number only includes those that primarily sell mineral specimens; it does not include information pages or those selling other earth materials, such as fossils. Although the quantitative value of the minerals on these sites is dynamic, one cursory 2012 assessment found that the total value of the mineral specimens for sale on the World Wide Web on a given day exceeded 90 million dollars. Nearly half of that value is commanded by fewer than 2,500 specimens and by only a few species—tourmaline (color required!), beryl, gold, silver, rhodochrosite, and pyromorphite.

The changes in the global mineral market at the start of the new milleniium were reflected in Arizona's institutions, collecting community, and mineral-specimen production. While it may be more difficult to find places in Arizona where mineral collecting is allowed, localities still exist and there are others still waiting to be discovered.

The University of Arizona Mineral Museum

In spite of funding struggles, waning interest on the part of the university, and constant pressure to sell specimens to an increasingly hungry market, the University of Arizona Mineral Museum continues to provide an opportunity to view and study the minerals of Arizona and Mexico.

In 2003 Dr. Spence Titley, a distinguished economic geologist, became the fifth curator of the Mineral Museum. By then most universities had moved away from maintaining and curating mineral collections, and the University of Arizona was no exception. Although the museum remains popular with collectors, the university has decided that it must be a self-supporting entity, ultimately moving off campus to free up real estate for research and instructional facilities.

In 2008 Dr. Robert Downs, a mineralogist, became the sixth curator of the museum. Downs has the lead on the RRUFF Project, an ambitious program to measure the properties of all minerals especially those properties which define or characterize. He is a researcher with the next Mars Rover mission for which he will X-ray minerals on Mars.

The museum today houses more than 19,000 specimens including a micromount collection boasting more than 7,000 specimens. Over 2,100 species are represented. In 2011 the Mineral Museum accepted a long-term loan of the important Flagg Mineral Foundation collection, which is currently on display at the museum.

The relationship between Phelps-Dodge and the Mineral Museum that was established during Bisbee's heyday has been renewed. Freeport-McMoRan Copper and Gold, which acquired Phelps-Dodge in 2007, has established an endowment to maintain the museum and fund a position for a permanent collection manager. Mark Candee is the first person to hold this position. The endowment signifies the promise of a healthy museum into the future, with a signature of aesthetics and scientific research.

The Sierrita Mountains

with contributions by Barbara Muntyan

While the museum offers an opportunity for the public to view important and historic mineral specimens, collectors continue to search Arizona's mountains and deserts for new and significant finds. The western slopes of the Sierrita Mountains southwest of Tucson yielded small beryl crystals in the past, but the locality was largely forgotten until collector Barbara Muntyan moved to Tucson and began to try to find it.

In September 2008 she obtained a cloudy, sky-blue aquamarine crystal associated with a minor amount of biotite. The crystal is 5.7 centimeters long. It was the size of the crystal that roused Muntyan's interest, as she likes

Dick Jones collected this 5 cm wide beryl on matrix in the Sierrita Mountains, Pima County in the 1960s. This and other beryl specimens with similar provenance prompted Barbara Muntyan to try to rediscover the locality.

Evan Jones collection
Jeff Scovil photo

Beryl on quartz with mica, repaired, 8.5 cm tall
Sierrita Mountains, Pima County
Collected in the 1960s by Dick Jones
Barbara Muntyan collection
Jesse La Plante photo

to collect "Colorado-style" Arizona minerals, meaning she likes large, well-formed crystals with associated species, preferably on matrix.

The aquamarine variety of beryl is not abundant in Arizona, especially not in large crystals. Muntyan's specimen was labeled, "Palo Verde claim, Sierrita Mountains, Pima County, Arizona." The late Dick Jones had collected the specimen in the 1960s, and it had been in his collection before passing to Tom McKee, then Les Presmyk, and finally to Muntyan, who loves field-collecting and decided to go out to the site and prospect. She noted:

> *It didn't take long to discover that the Sierrita aquamarine locale was "lost." A kind of urban myth alleges that a graduate student from the University of Arizona had found a small pocket of gem-quality aquamarine sometime in the 1950s and sold it to finance his studies. The tale ran that not only was this a "lost locality" but that the aquamarine was "all gone." Thus began my quest to find the locality.*

Muntyan began by familiarizing herself with aquamarine specimens from the Sierrita Mountains. The Arizona-Sonora Desert Museum has two specimens; Evan Jones owns a fine matrix aquamarine from the "Bella Donna Claim," which is about 3-centimeters long, deep blue, and quite gemmy; and Les Presmyk has two crystals: a single crystal from the same pocket as Jones's and a cloudy, medium-blue, 7.5-centimeter crystal on massive quartz labeled "Palo Verde Claim." Both of Presmyk's pieces were collected by Dick Jones and had been in the collection of the Tom McKee.

Muntyan then interviewed several field-collectors who were active in Arizona in the 1960s, 70s and 80s: George Stevens, Andy Anderson, Freddie Rhodes, Bill Hefferon, Paul Matt, and Dick Thomssen. All remembered finding aquamarine in the Sierritas, but they couldn't remember exactly where. After further research at the University of Arizona Library and several scouting trips, Munytan and her collecting partner thought she had finally found the area.

> *We stopped the vehicle on a hillside because the track we were following clearly cut a white quartz dike. As soon as we got out, I spotted telltale sky blue, small, weathered aquamarine crystals up to 1/3 inch on white massive quartz. They were crummy specimens, but they were aquamarine. A subsequent grid search of the hill produced about a dozen other examples of aquamarine, all massive and all filling the space between massive white quartz blocks.*

Over the next eighteen months, they found several scattered pegmatite outcrops. To date they have located some 300 separate quartz outcrops in twelve distinct pegmatite zones of the Sierrita Mountains. Ten of these contain aquamarine crystals.

Muntyan was certain that she had found the Palo Verde Claim when she was digging in a likely prospect when, "the quartz suddenly parted and a 15-centimeter-long crystal and its negative cast was in my hands. A totally mind-blowing sight—this crystal was the continuation of my specimen!"

Muntyan decided not to repair the crystal. The smaller section was awarded the 2010 trophy for the best Arizona Gem Mineral in competition at the Tucson Gem and Mineral Society show in February.

The Santa Teresa Mountains

Another Colorado collector, mining geologist Dean Misantoni, found beryl-bearing pegmatite in Arizona. In 2004 Misantoni, began systematically exploring the Santa Teresa Mountains in Graham County. A locked gate blocked access to the Iron Cap Mine, so he switched his focus to the mountains further south where five years earlier he had noted cavities in a few granite cobbles in some of the drainages.

The area was roadless and primitive, and the only way in was by foot. Misantoni had previously found cavities containing regular and smoky quartz, epidote, hematite, fluorite, and blue beryl. And on one trip, he encountered pockets and collected good specimens of fluorite, quartz, microcline, and blue beryl from miarolitic cavities and to a lesser extent from veins in a pegmatite.

Another find—and one of the most celebrated thus far in the twenty-first century—was of a vug containing a fabulous spray of blue beryl (pictured on page 340). The 10-centimeter spray is made up of lustrous divergent columnar crystals with flattened tops, consisting of what appear to be aligned multiple terminations. John Crowley found it in 2005 while collecting in the Santa Teresas with Misantoni.

Crowley inspected a pocket that Misantoni had seen previously, but Crowley noticed something that Misantoni had missed: an upturn at the back of the arm-length tube. Crowley searched with his fingers and felt a group of crystals, which were only loosely attached to the matrix. He carefully freed the group, pulled his arm from the tube and revealed the spray of blue beryl.

Other pockets in the area show a diversity of species. Fluorite occurs as deep purple, rough octahedrons to 6 centimeters on an edge on gray to smoky-colored quartz crystals. Quartz is found as slightly translucent, prismatic crystals in groups which can exceed 20 centimeters. Microcline, though, is less common. It occurs occasionally as groups of sharp glassy brown-gray crystals to 2.5 centimeters that are sometimes associated with quartz and/or fluorite.

Below: Quartz on microcline, 7 cm tall
Santa Teresa Mountains, Graham County
Evan Jones collection
Jeff Scovil photo

Right: Fluorite and quartz, 14.6 cm tall
Santa Teresa Mountains, Graham County
Evan Jones collection
Jeff Scovil photo

Circa 1910 view of the Gold Road townsite in the Oatman gold district

John Taylor collection

Oatman Mining District

with contributions by Mark Hay

It wasn't just the odd pegmatite that produced specimens in the twenty-first century. Collectors still found their way into old mines and discovered untouched specimen bonanzas. The Oatman gold district (see page 100) is home to a few famous old mines such as the **Tom Reed** and **Gold Road**, but it was not known for producing anything of interest to collectors. Geologists and others working in the Oatman district in the 1970s and 80s reported finding fluorite crystals in the mines in the area. In 1983 Carousel Minerals briefly advertised fluorite from the Oatman district, but the area still did not attract much in the way of collector attention until the late 1990s when collectors Dick Morris and Mark Hay were looking for a collectable fluorite occurrence in Arizona. Their research brought them to Oatman. Hay and Morris had poked around in many of Arizona's mining districts, but they were unfamiliar with Oatman. They were motivated to find a source for Arizona fluorite, however, so they made regulars treks from Phoenix to Oatman (4½ hours one way) to explore the district.

Over the next few years they checked or explored around 35 mines, dumps, and cuts. And there are many more that they did not check. The pieces that Hay and Morris collected in the district are among the best fluorite specimens ever to have come from Arizona. And the Oatman area mines are certainly the state's most prolific source of high-quality fluorite specimens.

Hay and Morris focused on the small outlying mines to the north along the sides of Hardy Mountain and Silver Creek. Hay observed, "With few exceptions, veins in the area often carry fluorite as a gangue mineral, and to the good fortune of mineral collectors, the veins are vuggy and the fluorite is well crystallized." Three properties treated them particularly well: the **Oregon**, **Hardy**, and **Homestake**. The latter yielded the greatest quantity and finest quality of specimens.

Fluorite on quartz, 6.5 cm wide
Hardy Mine, Oatman, Mohave County
Collected in 1998 by Mark Hay and Dick Morris

Mark Hay collection
Jeff Scovil photo

When they first visited the Homestake, it was being worked for gold by Randy McIntosh and Mike Van Stratten, who were mining in partnership with Bernie Siegel, the owner of the mine. McIntosh, Van Stratten, and Siegel were open and friendly from Day One. They had no interest in the fluorite crystals, and once they were comfortable that Hay and Morris had no intention of mining gold, they agreed to lease the tunnel above the one they were working to Hay and Morris.

> *The main tunnel of the Homestake followed the vein in for several hundred feet. Above it were two smaller, exploratory tunnels that had been driven many years earlier. A steep path led from the lower workings, up the mountain to the middle tunnel, which was about 100 feet above the main tunnel and was only about 20 feet long. The back two thirds of the tunnel were devoid of fluorite. But the first several feet were something else!*

Hay went on to describe,

> *Massive bull-quartz with pockets the size of a quarter up to maybe seven centimeters, many with sweet green octahedrons to about one centimeter*

looking like eggs in a nest. In the ceiling, there was a large open pocket that had been hit by the original miners. What remained was about a foot and a half long and went more than 18 inches into the ceiling. Heaven knows how far it had extended down into what was now the open void of the tunnel. The walls of the pocket were studded with fluorite crystals to 5 centimeters! Of course, the vein quartz was spectacularly hard and recovering any of them intact was going to be difficult but the potential was incredible! We had found our fluorite dig.

The fluorite crystals were always octahedral and usually 2.5 centimeters or less, although they did occasionally find crystals exceeding 5 centimeters on edge. The largest was a single, loose rounded crystal almost 11 centimeters on a side. The crystals were always green, varying from a rich lime green to pale watery green. The occasional crystal shows minor purple zoning. The vast majority occur on massive, white, vein quartz, which was usually coated with small drusy, milky, or clear quartz crystals.

Morris and Hay worked the middle tunnel of the Homestake Mine for about five years. They finally stopped renewing their lease, not because the fluorite had played out, but because they were ready to move on.

Below left: Fluorite and quartz, 5.6 cm tall
Homestake Mine, Mohave County
Collected in 2002 by Dick Morris and Mark Hay
Dick Morris collection
Jeff Scovil photo

Below right: Fluorite and quartz
5.8 cm tall
Homestake Mine, Mohave County
Collected in summer 1999 by Dick Morris and Mark Hay
Dick Morris collection
Jeff Scovil photo

Rowley Mine

with contributions by Jim Ricker and Evan Jones

The Rowley Mine did not produce many specimens in the late 1990s. That was due in part to the fact that collectors had buried major areas of the mine in muck, but collecting at the mine also dropped off because the collector market was losing its taste for fragile mineral specimens. The internet had taken off as a means for selling specimens, and Rowley wulfenite does not ship well.

In the early 2000s, collector Ed Davis renewed interest in the locality when he exposed a series of pockets around the Jobes shaft. Tucson-based Jim Ricker and Mike Joseph, a friend of Jim's from Montana, also worked this prolific zone. Ricker and Joesph had looked over a map of Arizona's wulfenite localities and decided to visit the Rowley Mine, as neither had ever been there before.

They arrived to find that there was no ladderway in any of the three shafts. The pair did a cursory inspection of the Main inclined shaft and then moved over to the Jobes. They dropped a rope down, descended, and walked the decline to the Jobes stope where they found a terrific surprise. Someone had exposed a fracture full of bright orange, yellow, and red wulfenite. The crack and the wulfenite crystals reached 1.5 centimeters in places. About 10 square feet of the mineralized area was exposed, and the back surface was still intact. The find made for a memorable collecting trip.

Wulfenite on mimetite and barite
5.1 cm tall
Jobes stope, Rowley Mine
Maricopa County
Collected in November 2005 by Jim Ricker and Ed Davis
Dick Morris collection
Jeff Scovil photo

Over the next several days, they collected from several pockets and took home perhaps 30 flats of medium- and high-quality specimens, with crystals exceeding 3 centimeters on edge. Colors varied from classic, bright red-orange to coffee-yellow. The latter are extremely thin and many were damaged during the collecting process.

However, as was pointed out by many other collectors, the hanging wall in this stope was frightening. It was approximately six feet thick with vertical stress cracks and a clear separation running along the top contact with the country rock. Ricker and Joesph were careful to stay clear of it. When they had moved as close to the hanging wall as their comfort level allowed, they retreated with hopes of returning at a later date after the wall had fallen naturally.

Some months later Ricker went by the mine to assess the situation. He was excited to find that the hanging wall had indeed come down. (He later learned that Ed Davis and James Dunn had been there a week or so before and had caused the collapse, but they felt it was too unstable for collecting).

All seemed safe, so he retrieved the only tools he had in his car: a small screwdriver and a jeweler's peening hammer. Within just a few minutes, he was opening small pockets of exceptional-quality crystals. The first pocket contained a perfectly placed wulfenite crystal of a habit he had rarely seen anywhere and never saw again at Rowley. It was exactly twice as tall as it was wide and just under 2.5 centimeters on edge. Ricker recovered this "Gum Stick" and left the mine. The next day he loaded his RV with all the tools and packing he thought he would need and got ready for an extended stay at the mine. However, he arrived the following morning to find four

Wulfenite and mimetite in a barite vug
14.5 cm wide
Jobes stope, Rowley Mine
Maricopa County
Collected in 2004 by Evan Jones and Jim Ricker

Evan Jones collection
Jeff Scovil photo

cars already parked at the mine:

> *A chain ladder was hanging in the shaft and I went down to see how the collectors were doing. Eight people were hammering on the walls. It was a group of day trippers who had no idea the value of the pretty orange chunks of rock they were knocking off the walls and rolling down the skree pile. For over an hour, I watched tens of thousands of dollars worth of exceptional wulfenite specimens being systematically destroyed. Thankfully they got hungry early and left to go to Burger King in Gila Bend, leaving the mine to me.*

Ricker collected 70 flats of material that day and was back several weeks later with collector Evan Jones. They had intended to collect for just one day but exposed a large pocket that took the two of them two days to extract. Jones recalls the experience:

> *After a couple of hours of collecting, I popped a small chunk of barite out of the ceiling where we were working and exposed a relatively large (for Rowley) pocket. It was about 10 by 12 centimeters and was lined with wulfenite and mimetite on bladed white barite. What a sight! The wulfenite crystals were atypical and almost too good to be true: large and very thick, with sharp, beveled edges and a distinct yellow-orange color. Jim and I just stood there gawking at this incredible pocket. We decided to remove it in its entirety.*

Ricker drove home to retrieve his compressor and hammer drill, and the two spent the better part of the next day carefully drilling and chiseling until they finally popped the huge piece of barite out with the pocket intact in the center. They hauled the 50-pound chuck up the ladder. The Collector's Edge in Colorado carved 47 pounds of the barite away. The specimen (pictured on page 352) is now in Jones's collection.

Ricker intermittently worked the Rowley over the next several months but with diminishing returns. He was back after a hiatus of several months to find Ed Davis emerging from the mine. The property had changed hands and the new owner had hired Davis as the caretaker. Ricker and Davis agreed to work the Rowley together for the season. They mined the same fracture on the Jobes stope for several months. Toward the end of August 2005, they hit what Davis called "the wall of a thousand pockets."

Just one week prior to Denver show in mid-September, they hit their best: a vug containing crystals more than 2 centimeters on edge. The vug was collected intact. It is a great specimen that could have been just a little better. When Davis and Ricker extracted the specimen, a silver wire was nestled in with the wulfenite crystals. In his rush to prepare all of the Rowley material for the looming Denver Show, Ricker mistook the silver for a dog hair. But the silver "hair" broke off his probe.

Collectors including James Dunn and John Cornish also enjoyed success in the Rowley Mine in the 2000s. The property is privately owned but is not currently being leased.

Lower left: Ed Davis climbing out of the Jobes shaft of the Rowley Mine

Steve Knox photo

Upper right: Vanadinite pseudomorph after wulfenite on mimetite
Field of view 1.5 cm wide
Jobes stope, Rowley Mine
Maricopa County
Collected in 2005 by Ed Davis and Jim Ricker

Steve Knox collection
Jeff Scovil photo

Lower right: Wulfenite pocket on the Jobes stope in the Rowley Mine

Steve Knox photo

The Carlota Mine

with contributions by Evan Jones

A few of miles southwest of the Pinto Valley Mine in the Globe-Miami mining district, the old Carlota copper mine was absorbed into the Cactus Pit operation after 2000. Collector Evan Jones had heard about the specimen potential of the Carlota. He had a contact at the mine who agreed to call if anything of interest was found during mining.

On a weekend in October 2009, Jones got a call in the middle of the night—an azurite vein had been exposed on the 3460 foot level. He heard his contact ask, "You know those azurite crystals you're looking for? We just hit a whole bunch of 'em. How much do you want?" "As much as you can pick up." Jones replied.

Jones remembers meeting at his apartment the following day. "I couldn't believe my eyes: laid out on the kitchen table were about 30 plates, up to palm-size, of sharp, lustrous and blocky azurite crystal groups to an inch." The find also included chatoyant, blocky malachite pseudomorphs after azurite up to 3 centimeters on edge. They were recovered as isolated crystals and as groups on a green malachite matrix.

"Is this the kind of stuff you want?" The miner asked. "Oh yes," Jones answered, "That's the stuff."

These were easily the finest azurite specimens found in the district in decades, but the rare accessory minerals were even more fascinating. Grains

Above: Azurite, 9.5 cm wide
3460 bench, Carlota Mine, Gila County
Collected in 2009
Evan Jones collection
Jeff Scovil photo

Facing page, upper: Azurite and prosopite
Azurite crystals to 8 mm
3460 bench, Carlota Mine, Gila County
Collected in 2009
Dick Morris collection
Jeff Scovil photo

Facing page, lower: Malachite
6.8 cm wide
3460 bench, Carlota Mine, Gila County
Collected in 2009
Mark Hay collection
Jeff Scovil photo

and small crystal pockets of powellite were admixed with the azurite. Light blue druses of prosopite were also present. This is the second, and by far the best occurrence of prosopite in the state. Joe Ruiz subsequently identified szenicsite on material from the mine that he had purchased. The szenicsite occurs as small yellow-green blebs in the azurite veins. The find at the Carlota Mine is the first North American occurrence of the species.

Unfortunately mine management did not respond well when collectors began soliciting employees for crystals and collecting privileges. Miners and staff were notified that any further mineral collecting would be grounds for instant discharge. No further specimens have come from the property since.

Santo Niño Mine

with contributions by Barb Muntyan

These final localities bring *Collecting Arizona* full circle—back to prospects not far from the site of the pre-territorial, Civil War-era Mowry Mine (see page 26). Collectors have been working the area for Japan-law twinned quartz as well as other quartz varieties and associations since the 1950s. And that tradition has been carried forward into the new millennium.

The Santo Niño Mine is located on the northeast slopes of Mount Washington, about 2.5 miles southwest of **Duquesne-Washington Camp**. Mount Washington is one of the highest peaks in the Patagonia Mountains, topping out at 7,200 feet; the collecting areas are at elevations of between 6,200 and 6,700 feet.

Quartz, 5.6 cm tall
Santo Niño Mine, Santa Cruz County
Barbara Muntyan collection
Jeff Scovil photo

Located on five patented mining claims within Coronado National Forest, the Santo Niño Mine was a small surface and underground operation that extracted modest amounts of copper, silver, molybdenum, gold, lead, and zinc. The ore vein dips at a steep angle, and the claims run almost straight up the hillside with portals on even-numbered levels and a shaft at the top of the vein. A rough trail zigzags down the hillside from portal to portal and ends at a large ore bin along the creek.

The deposit was discovered in 1908 by prospector Dennis Coughlin. Production was intermittent from 1912 until 1955. In comparison, some of the major mines of the Washington Camp-Duquesne district were known since Mexican times. By 1908 several big mines including the **Holland**, **Kansas**, and **Belmont** were in operation as was the Duquesne Mill.

Tony Potucek first visited the locality in 1975 with Bill and Millie Schupp. The trip was planned to collect rutile from the outcrops in the area. It was not long after arriving at the bottom of the gulch below the Santo Nino workings that they found red, reticulated rutile groups up to 5 centimeters wide.

Most collectors however are drawn to the Santo Niño Mine, or more accurately from an area west of the actual mine, for sceptered amethystine quartz crystals. The locality is a favorite among collectors, many of whom have had good luck digging for quartz and other minerals.

Eddie Granillo, who grew up in Washington Camp, has been among the most successful. Eddie and his brother roamed these hills as teenagers and young men; they have found numerous vugs containing amethyst scepters over the years. One of their most significant finds in the early 1990s contained amethyst crystals up to 10 centimeters in length and clusters 22.86 centimeters across.

The collecting area is accessed by following a poorly-maintained Jeep trail from the south end of Washington Camp through the national forest approximately two miles westward and then hiking up a steep, heavily vegetated, north-facing hillside beyond the mine's ore bin. About halfway to the top of the ridge, a huge excavation—the "Trench"—comes into view. Collectors have hand-excavated the roughly 50-foot-long, 6- to 8-foot wide, 4-foot deep trench, and evidence of vugs is everywhere along it. The quartz vein lies horizontally along the back wall of the excavation. There are numerous other, smaller prospect pits to the east, west, and above the Trench. These range from barely a foot across to 8-foot-wide pits.

Barbara Muntyan and the late Don Belsher excavated a large pocket from the Trench area in February 2003. Their efforts were rewarded with six flats of amethyst scepters and groups, white quartz clusters, and numerous other specimens. Most of the crystals were 2.5 to 3 centimeters long, although some were as large as 7.5 centimeters.

About three-quarters of the way up this hillside, another large prospect pit has yielded significant amounts of white, bladed barite; in fact, a variety of well-crystallized mineral species can be found at the Santo Niño locality in addition to quartz and barite. These include anatase, brookite, barite, microcline (Anthony 1995), molybdenite (Galbraith and Brennan 1959), muscovite, pyrite, pyromorphite, and rutile. Jeff Scovil reports having found

Right: Amethyst scepter, 6.9 cm tall
Santo Niño Mine, Santa Cruz County
University of Arizona collection
Jeff Scovil photo

Above: Quartz (variety amethyst) scepters *in situ* in the hillside near the Santo Niño Mine, Santa Cruz County
Steve Knox photo, May 2009

microcrystals of wulfenite on the Santo Niño dump when he was there as part of a club field trip.

This locality has been known to southern Arizona mineral collectors for many years. However, because of the remoteness, rugged terrain, and heavy vegetation on the hillside, persistence is required to find the best collecting zones. In addition, working this area requires the use of heavy tools: sledge hammers, long chisels, picks, pry bars, and shovels. Once a mineralized zone is identified, collecting becomes a careful and tedious removal exercise of scooping out the mud which fills the vug interior and which makes the underlying quartz crystals difficult to see and to extract without damaging them. Perhaps eighty percent of the time, a vug will contain only milk-white quartz clusters, but if amethystine sceptered crystals are found, it is worth all the effort.

AFTERWORD

The mineral heritage of Arizona is extraordinary: the state's ores have added copious amounts of wealth to the nation's coffers, and its spectacular portfolio of mineral specimens populate major collections around the world. In the hundred years since Arizona was admitted into the Union, thousands of collectors—amateur and professional—have scoured the state looking for bits of color and sharp crystals.

However, the state, the nation, in fact, the world has changed. Most of the contributors to *Collecting Arizona* are of a generation whose forefathers had opened the earth for them, and they in turn climbed in and retrieved colorful treasures. There is no one who does not long to share those glory days with their children, but lamentably, the next century will probably not compare to the prolific past.

Every collector who contributed to Arizona's mineral legacy—those that collected the rocks from holes in the ground and those that traded and bought those rocks—has wonderful memories of Arizona and its rich, wild country. Our intent was to place the specimens in historical context, to share the experiences, and illustrate the connections that Arizona collectors have with these specimens, as in the final analysis, it is those connections that determine a specimen's value. And it is that value—the great intangible legacy that these specimens represent—that we wish to pass on.

In 2012 Arizona has a passionate and dedicated (but aging) collector community, one that is willing to explore every nook and cranny of the state in pursuit of minerals. Their stories paint a picture like that of prospectors of old: give them a grubstake and they will endure hardship and danger to collect a few cents worth of copper, silver, or lead trapped in the delicate structure of a colorful mineral. And oh what minerals they have collected! Arizona's royalty—its wulfenite and azurite—are taking their rightful place among the classics in the world's mineral collections.

Although it is impossible to prognosticate, the great stories of the twenty-first century will likely focus on the genealogy of prized specimens as they pass from one collector to another. Arizona will always be the place of great minerals, and its specimens will carry a powerful mythology in time. It is up to another generation of prospectors to discover their treasure, their legacy. To be sure, there will be many wonderful minerals collected, mostly from new ground and occasionally from old, forgotten prospects, but the twentieth century will remain a grand time to have been Collecting Arizona.

Leadhillite is named after its type locality of Leadhills, Lanarkshire, Strathclyde, Scotland. The crystals are monoclinic but typically exhibit pseudohexagonal twinning that gives rise to tabular prisms.

The Mammoth-St. Anthony Mine in the ghost town of Tiger, Arizona, is the premiere locality for leadhillite. The finest specimens show a distinct sky-blue color and an adamantine luster, a combination only found at this much beloved, early to mid-twentieth century locality. Tiger leadhillite specimens are "must haves" for Arizona collectors.

Leadhillite, 7.9 cm tall
Mammoth-St. Anthony, Tiger, Pinal County

Flagg Mineral Foundation collection
Jeff Scovil photo

LITERATURE AND CITED WORKS

Anderson, C.A. and S.C. Creasey. 1955. "Geology and ore deposits of the Bagdad area, Yavapai County, Arizona." *U.S.G.S. Professional Paper No. 278.*

Anderson, C.A. and S.C. Creasey. 1958. "Geology and ore deposits of the Jerome Area, Yavapai County, Arizona." *U.S.G.S. Professional Paper No. 308.*

Anderson, C.A., E.H. Scholz, and J.D. Stroell. 1955. "Geology and ore deposits of the Bagdad Area, Yavapai County, Arizona." *U.S.G.S. Professional Paper No. 278.*

Anderson, R.E. 1971. "Thin skin distension in Tertiary rocks of southeastern Nevada." *Geological Society of America Bulletin* 82:43-58.

Anderson, R.E., C.R. Longwell, R.L. Armstrong, and R.F. Marvin. 1972. "Significance of K-Ar ages of Tertiary rocks from the Lake Mead region Nevada-Arizona." *Geological Society of America Bulletin* 83:273-288.

Anonymous. 1945. *Rocks and Minerals* 20:2.

Anthony, J.W. 1980. "The Mineral Museum of the University of Arizona." *The Mineralogical Record* 11:239-241.

Anthony, J.W., S.A. Williams, R.A. Bideaux, and R.A. Grant. 1995. *Mineralogy of Arizona* (3rd edition). Tucson: University of Arizona Press.

Arizona Bureau of Mines. 1950. "Arizona zinc and lead deposits, Part I." *Arizona Bureau of Mines Bulletin 156.*

——. 1951. "Arizona zinc and lead deposits, Part II." *Arizona Bureau of Mines Bulletin 158.*

Arizona Department of Mines and Mineral Resources. nd. "Files on various mineral specimen producing localities."

Bancroft, P. 1980. "Famous mineral localities: Tiger, Arizona." *The Mineralogical Record* 11 (No. 3):151-168.

Bancroft, P. 1984. *Gem and Crystal Treasures.* Fallbrook: Western Enterprises/The Mineralogical Record.

Bancroft, P. and G. Bricker. 1990. "Arizona's silver mining district." *The Mineralogical Record* 21:151-168.

Barnes, R., and M. Hay. 1983. "Famous mineral localities, The Magma Mine." *The Mineralogical Record* 14: 72-82.

Barnes, W.C. 1988. *Arizona Place Names.* Tucson: University of Arizona.

Bates, A.C. 1896. "News and Comments." *The Mineral Collector* 2 (No. 12):179.

Bideaux, R.A. 1952. "World news on mineral occurrences." *Rocks & Minerals* 27:594.

Bideaux, R.A. 1957. *The Mineral Explorer* (Mar-Apr).

Bideaux, R.A. 1959. "Wulfenite from the Glove mine, near Amado, Arizona." *The Mineral Collector* (Nov.-Dec.).

——. 1959. "Notable specimens from Duquesne, Santa Cruz County, Arizona." *The Mineral Collector* (Jan.-Feb.).

——. 1971. "The collector." *The Mineralogical Record* vol. 2: 185-186.

——. 1972. "The collector." *The Mineralogical Record* vol. 3: 148-150.

——. 1972. "The collector." *The Mineralogical Record* vol. 3: 198-201.

——. 1973. "The collector." *The Mineralogical Record* vol. 4: 4-5.

——. 1973. "The collector." *The Mineralogical Record* Vol. 4:140-142.

Bideaux, R.A. 1980. "Famous mineral localities: Tiger, Arizona." *The Mineralogical Record* 11:155.

Bideaux, R. A., and T. C. Wallace. 1997. "Arizona copper." *Rocks and Minerals* 72:10-27.

Bixby, M. 1897. Advertisements. *The Mineral Collector* iv:5:71.

Blake, W.P. 1881. "On the occurrence of vanadinites of lead at the Castle Dome mines in Arizona." *American Journal of Science* vol. 22.

——. 1883. *The Silver King mine of Arizona.* New Haven: Tuttle, Morehouse & Taylor.

——. 1909. *Minerals of Arizona: their occurrence and associations with notes on their composition: A Report to the Governor of Arizona.*

Bolton, H.E. 1919. *Kino's Historical Memoir of Pimeria Alta.* Cleveland: The Arthur H. Clark Company.

Brenton, H. 2011. Unpublished report.

Bricker, G. 2007-2010. "Red Cloud mine and Silver Mining district (various articles on)." *Lithosphere* (newsletter of the Fallbrook Gem and Mineral Society).

Brittain, R.L. 1954. *Geology and Ore Deposits of the Western Portion of the Hilltop Mine Area, Cochise County, Arizona.* Tucson: University of Arizona.

Browne, J.R. 1864. *Adventures in the Apache Country.* New York: Harper & Brothers.

Burminco. 1946. Advertisement. *Rocks & Minerals* (Dec.).

Butchart, H. 1997-2006. *Grand Canyon Treks.* Bishop (CA): Spotted Dog Press.

Butler, B.S. , and E.D. Wilson et. al. 1938. "Some Arizona ore deposits." *Arizona Department of Mines Bulletin 145.*

Butler, B.S., E.D. Wilson, and C.A. Rasor. 1938. "Geology and ore deposits of the Tombstone District, Arizona." *Arizona Bureau of Mines Bulletin 143.*

Butler, B.S. and Wilson, E.D. 1942. "Ore deposits at Tombstone, Arizona." In *Ore Deposits as Related to Structural Features.* W.H. Newhouse (ed.). New York and London: Hafner Publishing Company.

Calumet and Arizona Mining Company. 1902-1930. Company annual reports.

Chaffee, M.A. 1974. "Stratigraphic relations of the Bolsa Quartzite, Vekol Mountains, Pinal County, Arizona." *U.S.G.S. Journal of Research.* Vol. 2 (No. 2):143-146.

——. 1977. "Geochemical exploration techniques based on distribution of selected elements in rocks, soils, and plants, Vekol Porphyry Copper Deposit

Area, Pinal County, Arizona." *U.S.G.S. Bulletin 1278-E.*

Chappell, G. 1973. *Rails to Carry Copper, a History of the Magma Arizona Railroad.* Boulder (CO): Pruett Publishing Co.

Chicago Daily Tribune. 1882. "Mining Exposition." (Aug. 17):6.

Chicago Daily Tribune. 1893. "Wealth of Arizona Mountains." (Apr. 24):3.

Clark, A.M., and G.Fleck. 1980. "The Greyhorse mine, Pinal County, Arizona." *The Mineralogical Record* vol. 11: 231-233.

Coggin, H.M. 1987. "A History of Placer Mining in Arizona" in *History of Mining in Arizona.* Tucson: Mining Club of the Southwest Foundation.

Colquhoun, J. 1924. *The History of the Clifton-Morenci Mining District.* London: John Murray.

Conger, W.C. 1987. "History of the Clifton-Morenci District." in *History of Mining in Arizona* vol I, J.M. Canty and M.N. Greeley (eds.). Tucson: Mining Foundation of the Southwest.

Conklin, E. 1878. *Picturesque Arizona.* New York: The Mining Record Printing Establishment.

Creasey, S.C. 1967. "General geology of the Mammoth Quadrangle, Pinal County, Arizona." *U.S.G.S. Bulletin 1218.*

Creasey, S.C., and J.D. Pelletier. 1965. "Geology of the San Manuel Area, Pinal County, Arizona." *U.S.G.S. Professional Paper 471.*

Creasey, S.C., and G.L. Quick. 1955. "Copper deposits of parts of Helvetia Mining District, Pima County, Arizona." *U.S.G.S. Bulletin 1027-F.*

Criddle, A.J., Stanley, C.J., and Feyer, E.E. 1983. "Henryite, a new copper-silver telluride from Bisbee, Arizona." *Bulletin de Mineralogie* 106:511-517.

Cummings, J.B. 1946. "Exploration of New Planet Iron Deposit, Yuma County, Arizona." *U.S.G.S. Report of Investigations 3982.*

Dana, E.S. 1892. *The System of Mineralogy of James Dwight Dana 1837-1868.* 6th ed. New York: John Wiley & Sons, Inc.

Dana, J.D. 1868. *A System of Mineralogy.* 5th ed. New York: John Wiley & Sons, Publishers.

Dewitt, E. 1976. *Precambrian Geology and Ore deposits of the Mayer-Crown King Area, Yavapai County, Arizona.* Tucson: University of Arizona Press.

Dewitt, E., J.P. Thorson, and R.C. Smith. 1986. "Geology and gold deposits of the Oatman District, Northwestern, Arizona." *U.S.G.S. Open-File Report 86-0638.*

Dings, M.G. 1951. "The Wallapai Mining District, Cerbat Mountains, Mohave County, Arizona." *U.S.G.S. Bulletin 978E.*

Domitrovic, A.M., W.E. Wilson, and M. Hay. 1998. "Famous mineral localities: The Castle Dome District, Yuma County, Arizona." *The Mineralogical Record* vol. 29.

Douglas, J. 1881. Report on the Copper Queen Mine. Letter to Prof. Silliman dated February 8, 1881. Arizona Historical Society files (7).

——. 1909. Unpublished memoirs, Phelps-Dodge Corp. files.

Dowell, G.P. 1978. *The Total Wreck, Arizona's Forgotten "Bonanza" Mine.* Tucson: University of Arizona Press.

Drewes, H. 1971. "Geological Investigations Map 1-614." United States Geological Survey.

Drewes, H., and F. Williams. 1973. "Mineral resources of the Chiricahua Wilderness Area, Cochise Co." *Arizona Geological Survey Bulletin 1385-A.*

Dudley, F. 1957. "Oxidized minerals from the Silver Bill Mine, Gleeson, Arizona." *The Mineral Explorer* (Nov.-Dec.).

Dunning, C.H. 1959. *Rocks to Riches.* Phoenix: Southwest Publishing Co., Inc.

——. 1984. "Structure and mineralization-Kingman Area, Arizona." Arizona Geological Society.

Eastlick, J.T. 1968. "Geology of the Christmas Mine and vicinity, Banner Mining District, Arizona" in *Ore Deposits of the United States Vol II.* J.D. Ridge (ed.). New York: The American Institute of Mining, Metallurgical, and Petroleum Engineers, Inc.

Esbenshade, S., and W.A. Thompson. 2008. "The Morenci Mine" in *American Mineral Treasures.* G.A. Staebler and W.E. Wilson (eds.). East Hampton: Lithographie.

Edson, G.M. 1980. "Famous mineral localities: The Red Cloud Mine, Yuma County, Arizona." *The Mineralogical Record* 11:151-152.

Elevatorski, E.A. 1971. *Arizona Fluorspar.* Phoenix: Arizona Department of Mines and Mineral Resources.

Enders, M.S. 2000. *The Evolution of Supergene Enrichment in the Morenci Porphyry Copper Deposit, Greenlee County, Arizona.* University of Arizona Doctoral Thesis.

"Calumet & Arizona finds new body of carbonate ore." 1921. *Engineering and Mining Journal* vol. 111:406.

English, G.L. 1890. *Catalogue of Minerals for Sale by Geo. L. English & Co.* New York and Philadelphia: Geo. L. English Co.

Flagg, A.L. 1919. "U. S. vanadium development property near Kelvin, Arizona." Arizona Department of Mines and Mineral Resources files, unpublished report.

——. 1944. *Rockhounds & Arizona Minerals.* Phoenix: Whispering Wind Press.

——. 1949. "World news on mineral occurrences." *Rocks & Minerals* 24:6.

——. 1958. *Mineralogical Journeys in Arizona.* Scottsdale: Fred H. Bitner.

Fleischer, M. 1972. "New mineral names." *American Mineralogist* 57:1003-1006.

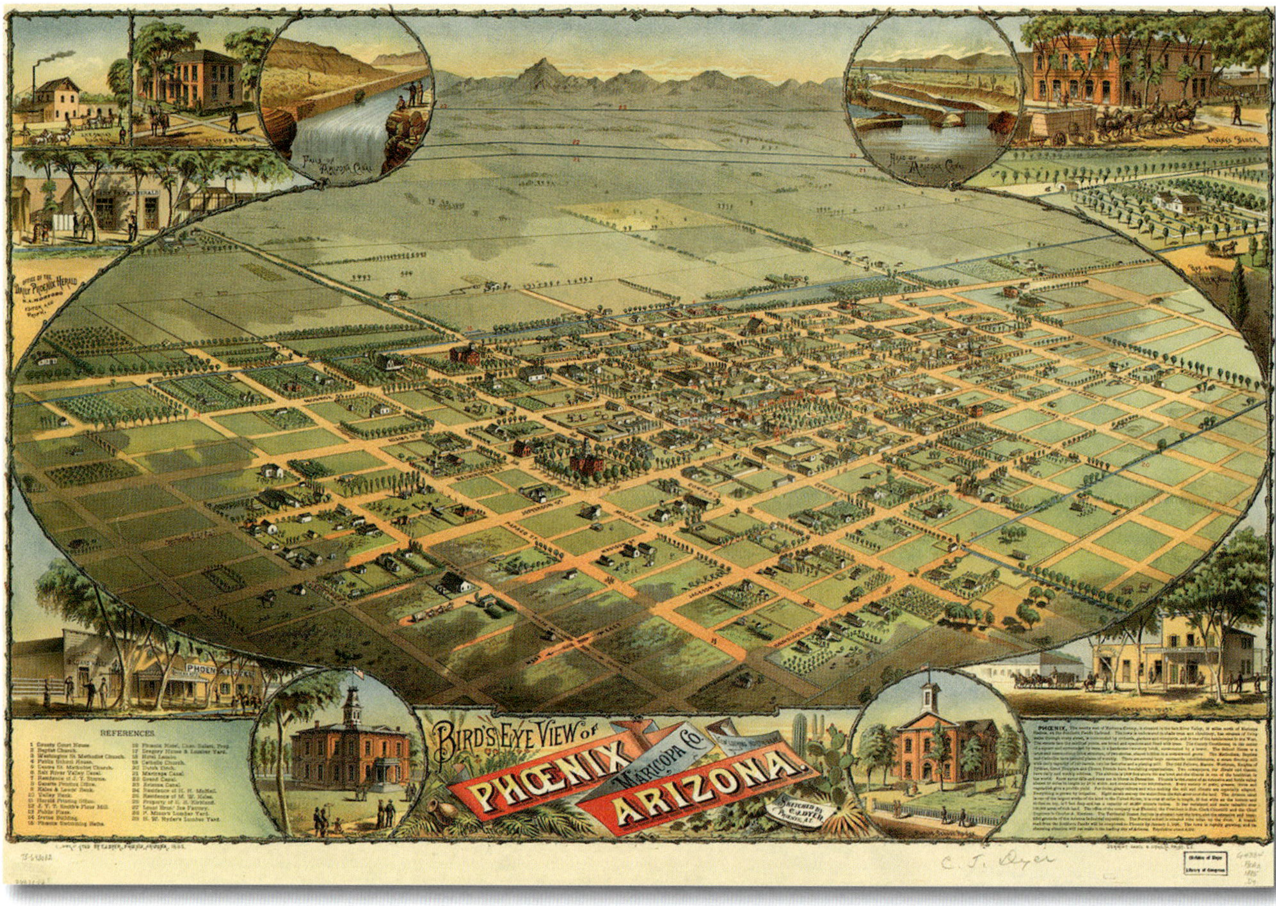

1885 "Bird's Eye View, Phoenix, Arizona" The fifth most populous city in the United States, Phoenix was founded in 1868.

Foote, A.E. 1892. *The Naturalist's Leisure Hour and Monthly Bulletin* (February).

Ford, W.E. 1932. *A Textbook of Mineralogy with an Extended Treatise on Crystallography and Physical Mineralogy.* New York: John Wiley & Sons, Inc.

Ford, W.E., and W.M. Bradley. 1915. "On the identity of footeite with connellite together with the description of two new occurrences of the mineral." *American Journal of Science* 4th series, 39:181-187.

Foshag, W.F. 1919. "Famous mineral localities: Yuma County, Arizona." *American Mineralogist* Vol. 4:149-150.

Friedrich, J.J. 1888. "On copper minerals from the Bisbee district, Arizona." New York Academy of Science 8:45-46.

Galbraith, F.W., and D.J. Brennan. 1959. "Minerals of Arizona" (3rd ed.). *University of Arizona Physical Science Bulletin* 4: 116.

Getsinger, F.R. "The Woolery Mineral Collection." *Arizona Highways* XLI: 5

Gilluly, J. 1946. "The Ajo Mining District, Arizona." *U.S.G.S. Professional Paper 209.*

Gilmour, P., and A.R. Still. 1968. "Geology of the Iron King mine" in *Ore deposits of the United States* vol II. J.D. Ridge (ed.). Society of Mining Engineers.

Goudy, K. 1987. "Oatman, Arizona" in *History of Mining in Arizona Vol. I.* J. M. Canty and M.N. Greeley (eds.). Tucson: Mining Foundation of the Southwest.

Graeme, R.W. 1981. "Famous mineral localities: Bisbee, Arizona." *The Mineralogical Record* 12:258-319.

Grant, R.W. 1986. *Rockhound Record* Apr.:7.

——. 1989. "The Blue Ball mine, Pinal County, Arizona." *The Mineralogical Record* 20:6.

Guild, F.N. 1910. *The Mineralogy of Arizona.* Easton: The Chemical Publishing Co.

Hammer, D.F., and D.W. Peterson. 1968. "Geology of the Magma Mine Area, Arizona" in *Ore Deposits of the United States Vol II*, J.D. Ridge (ed.). New York: The American Institute of Mining, Metallurgical, and Petroleum Engineers.

Harris, D.C., Roberts, A.C., Thorpe, R.I., Criddle, A.J., and Stanley, C.J. 1984. "Kiddcreekite, a new mineral species from the Kidd Creek Mine, Timmins, Ontario and from the Campbell orebody, Bisbee, Arizona." *Canadian Mineralogist* 22:227-232.

Haxel, G., and J. Dillon. 1978. "The Pelona-Orocopia schist and Vincent-Chocolate Mountain thrust system, Southern California" in *Mesozoic Paleogeography of the Western U. S.* D. G. Howell and K. A. McDougall (eds.). Pacific Section, SEPM Pacific Coast Paleogeography Symposium No. 2.

Hay, M. 1997. "The Western Union Mine,

Cerbat Mountains, Mohave County, Arizona." *The Mineralogical Record* 28:3.

Hay, M., and R. Morris. 2007. Minerals of Arizona in *Annual Arizona Mineral Symposium*, March 10, 2007.

Hills, R.C. 1882. "Dioptase from Arizona." *American Journal of Science*. 3rd series, 23.

——. 1891. "Pseudomorphous crystals of malachite after azurite from Bisbee, Arizona." *Colorado Scientific Society Proceedings* 3:257.

Hinton, R.J. 1878. *The Handbook to Arizona*. San Francisco: Payot, Upham & Company.

Holden, E.F. 1922. "Ceruleofibrite, a new mineral (Bisbee)." *American Mineralogist* 7:80-83.

——. 1924. "'Ceruleofibrite' is connellite." *American Mineralogist* 9:55-56.

Jones, R.L. 1957. *Geology of the Cerro Colorado Mining District*. Tucson: University of Arizona Press.

——. 1957. "Famous mineral localities: The Old Yuma Mine." *The Mineralogical Record* 14:95.

——. 1980. "The Grand Reef Mine, Graham County, Arizona." *The Mineralogical Record* 11:4.

Jones, R.L., and Wilson W.E. 1983. "Famous mineral localities: The Ray Mine." *The Mineralogical Record* 14: 5.

Joralemon, I.B. 1934. *Romantic Copper*. Berkeley: Howell-North Books.

Kalt, W.D. 1968. *Awake the Copper Ghosts! The History of Banner Mining Company and the Treasure of Twin Buttes*. Tucson: Banner Mining Company.

Keith, S.B. 1974. "Index of mining properties in Pima County, Arizona." *Arizona Bureau of Mines Bulletin 189*.

——. 1975. "Index of mining properties in Santa Cruz County, Arizona." *Arizona Bureau of Mines Bulletin 191*.

Keith, S.B. 1972. "Mineralogy and paragenesis of the 79 Mine lead-zinc-copper deposit." *The Mineralogical Record* 3:6.

Keith, S.B., D.E. Gest, E.Dewitt, N.W. Toll, and B.A. Everson. 1983. "Metallic mineral districts and production in Arizona." *Arizona Bureau of Geology and Mineral Technology, Geological Survey Branch, Bulletin 194*.

Keithley, H.S. 1945. "The Apache Vanadium Mine." *Rocks & Minerals* 20:12.

Kerry, R. 1995. "A rock climber's Shangri-La." *Arizona Highways* 16.

King, J.B. 1940. "Occurrence of wulfenite in Arizona." *The Mineralogist* 8:6, 261-262.

Knudsen, H.P. 1983. "The Silver Bell Mine, Gleeson, Arizona." *The Mineralogical Record* 14:2.

Koenig, G.A. 1891. "Über Paramelaconit und footiet." *Zeitschrift für Krystallographie und Mineralogie* 5:79-83.

Koutz, K.R. 2001. "Geology and Mineralization of the Hardshell Silver, Base Metal, Manganese Oxide deposits, Hardshaw Mining District, Patagonia Mountains, Santa Cruz Co., Arizona." *Arizona Geological Society Newsletter* (October).

Kunz, G.F. 1885. "On the agatized woods and the malachite, azurite, etc. from Arizona." *New York Academy of Science Transactions* 5:9-11.

Lacy, J.C. 1991. "Pursuing segments of the Glass Snake; The battle for the Big Jim" in *History of Mining in Arizona Vol. II*. J. M Canty and M.N. Greeley (eds.). Tucson: Mining Foundation of the Southwest.

Langton, J.M. 1973. "Ore genesis in the Morenci-Metcalf district." *Society of Mining Engineers, AIME, Transactions* 254:244-257.

Larsen, E.S., and H.E. Vassar. 1925. "Chalcoalumite, a new mineral from Bisbee, Arizona." *American Mineralogist* 10:79-83.

Lassiter, W.E. 1965. "World news on mineral occurrences." *Rocks & Minerals* 40:16.

Lausen, C. 1928. "Hydrous sulphates formed under fumerolic condition at the United Verde Mine." *American Mineralogist* 13:203-229.

——. 1931. "Geology and ore deposits of the Oatman and Katherine districts, Arizona." *Arizona Bureau of Mines Bulletin 131*.

Lehman, N.H. 1978. *The Geology and Pyrometasomatic Ore Deposits of the Washington Camp-Duquesne district, Santa Cruz County, Arizona*. Tucson: University of Arizona Press.

Leicht, Wayne. 1971. "Minerals of the Grandview Mine." *The Mineralogical Record* 2:5.

Lenon, R. 1999. "The Patagonia Area Mining Districts, Santa Cruz County, Arizona—1539-1930" in *History of Mining in Arizona, Vol. III*. J.M Canty, H.M. Coggin and M.N. Greeley (eds.). Tucson: Mining Foundation of the Southwest.

Lindberg, P.A. 1989. "Precambrian ore deposits of Arizona" in Geologic Evolution of Arizona. J.P. and S.J. Reynolds Jenny (eds.). *Geological Society Digest* 17.

Lindgren, W. 1905. "The copper deposits of the Clifton-Morenci district, Arizona." *U.S.G.S. Professional Paper 43*.

Lindgren, W. 1926. "Ore deposits of the Jerome and Bradshaw Mountains Quadrangle, Arizona." *U.S.G.S. Bulletin 782*.

Lowell, J. 1976. "Mineralization of the Four Peaks amethyst deposit, Maricopa County, Arizona." *The Mineralogical Record* 7:2.

Lowell, J.D. 1987. "Thirty-year evolution of porphyry copper exploration in southwest U.S.A. Part 2, case histories of discoveries." *Arizona Geological Society Digest* 11:175-178.

Lowell, J.D., and J.M. Guilbert. 1970. "Lateral and vertical alteration—mineralization zoning in porphyry copper deposits." *Economic Geology*, 65:373-408.

McClintock, J.H. 1916. *Arizona, Prehistoric-Aboriginal-Pioneer-Modern, The Nation's Youngest Commonwealth within a Land of Ancient Culture*. Chicago: The S.J. Clarke Publishing Co.

McClymonds, N.E. 1952. "Stratigraphy and structure of the southern portion of the Waterman Mountains, Pima County, Arizona." University of Arizona Doctoral thesis.

Metz, R.A., and A.W. Rose. 1966. "Geology of the Ray copper deposit, Ray, Arizona" in Geology of the Porphyry Copper Deposits, Southwestern North America. S.R. Titley and C.L. Hicks (eds.). Tucson: University of Arizona Press.

Miller, D.K., and Wilson W.E. 1983. "Famous mineral localities: The Hilltop mine." *The Mineralogical Record* 14:2.

Mills, C.E. 1958. "Notations from annual reports" (Copper Queen Consolidated Mining Company, Phelps Dodge Company, and Phelps Dodge

Corporation) years 1909-1950. Unpublished, Phelps Dodge Corp. files.
"Mineral Appraisal of Coronado National Forest Part 7"in *Mineral Land Assessment Open File Report* 22-94. 1994. Washington, D.C.: United States Bureau Mines.
Mitchell, G.J. 1920. "Vertical extent of copper minerals in the Junction Mine, Warren district." *Engineering and Mining Journal* 109:1411-1412.
Mowry, S. 1863. *The Geography and Resources of Arizona and Sonora.* An Address before the American Geographical and Statistical Society. San Francisco: A. Roman & Co.
Murbarger, N. 1964. *Ghosts of the Adobe Walls: Human Interest and Historical Highlights from 400 Ghost Haunts of Old Arizona.* Los Angeles: Westernlore Press.
Myrick, D.F. 1984. *Railroads to Carry Copper, Vol. III.* Glendale: Trans-Anglo Books.
Nations, D., and E. Stump. 1981. *Geology of Arizona.* Iowa: Kendell/Hunt Publishing Company.
North, D.M.T. 1980. *Samuel Peter Heintzelman and the Sonora Exploring & Mining Company.* Tucson: University of Arizona Press.
Novak, G., and W.W. Besse. 1986. "Vanadinite from the J. C. Holmes Claim, Santa Cruz County, Arizona." *The Mineralogical Record* 17:2.
Olson, H. J. 1961. "The geology of the Glove Mine, Santa Cruz County, Arizona." Doctoral thesis. University of Arizona, Tucson, Arizona.
Oosterwyck-Gastuche, M.V. 1967. "Etude des silicates de culver de Katanga." *Museum Royal Africa Central Ann. Ser.* 8:58.
Palache, C.I. 1939a. "Antlerite (Bisbee)." *American Mineralogist* 24:293-302.
——. 1939 b. "Brochantite (Bisbee)." *American Mineralogist* 24:463-481.
Palache, C.I., and H.E. Merwin. 1909. "On connellite and chalcopyllite from Bisbee, Arizona." *American Journal of Science* 4th Series. 28:537-540.
Palache, C.I., and E.V. Shannon. 1920. "Higginsite, a new mineral of the olivene group." *American Mineralogist* 5:155-157.
Panczner, W.D. 1976. "The Mammoth-St.Anthony Mine, Tiger, Arizona, its history, geology and mineralogy." Mineralogical Society of America-Friends of Mineralogy 2nd biennial symposium field trip.
Panczner, W.D., and C.S. Panczner. 1983. "Terra incognita—the search for minerals in Spanish Arizona." *The Mineralogical Record*, 14:2.
Papke, K.G. 1952. "Geology and ore deposits of the eastern portions of the Hilltop mine area, Cochise County, Arizona." Doctoral thesis. University of Arizona, Tucson.
Parker, B. 1986. *Einstein's Dream: The Search for a Unified Theory of the Universe.* New York: Plenum Press.
Parker, F.Z. 1966. "The geology and ore deposits of the Silver district, Trigo Mts., Yuma Co., Arizona." Doctoral thesis. San Diego State University, San Diego, CA.
Peacor, D., P.J. Dunn, G. Schnorrer-Kohler, and R.A. Bideaux. 1985. "Mammothite, a new mineral from Tiger, Arizona and Laurium, Greece." *The Mineralogical Record* 15:117-120.
Penfield, S.L. 1890. "On spangolite, a new copper mineral." *American Journal of Science* 28:537.
Peterson, N.P. 1938. "Geology and ore deposits of the Mammoth Mining Camp area, Pinal County, Arizona." *Arizona Bureau of Mines Bulletin 144.*
——. 1962. "The geology and ore deposits of the Globe-Miami district, Arizona." *U.S.G.S. Professional Paper 342.*
Peterson, N.P., C.M. Gilbert, and G.L. Quick. 1946. "Geology and ore deposits of the Castle Dome area, Gila County, Arizona." *U.S.G.S. Bulletin 971.*
Peterson, N.P., and R.W. Swanson. 1956. "Geology of the Christmas Copper Mine, Gila County, Arizona." *U.S.G.S. Bulletin 1027-H.*
Pietenpol, D.J. 1983. "Structure and ore deposits of the Silver district, La Paz County, Arizona." Doctoral Thesis. University of Arizona, Tucson, Arizona.
Powell, J.W. 1875. "Exploration of the Colorado River of the west and its tributaries. Explored in 1869, 1870, 1871, and 1872, under the direction of the Secretary of the Smithsonian Institution." Washington D.C.: Government Printing Office.
Preece, R.K., R.J. Stegen, and T.A. Weiskopf. "Hydrothermal alteration and mineralization of the Morenci Porphyry copper deposit Greenlee County, Arizona." Undated Phelps-Dodge Morenci, Inc. document.
Pumpelly, R. 1870. *Across America and Asia* (3rd ed.). New York: Leypoldt & Holt.
Ransome, F.L. 1903. "Geology of the Globe copper district, Arizona." *U.S.G.S. Professional Paper 12.*
——. 1913. "The Turquoise Copper Mining district, Arizona." *U.S.G.S. Bulletin 530.*
——. 1914. "Copper deposits near Superior, Arizona." *U.S.G.S. Bulletin 540-D.*
——. 1919. "The copper deposits of Ray and Maimi." *U.S.G.S. Professional Paper 115.*
——. 1922. "Ore deposits of the Sierrita Mountains, Pima County, Arizona." *U.S.G.S. Bulletin 725-J.*
——. 1923. "Geology of the Oatman Gold district, Arizona, a preliminary report." *U.S.G.S. Bulletin 743.*
——. 1923. "Geology of the Oatman gold district, Arizona" *U.S.G.S. Bulletin 743.*
Record, D.P. 1951. "World news on mineral occurrences- Arizona." *Rocks & Minerals* 26:5-6.
Reynolds, J.R. 1988. Arizona state geological map, based on Arizona Geological Survey Map 26. Arizona Geological Survey and Arizona Geological Society.
Rickard, F.R. 1996. *Exploring, Mining, Leaching, and Concentrating of Copper Ores as Related to the Development of Ajo, Arizona.* Self-published.
——. 1999. "The copper ores at Ajo, Arizona—Discovery to World War II" in *History of Mining in Arizona, Vol. III.* J.M. Canty, H.M. Coggin and M.N. Greeley (eds.). Tucson: Mining Foundation of the Southwest.
Robinson, G., and V. King. 1988. "What's new in minerals." *The Mineralogical Record* 19:5.
Rogers, A.F. 1910. "Notes on some pseudomorphs, petrifactions and alterations." *Proceedings of the American Philanthropic Society* 49:17-23.
——. 1913. "Delfossite, a cuprous metaferrite from Bisbee, Arizona." *American Journal of Science 4th Series,* 36:290-294.
Ross, C.P. 1925. "Geology and ore deposits of the Aravaipa and Stanley Mining districts, Graham County, Arizona." *U.S.G.S. Bulletin 763.*
Rytuba, J.J. et. al. 1978. *Reconnaissance*

Geologic Map of the Quijotoa Mountains Quadrangle, Pima County, Arizona. U.S.G.S.Misc, Field Studies Map MF-937.

Schaller, W.T. 1915. "Four new minerals [includes preliminary description of shattuckite and bisbeeite]." *Journal of the Washington Academy of Science* 5:7.

Schmeltz, F.W. 1931. "Interesting localities and how to reach them; the Yuma Mine." *Rocks & Minerals* 6:180.

Schrader, F.C. 1909. "Mineral deposits of the Cerbat Range, Black Mountains, and Grand Wash Cliffs, Mohave County, Arizona." *U.S.G.S. Bulletin 397.*

——. 1915. "Mineral deposits of the Santa Rita and Patagonia Mountains, Arizona." *U.S.G.S. Bulletin 582.*

Scovil, J.A., and L. Wagner, Jr. 1991. "The Fat Jack Mine, Yavapai County, Arizona." *The Mineralogical Record* 22:1.

Shackelford, T.J. 1976. "Structural Geology of the Rawhide Mountains, Mohave County, Arizona." Doctoral thesis. USC, Los Angeles.

Shannon, D.M. 1983. "Ferroan dolomite from the Vekol Mine." *The Mineralogical Record* 14:2.

——. Unpublished. "Rawhide Mine Artillery Peak mining district, Mohave County, Arizona."

Sheridan, T.E. 1995. *Arizona: A History.* Tucson: University of Arizona Press.

Short, M.N., F.W. Galbraith, E.N. Harshman, and T.H. Kuhn. 1943. "Geology and ore deposits of the Superior mining area, Arizona." *Arizona Bureau of Mines Bulletin 151.*

Silliman, B. 1881. "Mineralogical notes." *American Journal of Science* 22:198-205.

Simons, F.S. 1964. "Geology of the Klondyke Quadrangle, Graham and Pinal Counties, Arizona." *U.S.G.S. Professional Paper 461.*

Smith, D.S. 1951. "World news on mineral occurrences." *Rocks & Minerals* 26:11-12.

Smith, K.D. 1995. "The Finch Mine and vicinity, Gila County, Arizona." *The Mineralogical Record* 26:5.

Spencer, J.E., and S.J. Reynolds. 1989. "Geology and mineral resources of the Buckskin and Rawhide Mountains, West-Central Arizona." *Arizona Geology* 19:2.

Sprunger, M. 1980. "Hematite from the Dome Rock Mountains near Quartzite" Arizona." *The Mineralogical Record* 11:4.

St. Cyr, E.A. 1963. "World news of mineral occurrences." *Rocks & Minerals* 38:142.

Staebler, G.A., and W. E. Wilson, eds. 2008. *American Mineral Treasures*. East Hampton: Lithographie.

Sterrett, G.A. 1907. "Precious stones." *U.S.G.S. Mineral Resources*, 802-803.

——. 1909. "Precious stones." *U.S.G.S. Mineral Resources.*

——. 1911. "Precious stones." *U.S.G.S. Mineral Resources.*

Stewart, L.A., and A.J. Pfister. 1960. "Barite deposits of Arizona." *United States Bureau of Mines Report of Investigations 5651.*

Stolberg, C.S. 1988. "Osarizawaite from the Silver Hill mine, Arizona." *The Mineralogical Record* 19:5.

Szenics, T. 1975. "Ed Over at the Red Cloud Mine: the story of one man's discovery." *The Mineralogical Record* 6:4.

Tenney, J.B. 1934. *Economic Geological Reconnaissance of Casa Grande Mining District, Pinal County, Arizona.* Casa Grande: Casa Grande Chamber of Commerce.

Thomas, R.J., and R.B. Gibbs. 1983. "Famous mineral localities: The New Cornelia Mine, Ajo, Arizona." *The Mineralogical Record* 14:283.

Thompson, J.R. 1980. "The Defiance Mine and vicinity, Cochise County, Arizona." *The Mineralogical Record* 11:4.

——. 1983. "Camp Verde evaporates." *The Mineralogical Record* 14:2.

Thomssen, R. 1957. "Micromounts from the Apache Mine, near Globe, Gila County, Arizona." *The Mineral Explorer* 1:1.

Thorson, J.P. 1971. "Igneous petrology of the Oatman district, Mohave County, Arizona." Doctoral thesis. University of California at Santa Barbara, Santa Barbara, California.

Titley, S.R. 1982. *Advances in the Geology of the Porphyry Copper Deposits.* Tucson: University of Arizona Press.

Titley, S.R., and C.L. Hicks. 1966. *Geology of the Porphyry Copper Deposits, Southwestern North America.* Tucson: University of Arizona Press.

Tombstone Epitaph, 86:3. 1882. Local Personals 2.

——. 1899. Copperings 16:2.

Tovote, W.L. 1911. "Bisbee, Arizona: a geological sketch." *Mining and Scientific Press* 102:203-208.

Trimble, M. 2001. Arizona: A Cavalcade of History (2 nd ed.). Rio Nuevo Publishers.

Trimble, M. 2004. *Roadside History of Arizona* (2nd ed.). Missoula: Mountain Press Publishing Company.

Trischka, C. 1928. "The silica outcrops of the Warren mining district,." *Engineering*

David Deathridge at the 79 Mine, Gila County, Arizona

Casey Jones photo 1972

and Mining Journal 125:1045-1050.
Tupper, C.A. 1915. "The Bisbee-Warren district-Copper Queen Mine." *Mining and Engineering World* 43:515-522.
United States Bureau of Mines. 1994. "Mineral Appraisal of Coronado National Forest, Part 7." *Mineral Land Assessment Open File Report 22-94.*
Urban, J. 1957. "Phantom quartz crystals from the Mazatzals." *The Mineral Explorer* (January-March).
Vega, L, and W. Wilkinson. 1984. "Structure and mineralization-Kingman area, Arizona, Mineral Park Mine information." Arizona Geological Society.
Ward, M.B. 1977. "The volcanic geology of the Castle Hot Springs area, Yavapai County, Arizona." Doctoral thesis. Arizona State University, Tempe, AZ.
Weed, W.H. 1918. *The Mines Handbook, 13.* New York: Walter Harvey Weed.
Wells, R.C. 1913. "A new occurrence of cuprodescloizite (Bisbee)." *American Journal of Science* 4th series, 36:43-47.
Welty, J.B., S.J. Reynolds, D.E. Gest, R.A. Trapp, and E. Dewitt. 1985. "Mine index for metallic mineral districts of Arizona." *Arizona Bureau of Geology and Mineral Technology, Geological Survey Branch Bulletin 196.*
Western States Minerals Corporation. nd. On Portland Mine, file data.
White, J. 1974. "What's new in minerals." *The Mineralogical Record* 5:233.
White, J.S. 1992. "An occurrence of bixbyite, spessartine, topaz and pseudobrookite from Ash Creek near Hayden, Arizona." *The Mineralogical Record* 23:6.
Wilkinson Jr., W.H., R.W. Algood, G.M. Algood, and C. Williams. 1983. "Ramsdellite from the Mistake Mine." *The Mineralogical Record* 14:5.
Wilkinson, W.H. Jr., L.A. Vega, and S.R. Titley. 1982. "Geology and ore deposits at Mineral Park, Mohave County, Arizona" in *Advances in the Geology of the Porphyry Copper Deposits, Southwestern North America.* S.R Titley (ed.) Tucson: University of Arizona Press.
Williams, B. 1884, 1899. "Notations from the personal journals of Ben Williams for the years noted." Originals in the William Larson Collection.
Williams, S.A., and B. Khin. 1971. "Chalcoalumite from Bisbee, Arizona." *The Mineralogical Record* 2:126-127.
Williams, S.A., and P. Matter III. 1975. "Graemite, a new Bisbee mineral." *The Mineralogical Record* 6:32-34.
Wilson, E.D. 1927. "Geology and ore deposits of the Courtland-Gleeson region, Arizona." *Arizona Bureau of Mines Bulletin 123.*
——. 1933. "Geology and mineral deposits of Southern Yuma County, Arizona." *Arizona Bureau of Mines Bulletin 134.*
——. 1962. "A resume of the geology of Arizona." *Arizona Bureau of Mines Bulletin 171.*
Wilson, E.D., J.B. Cunningham, and G.M. Butler. 1934. "Arizona lode gold mines and gold mining." *Arizona Bureau of Mines Bulletin 137.*
Wilson, E.D., G.R. Fansett, C.H. Johnson, and G.H. Roseveare. 1961. "Gold placers and placering in Arizona." *Arizona Bureau of Mines Bulletin 168.*
Wilson, E.D. et. al. 1950. "Arizona zinc and lead deposits, Part 1." *Arizona Bureau of Mines Bulletin 156.*
——. 1951. "Arizona zinc and lead deposits, Part II." *Arizona Bureau of Mines Bulletin 158.*
——. 1962. "A resume of the geology of Arizona." *Arizona Bureau of Mines Bulletin 171.*
Wilson, W.E., and R.A. Bideaux. 1983. "Famous mineral localities: The Glove Mine." *The Mineralogical Record* 14:299-306.
Wilson, W.E. 1971. "Classic locality: The Apache Mine." *The Mineralogical Record* 2:6.
——. 1978. "What's new in minerals." *The Mineralogical Record* 7:131-133.
——. 1979. "What's new in minerals." *The Mineralogical Record* 10:35.
——. 1981. "What's new in minerals." *The Mineralogical Record* 12:183.
——. 1987. "What's new in minerals." *The Mineralogical Record* 18:241-246, 299-300.
——. 1988. "The Iron Cap Mine, Graham County, Arizona." *The Mineralogical Record* 19:81.
——. 1988. "What's new in minerals." *The Mineralogical Record* 19:117.
——. 1996. "Bonanza at the Red Cloud Mine." *The Mineralogical Record* 27: 5.
——. 2006. "Wayne A. Thompson; a half-century of collecting and dealing in minerals." *The Mineralogical Record* 37:6.
——. 2008. "Wulfenite and associated minerals from the Glove Mine, Santa Cruz County." *The Mineralogical Record*
Wilson, W.E., and G. Godas. 1996. "The North Geronimo Mine, La Paz County, Arizona." *The Mineralogical Record* 27:363.
Wilson, W.E., and D.K Miller. 1974. "Minerals of the Rowley Mine." *The Mineralogical Record* 5:1.
Wilson, W.E., and G. Schlepp. 2008. "Old Yuma Mine" in *American Mineral Treasures.* G.A. Staebler and W.E. Wilson (eds). East Hampton: Lithographie.
Wise, F. 1943. "Notes on Mammoth, Arizona." *Rocks & Minerals.* 18:267.
Young, O. 1964. *Western Mining.* Norman: University of Oklahoma Press.
Zeitner, J.C. 1953. "A Pilgrimage to Bisbee." *The Mineralogist* 21:339-340.

Malachite
Field of view 7.9 cm wide
Czar Mine, Bisbee, Cochise County
Graeme collection
Jeff Scovil photo